AF862321

Variational Methods in Image Processing

CHAPMAN & HALL/CRC MATHEMATICAL AND COMPUTATIONAL IMAGING SCIENCES

Series Editors

Chandrajit Bajaj
Center for Computational Visualization
The University of Texas at Austin

Guillermo Sapiro
Department of Electrical
and Computer Engineering
Duke University

Aims and Scope

This series aims to capture new developments and summarize what is known over the whole spectrum of mathematical and computational imaging sciences. It seeks to encourage the integration of mathematical, statistical and computational methods in image acquisition and processing by publishing a broad range of textbooks, reference works and handbooks. The titles included in the series are meant to appeal to students, researchers and professionals in the mathematical, statistical and computational sciences, application areas, as well as interdisciplinary researchers involved in the field. The inclusion of concrete examples and applications, and programming code and examples, is highly encouraged.

Published Titles

Image Processing for Cinema
by Marcelo Bertalmío

Image Processing and Acquisition using Python
by Ravishankar Chityala and Sridevi Pudipeddi

Statistical and Computational Methods in Brain Image Analysis
by Moo K. Chung

Rough Fuzzy Image Analysis: Foundations and Methodologies
by Sankar K. Pal and James F. Peters

Variational Methods in Image Processing
by Luminita A. Vese and Carole Le Guyader

Theoretical Foundations of Digital Imaging Using MATLAB®
by Leonid P. Yaroslavsky

Proposals for the series should be submitted to the series editors above or directly to:
CRC Press, Taylor & Francis Group
3 Park Square, Milton Park, Abingdon, OX14 4RN, UK

CHAPMAN & HALL/CRC
MATHEMATICAL AND COMPUTATIONAL IMAGING SCIENCES

Variational Methods in Image Processing

Luminita A. Vese
University of California
Los Angeles, California, USA

Carole Le Guyader
National Institute of Applied Sciences (INSA)
Rouen, France

CRC Press
Taylor & Francis Group
Boca Raton London New York

CRC Press is an imprint of the
Taylor & Francis Group an **informa** business
A CHAPMAN & HALL BOOK

MATLAB® is a trademark of The MathWorks, Inc. and is used with permission. The MathWorks does not warrant the accuracy of the text or exercises in this book. This book's use or discussion of MATLAB® software or related products does not constitute endorsement or sponsorship by The MathWorks of a particular pedagogical approach or particular use of the MATLAB® software.

Reprint 2019

CRC Press
Taylor & Francis Group
6000 Broken Sound Parkway NW, Suite 300
Boca Raton, FL 33487-2742

© 2016 by Taylor & Francis Group, LLC
CRC Press is an imprint of Taylor & Francis Group, an Informa business

ISBN: 978-0-367-24176-6 (hbk)

Digitally Printed at Replika Press Pvt. Ltd.

This book contains information obtained from authentic and highly regarded sources. Reasonable efforts have been made to publish reliable data and information, but the author and publisher cannot assume responsibility for the validity of all materials or the consequences of their use. The authors and publishers have attempted to trace the copyright holders of all material reproduced in this publication and apologize to copyright holders if permission to publish in this form has not been obtained. If any copyright material has not been acknowledged please write and let us know so we may rectify in any future reprint.

Except as permitted under U.S. Copyright Law, no part of this book may be reprinted, reproduced, transmitted, or utilized in any form by any electronic, mechanical, or other means, now known or hereafter invented, including photocopying, microfilming, and recording, or in any information storage or retrieval system, without written permission from the publishers.

For permission to photocopy or use material electronically from this work, please access www.copyright.com (http://www.copyright.com/) or contact the Copyright Clearance Center, Inc. (CCC), 222 Rosewood Drive, Danvers, MA 01923, 978-750-8400. CCC is a not-for-profit organization that provides licenses and registration for a variety of users. For organizations that have been granted a photocopy license by the CCC, a separate system of payment has been arranged.

Trademark Notice: Product or corporate names may be trademarks or registered trademarks, and are used only for identification and explanation without intent to infringe.

Visit the Taylor & Francis Web site at
http://www.taylorandfrancis.com

and the CRC Press Web site at
http://www.crcpress.com

For sale in India, Pakistan, Nepal, Bhutan, Bangladesh and Sri Lanka only.

Contents

List of Figures

List of Tables

Preface

This manuscript is devoted to variational models, their corresponding Euler–Lagrange equations and numerical implementations for image processing. Such techniques allow us to solve many inverse problems by minimization and regularization using rigorous tools from function spaces, calculus of variations, numerical analysis, and scientific computing. The most important problems in image processing are studied here, such as image restoration and image segmentation. Other related problems and applications are also presented and analyzed in detail. The variational approach offers an optimal and elegant solution in many cases, given knowledge about the image formation model, constraints, and a priori information. The variational method by regularization has been proven to be one of the most powerful techniques for solving many image processing tasks. This book covers numerous methods and applications, with accompanying tables, illustrations, algorithms, exercises, and online electronic material. It seeks to balance the theory with practice and the use of computational approaches.

Our goal is to offer a general textbook on variational approaches for image processing. Several topics are discussed in detail and may appeal to a larger audience. Instead of minimizing overlap with existing textbooks, the aim is for a more comprehensive and up-to-date presentation.

Each chapter includes the presentation of the problem, its mathematical formulation as a minimization operation, discussion and analysis of its mathematical well-posedness, derivation of the associated Euler–Lagrange equations, numerical approximations and algorithm descriptions, several numerical results, and a list of exercises.

In line with a desire for accessibility, this book attempts to be a self-contained guide to variational models in image processing, providing a synopsis of the required mathematical background necessary to understand the presented methods.

There are a number of successful advanced texts, including textbooks, on variational models and partial differential equations for image processing and related topics from image analysis and computer vision. This is a proof of the strong and continuing interest in these areas. This textbook focuses specifically on the principles and techniques for variational image processing and applications, balancing the traditional computational models with the more modern techniques developed to answer new challenges introduced by the new image acquisition devices.

The audience

This text is intended primarily for advanced undergraduate and graduate students in applied mathematics, scientific computing, medical imaging, computer vision, computer science, engineering and related fields and for engineers, professionals from academia, and the image processing industry. The manuscript can be used as a textbook for a graduate course or for a graduate summer school. It will serve as a self-contained handbook and detailed overview of the relevant variational models for image processing. The general area of image processing and its state-of-the-art methods have become essential in many fields, including medical imaging, defense, surveillance, Internet, television, image transmission, special effects, physics, astronomy, and other fields that require image acquisition for further processing and analysis.

Topics not covered

There are many other important topics in variational image processing that could not be covered here. These include higher-order models, wavelet and statistical methods, convexification algorithms for image segmentation and partition, proximal point methods, and other relevant applications from computer vision and medical imaging.

Additional resources

Electronic resources accompany the manuscript. These can be found at the online link given below, and include MATLAB® codes for the main models and algorithms presented in the manuscript. The electronic resources will be updated when necessary. Other useful resources to students, instructors, researchers and practitioners will be found as well at this link:

http://www.math.ucla.edu/~lvese/VMIP

Acknowledgments

This text wouldn't have been possible without the tremendous support, work, inspiration and dedication of all our mentors, students, collaborators, and families.

Luminita Vese thanks all her mentors: Dorel Miheţ from West University of Timişoara who introduced her to mathematics and its rigor; her Ph.D. advisors Gilles Aubert and Michel Rascle, from the University of Nice-Sophia Antipolis, who introduced her to variational models in image processing, the topic of this text; and finally, her postdoctoral mentors Tony Chan and Stanley Osher, from the University of California, Los Angeles, who introduced her to many interesting and challenging problems in image processing and guided her throughout the years.

Luminita Vese thanks all her Ph.D. students for their hard work, talent, and dedication: Triet Le, Linh Lieu, Ginmo Chung, Nicolay Tanushev, Chris Elion, Yunho Kim, Miyoun Jung, Tungyou Lin, Pascal Getreuer, Carl Lederman, Ming Yan, Melissa Tong, and Hayden Schaeffer. This text would not have been possible without their enormous contributions to variational methods in image processing. The text contains numerous results obtained by Miyoun Jung, Ginmo Chung, Triet Le, Tungyou Lin, Nicolay Tanushev, and Ming Yan; she is very grateful to all of them.

Luminita Vese is grateful to her colleagues in the Mathematics Department from the University of California, Los Angeles for their constant support, especially to Andrea Bertozzi, John Garnett, and Stanley Osher.

Also, special thoughts go to her dear family for all the love, support and sacrifice throughout the years.

Carole Le Guyader thanks her mentors: Christian Gout from the National Institute of Applied Sciences of Rouen, who trusted her and guided her throughout the years, and Luminita Vese from the University of California, Los Angeles, for her constant support and kindness, for providing her with appealing new challenges in imaging sciences, and for her strong commitment.

Carole Le Guyader also thanks her colleague Nicolas Forcadel from the National Institute of Applied Sciences of Rouen for the enlightening mathematical discussions about the theory of viscosity solutions and calculus of variations and for his friendship.

She also thanks her Ph.D. students Ratiba Derfoul and Solène Ozeré for their involvement, diligence, and skills. A special thought goes to her grandfather André Le Corff for his unfailing love and to Vincent Bonamy.

This text could not have been written without our collaborators who, through their work, greatly contributed to many sections and chapters. We would especially like to thank Gilles Aubert, Xavier Bresson, Tony Chan, Ginmo Chung, Nicolas Forcadel, John Garnett, Peter Jones, Miyoun Jung, Triet Le, Tungyou Lin, Yves Meyer, Jean-Michel Morel, Stanley Osher, Nicolay Tanushev, Ming Yan, among many others.

We also thank Chapman & Hall/CRC Mathematical and Computational

Imaging Sciences Series and its editors Chandrajit Bajaj and Guillermo Sapiro for supporting this project and for giving us the opportunity to contribute to the series with this text.

Finally, we thank our editor, Sunil Nair, for his enormous patience and understanding, his advice and guidance, and for not giving up on the project. Also, we thank his assistants, Rachel Holt, Sarah Morris, Sarah Gelson, Sarfraz Khan, and Alexander Edwards, for their constant interest in the progress of the manuscript and for all their help and support throughout the process. We would also like to express our deep gratitude to Robin Starkes for contributing to the improvement of the writing of the book.

Chapter 1

Introduction and Book Overview

1.1 Introduction

Image processing is an essential field in many applications, including medical imaging, astronomy, astrophysics, surveillance, video, image compression and transmission, just to name a few. In one dimension, images are called signals; these are formed of samples defined at discrete points in time. In two dimensions we work with planar images that are formed of pixels. In three dimensions we work with volumetric images that are formed of voxels (such as magneto-resonance images). Images can be grayscale (represented by single-valued functions), or color (represented by vector-valued functions). Noise, blur and other types of imperfections often degrade acquired images. Thus such images have to be pre-processed before any further analysis.

Images are processed by computers as two-dimensional pictures. In image processing and analysis, the input is an image, such as a photograph or a video frame. The output of image processing and analysis is either an image (after processing), or a set of image attributes (after analysis). An image is often defined as a function of two variables, for example $f(x, y)$, with f as the amplitude (e.g. brightness) of the image at the pixel position (x, y). Modern digital technology allows for the acquisition and processing of multi-dimensional and vector-valued images.

The manipulation of images can be divided into three categories [159]:

(1) *Image processing*: the input is an observed image, while the output is a processed image. Image processing tasks include denoising, deblurring, restoration, or inpainting (such as filling in the missing information).

(2) *Image analysis*: the input is an image, while the output is given by image attributes. Image analysis tasks include segmentation or partition of an image into its constituent objects and their boundaries.

(3) *Image understanding*: the input is given by image attributes, while the output is given by high-level description of the scene. Image understanding tasks include making sense of a scene and assigning a label to a region or object.

As described below, image processing tasks (1) are considered part of low-level vision. Image analysis tasks (2) are part of mid-level vision, while image understanding tasks (3) are part of high-level vision.

1.1.1 Fundamental steps in image processing

Following Gonzalez and Woods [159], we give here the main steps in image processing and analysis.

Image acquisition (*output: digital image*) is the initial step before any processing, analysis, and understanding task. This process is also called sensing and requires two elements for acquiring digital images: a physical device that is sensitive to the energy radiated by the object we wish to image, and a digitizer for converting the output of the sensing device into digital form. Although it is sufficient to be given an image in digital form, it is important to have as much information about the acquisition process as possible, since such information can be incorporated in subsequent processing tasks. As we will see, in variational models for image processing with regularization (when minimizing the sum of a data fidelity term and a regularizing one), the data fidelity term usually contains information about the acquisition process. Moreover, the acquisition process often produces raw data in a transform domain (such as the Fourier domain when acquiring a magneto-resonance image, or the Radon domain when acquiring a computed tomography image). In this case, another step is needed before processing, which is called "image reconstruction". This can be seen as part of the acquisition process. Recently, much work has been done in compressive sensing in which the reconstruction of the image in the spatial domain is obtained from a limited number of measurements in a transform domain.

Low-level vision (*input: image; output: image*) includes image enhancement, image restoration, image compression, superresolution, and wavelet representation. Some of these tasks are well suited for variational models, such as the restoration (denoising, deblurring, or inpainting), which is a process for improving the quality of the observed data through noise and blur removal, and filling in missing information. When only low-resolution image data is available, variational models can be formulated to produce images of higher resolution and better visual quality.

Mid-level vision (*input: image; output: image components*) includes morphological processing (extracting image components useful in the representation and description of shapes) and image segmentation (partitioning an image into its constituent parts or objects). We will see that variational methods are well suited for formulating image segmentation and partition models. The output is given by a set of boundaries and/or regions.

High-level vision (*input: boundaries and regions; output: image attributes*) follows image segmentation and first includes representation of boundaries and regions and their description by a set of object attributes or features. The last step is recognition or understanding, by which we assign a label (e.g., "vehicle") to an object based on its descriptors and features, thus making sense of the scene.

This text deals with variational models, their associated Euler–Lagrange equations and numerical algorithms for solving several low-level vision and mid-level vision tasks.

1.1.2 Simple image formation model

In the continuous case, a planar monochromatic image is represented by a two-dimensional single-valued function $(x, y) \mapsto f(x, y)$. The value of f at the spatial coordinates (x, y) is positive and it is determined by the source of the image. We follow here the description presented by Gonzalez and Woods in [159]. If the image is generated from a physical process, its intensity values are proportional to the energy radiated by a physical source. Therefore, $f(x, y)$ must be nonzero and finite:

$$0 < f(x, y) < \infty.$$

The image-function f may be characterized by two components:

(1) The amount of source illumination $i(x, y)$ incident on the scene (called illumination).

(2) The amount of illumination $r(x, y)$ reflected by the objects (called reflectance).

We have

$$f(x, y) = i(x, y)r(x, y),$$

where

$$0 < i(x, y) < \infty \text{ and } 0 < r(x, y) < 1.$$

Reflectance is bounded below by 0 (total absorption) and above by 1 (total reflectance). The nature of $i(x, y)$ is determined by the illumination source, while the reflectance $r(x, y)$ is determined by the characteristics of the imaged objects.

From the above construction, we have

$$L_{min} \le l = f(x, y) \le L_{max},$$

where $l = f(x, y)$ is the gray level at coordinates (x, y). It is common to shift the grayscale (or intensity scale) from the interval $[L_{min}, L_{max}]$ to the interval $[0, L-1]$. Then $l = 0$ is considered black and $l = L-1$ is considered white on the grayscale. The intermediate values are shades of gray varying from black to white.

Finally, we need to convert the continuous sensed data $f(x, y)$ into digital form via two processes called *sampling* and *quantization*. This can be done using a digitizing device. Sampling is the process of digitizing the coordinate values (x, y), while quantization is the process of digitizing the amplitude values $l = f(x, y)$. The output is a digital image that can be represented as a matrix $\left(f_{i,j}\right)_{1\le i\le M, 1\le j\le N}$ of dimension $M \times N$, with at most L quantized intensity levels $f_{i,j}$. Usually, $f_{i,j} \in \{0, 1, 2, ..., L-1\}$, with L a power of 2. The standard choice is $L = 256$, with gray level 0 representing black and gray level 255 representing white. We show in Figure 1.1 a grayscale image with several shades of gray varying from black (intensity 0) to white (intensity 255).

FIGURE 1.1: A grayscale image with several shades of gray varying from black (intensity 0) to white (intensity 255).

Vector-valued functions are used to represent color images and other types of multivalued data. There are several ways of representing color images. The simplest one is the RGB color model. In this representation, we work with images $f = (f_R, f_G, f_B)$ that have three components corresponding to the red, green and blue channels, respectively. A pixel taking the value $(255, 0, 0)$ will be visualized in red, a pixel taking the value $(0, 255, 0)$ will be visualized in green, while a pixel taking the value $(0, 0, 255)$ will be visualized in blue. Similarly, $(0, 0, 0)$ corresponds to the black color and $(255, 255, 255)$ corresponds to the white color. Sometimes it is also useful to use the chromaticity-brightness (CB) color model. In this representation, the brightness B is defined as the vector magnitude $B = |f| = \sqrt{(f_R)^2 + (f_G)^2 + (f_B)^2}$, while the chromaticity C is given by directions $C = \frac{f}{|f|}$. As we will see, several sections and chapters illustrate variational models and algorithms for processing color images.

1.2 Overview

This introductory chapter gives an overview of the topics discussed, a brief history of the main variational methods in image processing, and examples of applications.

Chapter 2 is devoted to the mathematical background, definitions and terminology necessary for the formulation and analysis of the models presented in the other chapters. The chapter includes brief review of Tikhonov regularization of ill-posed inverse problems, of maximum a posteriori (MAP) estimate, and of many other essential mathematical notions for variational image processing, such as convolution and Fourier transform, Sobolev spaces and functions of bounded variation, together with important notions in calculus of variations, definition and computation of the Euler–Lagrange equation, variational level set methods for curve evolution, the gradient descent method, and finite difference formulas.

Part I, containing several chapters, is devoted to an important problem in image processing, namely, image restoration. It explains how inverse problems arising in image restoration (such as denoising and/or deblurring) are solved by variational methods with regularization, in particular how to choose the data fidelity term function of the noise model and the regularization function of a priori constraints, among other related techniques. Two other topics are presented: the decomposition of an image into cartoon and texture by variational methods and image reconstruction in computerized tomography.

Chapter 3 gives a detailed presentation and analysis of the total variation minimization model introduced in image processing by Rudin, Osher and Fatemi [277, 276], its properties, numerical approximations, and its applications to image denoising and deblurring. More precisely, given an observed noisy and/or blurry image function $f \in L^2(\Omega)$ (defined on a planar domain Ω) as a result of the linear degradation process $f = Ku + n$, where $K : L^2(\Omega) \to L^2(\Omega)$ is a continuous linear and smoothing operator representing the blur and n is additive Gaussian noise of zero mean and variance σ^2, the problem is to find the original image u given some a priori information. Figure 1.2 shows an original image u, its blurry version Ku (obtained using a convolution operator $u \mapsto Ku = k * u$ with k a Gaussian kernel), and its blurry and noisy version $f = Ku + n$ obtained from Ku by adding Gaussian noise of zero mean. The quadratic Tikhonov regularization amounts to finding u as the minimizer among Sobolev $H^1(\Omega)$ functions of

$$E(u) = \int_\Omega |Du|^2 dx + \lambda \int_\Omega (Ku - f)^2 dx, \tag{1.1}$$

where $\lambda > 0$ is a weighting parameter and Du is the distributional gradient vector of u.

In the expression of functional $E(u)$ (equation (1.1)), the first term

FIGURE 1.2: Top: an original image u. Bottom left: its blurry version Ku obtained using a convolution operator $Ku = k * u$ with k a Gaussian kernel. Bottom right: its blurry and noisy version $f = Ku + n$ obtained from Ku by adding Gaussian noise of zero mean.

$\int_\Omega |Du|^2 dx$ is a regularization imposing a priori information on the unknown minimizer u, while the second term $\int_\Omega (Ku - f)^2 dx$ is called a data fidelity term and takes into account the image formation model. If $u \in H^1(\Omega)$ is a minimizer of (1.1), then it must solve in the distributional sense the associated Euler–Lagrange equation,

$$\lambda K^*(Ku - f) = \triangle u \text{ in } \Omega \tag{1.2}$$

combined with homogeneous Neumann boundary conditions $\frac{\partial u}{\partial \vec{n}} = 0$ on $\partial\Omega$, where $\vec{n}$ is the exterior unit normal to $\partial\Omega$. The Laplacian $\triangle u$ is isotropic and strongly smoothing in all directions. At least for smooth functions $u \in C^2(\Omega)$, it can be easily checked that

$$\triangle u = u_{\vec{T}\vec{T}} + u_{\vec{N}\vec{N}},$$

where $\vec{T}$ and $\vec{N}$ are unit orthogonal vectors, such that $\vec{N}$ is in the same direction with the gradient vector Du (see Figure 1.3). Here, $u_{\vec{T}\vec{T}} = \vec{T}^t D^2 u \vec{T}$ and $u_{\vec{N}\vec{N}} = \vec{N}^t D^2 u \vec{N}$ are the second order derivatives of u in the $\vec{T}$ and $\vec{N}$ directions, respectively. The Laplacian $\triangle u$ does not distinguish between sharp

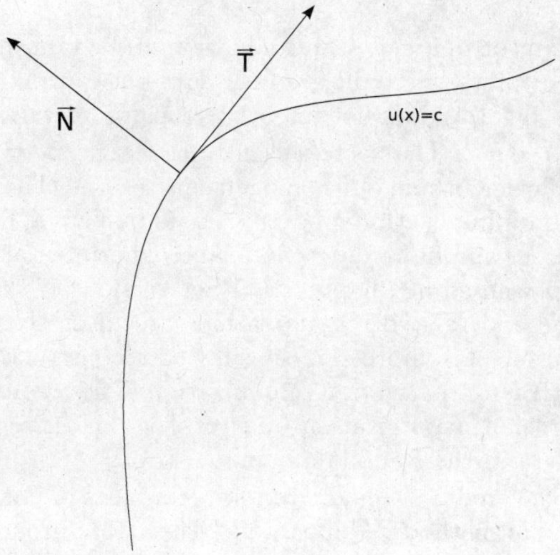

FIGURE 1.3: A level line $u(x) = c$ of the image function u and the unit normal and tangent vectors $\vec{T}$ and $\vec{N}$ to the level line.

edges and homogeneous regions. On edges, this operator will smooth the image u in both $\vec{T}$ and $\vec{N}$ directions. Thus, sharp edges cannot be preserved under the diffusion equation (1.2), as we know that the Sobolev space $H^1(\Omega)$ does not include functions with jumps along curves.

The quadratic regularization model (1.1) is related with constrained least squares filtering (introduced by Phillips [262] in one dimension; see also [159]), where the first regularizing term is instead taken to be $\int_\Omega |\triangle u|^2 dx$, with $\triangle u$ the Laplacian of u. In this case, a minimizer u would belong to the smoother Sobolev space $H^2(\Omega)$.

In the seminal work of Rudin, Osher and Fatemi [277, 276] for two dimensional images, it is proposed to work instead with functions of bounded variation and minimize the functional

$$F(u) = \int_\Omega |Du| + \lambda \int_\Omega (Ku - f)^2 dx. \tag{1.3}$$

A minimizer u of (1.3) must solve (in some sense) the associated Euler–Lagrange equation

$$2\lambda K^*(Ku - f) = \operatorname{div}\Big(\frac{Du}{|Du|}\Big) \text{ in } \Omega \tag{1.4}$$

with corresponding boundary conditions. Note that the divergence operator in the right-hand side of (1.4) is the mean curvature of level lines of u and can be expressed as

$$\mathrm{div}\Big(\frac{Du}{|Du|}\Big) = \frac{1}{|Du|} u_{\vec{T}\vec{T}}.$$

Thus (1.4) is an anisotropic smoothing equation in the tangent direction to the level lines of u, at points where the gradient does not vanish. The above partial differential equation (1.4) does not smooth in the gradient direction (the direction $\vec{N}$ normal to edges). Thanks to this geometrical property, sharp edges can be preserved while smoothing out homogeneous regions. Theoretical results of the total variation minimization model (1.3), its numerical discretizations using gradient descent and finite differences, and experimental results for image restoration are presented in Chapter 3. Other versions of the model function of the noise type and related regularizations are discussed. Chapter 3 ends with an application of compressive sensing to reconstruction in computerized tomography by total variation minimization. The reconstruction method combines expectation maximization and total variation regularization from undersampled data in the Radon domain.

Chapter 4 is devoted to the more recent techniques of image restoration by nonlocal variational methods. Gilboa and Osher [155] proposed the nonlocal total variation for image restoration by combining ideas from the nonlocal means filter of Buades, Coll, and Morel [72] with the total variation regularization [277, 276]. They have replaced the standard gradient magnitude in the Rudin, Osher, and Fatemi model [277, 276] by the nonlocal gradient. This modification is numerically more costly, but gives impressive image restoration results, especially for images with texture. The case of the nonlocal total variation regularization for image restoration will be discussed in detail.

Chapter 5 is devoted to the problem of image decomposition into cartoon and texture by variational methods. Following ideas of Meyer [236], the main problem is to separate a given planar image f into a cartoon component u (modeled, for instance, by a function of bounded variation) and a texture component v (modeled, for instance, by an oscillatory function in a dual space of distributions). This can be solved by a variational minimization model of the form

$$\inf_{u\in X_1, v\in X_2, f=u+v} F(u,v) = \|u\|_{X_1} + \lambda\|v\|_{X_2},$$

assuming that $f \in X_1 + X_2$. Here, X_1 can be taken as the space of functions of bounded variation, $BV(\Omega)$, while X_2 is usually a larger space of rougher functions and containing X_1. For example, X_2 could be $L^2(\Omega)$, $L^1(\Omega)$, or a dual space of distributions. The presented image decomposition models can be seen as extensions and generalizations of the Rudin, Osher and Fatemi model (1.3) when K is the identity. Indeed, the $BV - L^2$ model (1.3) with $K = I$ can be seen as a variational model for decomposing f into a cartoon component $u \in BV(\Omega)$ and an oscillatory component (noise or texture) $v = f - u \in$

$L^2(\Omega)$. An example of an image decomposition into cartoon and texture by minimization is shown in Figure 1.4, obtained by Le [205].

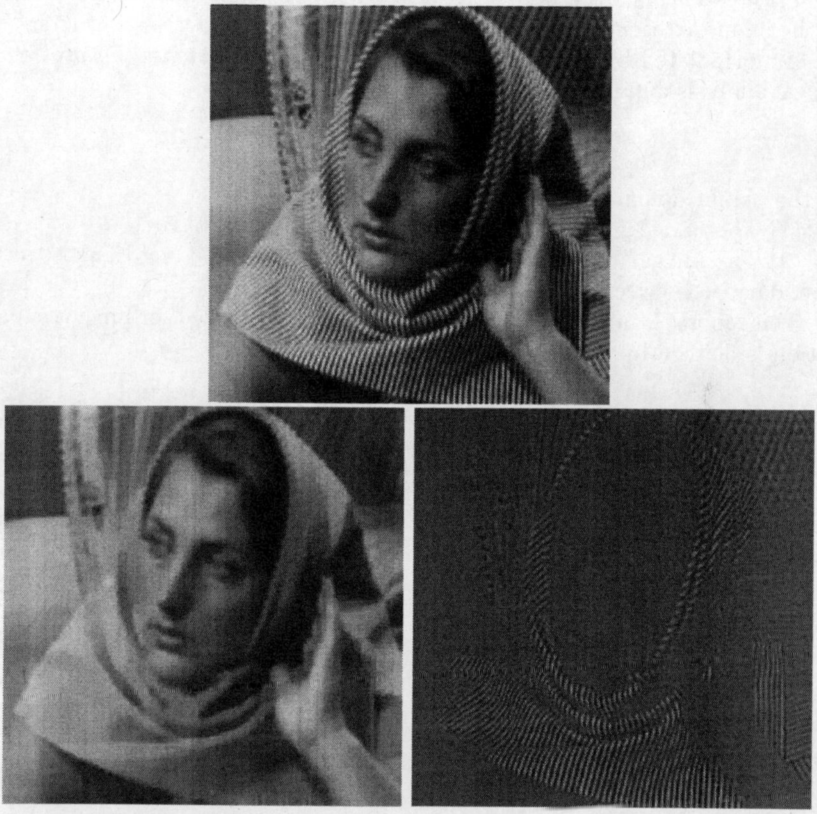

FIGURE 1.4: An example of image decomposition $f = u + v$ into cartoon and texture. Top: real image f. Bottom left: cartoon component u. Bottom right: texture component v.

Part II is devoted to the image segmentation problem by variational models and curve evolution. As mentioned earlier in this chapter, the aim is to partition an image into its main constituent parts or objects for further analysis and understanding. Another related problem is called active contours or snakes; here, the idea is to evolve a curve near an object for detecting the boundary of that object. Image segmentation and active contours are much used in medical imaging when an anatomical region or a tumor must be detected and then analyzed for its shape, area (or volume), perimeter (or surface area), or tissue characteristics.

Part II starts with a presentation of the Mumford and Shah segmentation model in Chapter 6, its phase-field approximations in Chapter 7 and its level set approximations and region-based active contours in Chapter 8. Chapters 7 and 8 also include applications to image restoration. Chapter 9 is devoted to the more classical (variational) edge-based active contour models, but updated with the most recent ideas such as topology-preserving active-contours.

As formulated by Mumford and Shah [247], the *segmentation problem* in image analysis aims to compute a partition

$$\Omega = \Omega_1 \cup \Omega_2 \cup ... \cup \Omega_i \cup ...$$

of the planar domain Ω of the image f such that

(1) The image f varies smoothly and/or slowly *within* each Ω_i.

(2) The image f varies discontinuously and/or rapidly across most of the boundary K between different Ω_i.

The general Mumford–Shah functional to be minimized for image segmentation is defined by [245, 247; 246],

$$E(u, K) = \mu^2 \int_\Omega (u - f)^2 dx + \int_{\Omega\setminus K} |\nabla u|^2 dx + \nu|K|, \tag{1.5}$$

where μ and ν are positive parameters. In two dimensions, $|K|$ stands for the length measure of the set of curves making up K. Note that (1.5), unlike the total variation functional (1.3), is not convex. Minimizers consist of pairs (u, K) with $u : \Omega \to \mathbb{R}$ smooth inside each Ω_i (in fact $u \in H^1(\Omega_i)$), but may be discontinuous across K.

An important specific case is obtained when we restrict E to piecewise-constant functions u, i.e., u = constant c_i in each open set Ω_i. This is now called the minimal partition problem and amounts to minimize

$$E(u, K) = \sum_i \int_{\Omega_i} (f - c_i)^2 dx + \nu_0|K|, \tag{1.6}$$

where $\nu_0 = \nu/\mu^2$.

It is not easy to minimize the above Mumford and Shah functionals due to the presence of the lower-dimensional object K as unknown. Also, these are not the lower semi-continuous envelopes of some functionals defined on the Sobolev spaces $W^{1,1}(\Omega)$ or $H^1(\Omega) = W^{1,2}(\Omega)$. Part II discusses two approaches to circumvent this difficulty: one uses elliptic approximations within the phase-field theory (Chapter 7), while the other uses variational level sets (Chapter 8). As we will see, the second approach provides a common link between Mumford–Shah segmentation and active contours via curve evolution.

Ambrosio and Tortorelli proposed two elliptic approximations [21], [22] to the weak formulation of the Mumford and Shah problem. The second one [22], being simpler, is commonly used in practice. Let

$$AT_\epsilon(u, v) = \int_\Omega (u-f)^2 dx + \beta \int_\Omega (o(\epsilon)+v^2)|\nabla u|^2 dx + \alpha \int_\Omega \left(\epsilon|\nabla v|^2 + \frac{(v-1)^2}{4\epsilon}\right) dx \tag{1.7}$$

if $(u, v) \in H^1(\Omega)^2$, $0 \le v \le 1$, and $AT_\epsilon(u, v) = +\infty$ otherwise. Here $o(\epsilon)$ is such that $0 < o(\epsilon) << \epsilon$. Minimizers (u_ϵ, v_ϵ) of AT_ϵ as $\epsilon \to 0$ will approximate the minimizer u of the weak formulation of the Mumford and Shah problem and the edge set or jump set $K = K_u$ is approximated by the regions where $v < 1$. Basically, the minimizer v with $0 \le v \le 1$ will act as an edge detector, being small (close to zero) around edges and close to 1 in the smoother regions away from edges. Ambrosio and Tortorelli [22] also proposed elliptic phase-field approximations to the minimal partition problem (1.6). Figure 1.5 shows an edge detection result obtained using the Ambrosio–Tortorelli approximation AT_ϵ to the Mumford–Shah functional.

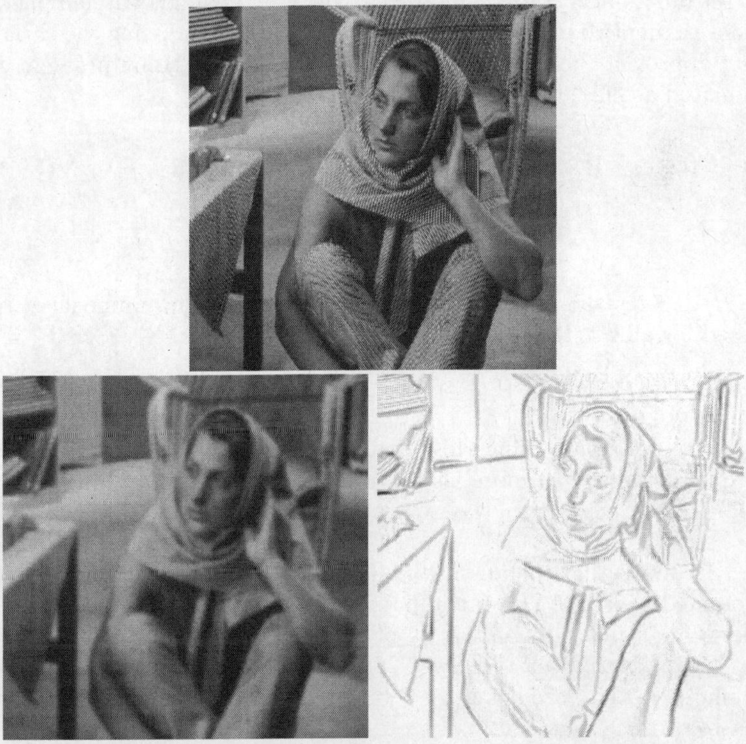

FIGURE 1.5: Segmentation and edge detection results using the Ambrosio–Tortorelli approximation AT_ϵ of the Mumford–Shah energy (1.5) in the weak formulation. Top: real image f. Bottom left: piecewise-smooth approximation u. Bottom right: edge function v.

Shah [288] and Alicandro, Braides and Shah [7] modified the above AT_ϵ approximation and computed a limiting functional that is non-convex, but similar to the total variation. We will see that such approximation (or the original AT_ϵ approximation) can be used for image restoration.

The connection between (region-based) active contours, the level set method of Osher and Sethian [254], and the Mumford and Shah segmentation model has been made in the work of Samson, Blanc-Féraud, Aubert and Zerubia [280, 279], Amadieu et al. [17], Chan and Vese [92, 94, 335], and Tsai, Yezzi and Willsky [328]. Chapter 8 presents this approach in detail following the work of Chan and Vese. Also, snakes or active contours combined with Mumford and Shah segmentation appear in the work of Cohen et al. [109, 108].

Let us assume that the unknown set of edges K is the boundary of an open and bounded subset of Ω; thus K can be represented by $K = \{x \in \Omega : \phi(x) = 0\}$ for some (unknown) Lipschitz continuous function $\phi : \Omega \to \mathbb{R}$, called a level set function. In a first case, we can restrict the unknown minimizer u to functions taking two unknown constant values c_1 and c_2. Using the variational level set approach [352], the corresponding minimal partition problem can be reformulated as [92, 94]

$$\begin{aligned}\inf_{c_1,c_2,\phi} E(c_1,c_2,\phi) &= \int_\Omega (f(x)-c_1)^2 H(\phi)dx + \int_\Omega (f(x)-c_2)^2 H(-\phi)dx \\ &+ \nu_0 \int_\Omega |DH(\phi)|, \qquad (1.8)\end{aligned}$$

where H denotes the Heaviside function. The unknown minimizer u is thus expressed for all $x \in \Omega$ as

$$u(x) = c_1 H(\phi(x)) + c_2(1 - H(\phi(x))) = c_1 H(\phi(x)) + c_2 H(-\phi(x)).$$

A segmentation result using this model is shown in Figure 1.6. The model has been generalized to more than two constant regions and more complex topologies of the edge set in the work of Chan, Chung, and Vese [93, 335, 103, 104].

Still assuming that the unknown set of edges K is the boundary of an open and bounded subset of Ω, we can have a level set formulation of the general Mumford–Shah functional (1.5), as proposed by Chan and Vese in [95, 335] and by Tsai, Yezzi, and Willsky [328]. Indeed, the functional (1.5) can be reformulated in this case using the level set method as the minimization with respect to u^+, u^-, and ϕ of

$$\begin{aligned}E(u^+,u^-,\phi) &= \mu^2 \int_\Omega (u^+ - f)^2 H(\phi)dx + \mu^2 \int_\Omega (u^- - f)^2 H(-\phi)dx \\ &+ \int_\Omega |\nabla u^+|^2 H(\phi)dx + \int_\Omega |\nabla u^-|^2 H(-\phi)dx + \nu \int_\Omega |DH(\phi)|,\end{aligned}$$

and the unknown minimizer u is expressed as $u = u^+ H(\phi) + u^- H(-\phi)$ in Ω.

As we can see, in both phase-field and level set approaches, the unknown set of edges K has been substituted by the function v or ϕ, respectively. For the reformulated functionals, their corresponding Euler–Lagrange equations can be easily computed and discretized.

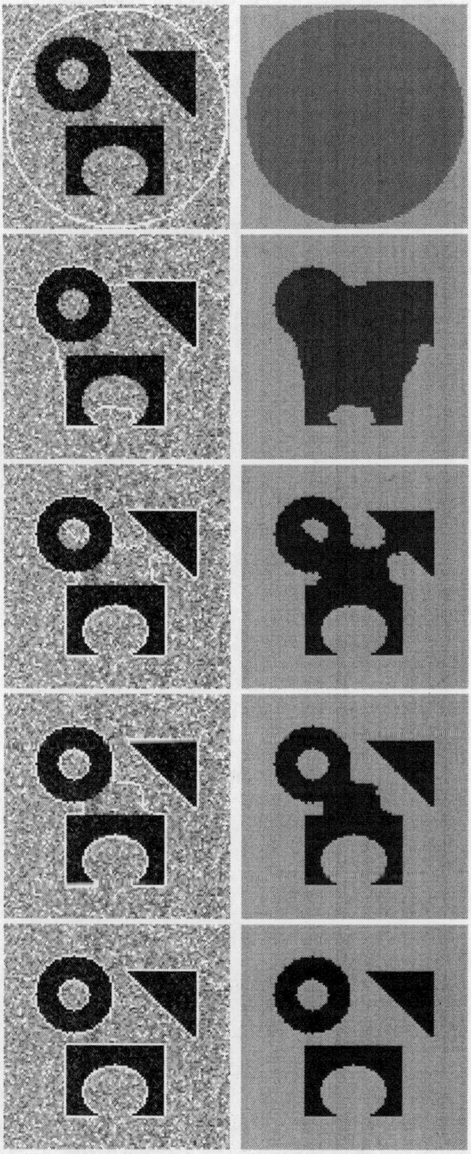

FIGURE 1.6: Active contour evolution and detection of different objects in a synthetic noisy image f, with various convexities and with an interior contour that is automatically detected using only one initial curve. Left column: active contour superposed over the image f. Right column: piecewise-constant images of averages as segmentations.

Part II ends with Chapter 9 dedicated to the more classical edge-based active contour models by variational methods that can be divided into two categories; parametric active contours in which the evolving contour is represented explicitly using splines as in the seminal work of Kass, Witkin and Terzopoulos [187], and geometric active contours in which the curve is represented implicitly as the zero level line of a function as in the seminal work of Caselles, Kimmel and Sapiro [80] and Kichenassamy et al. [188]:

(1) Methods that fall within the first category consist in evolving a parameterized curve subjected to internal regularization constraints and external data-driven forces so that it matches the object boundary. These models exhibit limitations: first, the involved energy is not intrinsic, that is, not invariant to a change of parameterization of the curve. It is thus not related to the geometry of the object. Second, through the evolution process, a reparameterization of the curve might be needed to obtain a more uniform distribution of the parameterization nodes. Third, topological changes are not automatically handled, which means that without additional artifices, multiple objects cannot be detected simultaneously. At last, the model is sensitive to initialization and shows slow/poor convergence near boundaries with strong concavities. Xu and Prince [346] address these two latter issues by introducing a new static external force called gradient vector flow (GVF). The main idea is to increase the capture range of the edge-map-related force field. In [177], Jifeng et al. propose improving the diffusion properties of the GVF force field by replacing the Laplacian operator used in the GVF model by its diffusion term in the normal direction, that is, the normalized infinity Laplacian operator.

(2) Methods included in the second category are still grounded on local edge detectors but the evolving contour is modelled as the zero level line of a function, so that the problem is discretized on a fixed rectangular grid. In [80, 188], Caselles et al. and Kichenassamy et al. prove that under certain assumptions, the classical active contour model is equivalent to searching for a geodesic curve in a Riemann space whose metric is related to the image content. The evolving contour is embedded in a level-set function and its evolution is phrased in terms of a Eulerian formulation. The ability of this intrinsic method to handle topological changes (merging and breaking) makes it useful in a wide range of applications. Nevertheless, this topological flexibility turns out to be undesirable in some specific contexts: when the shape to be detected has a known topology or when the resulting shape is homeomorphic to the initial one. The necessity of designing topology-preserving processes arises in medical imaging, for example, in the human cortex reconstruction. It is known that the human cortex has a spherical topology so throughout

the reconstruction process this feature must be preserved. This issue is the scope of the last part of Chapter 9.

Part III presents real-world applications of the variational image restoration and segmentation methods presented in Parts I and II. More precisely, Part III contains other related topics of interest: variational models for color image restoration (Chapter 10), variational models for image registration (Chapters 11 and 12), a variational model for electrical impedance tomography (Chapter 13), variational segmentation models for images corrupted by additive or multiplicative noise (Chapter 14), and minimization problems formulated on the unit sphere in $\mathbb{R}^N$, S^{N-1} (Chapter 15).

In addition to the programming exercises provided, some more theoretical applications are given at the end of the chapters.

Chapter 2

Mathematical Background

We will briefly review in this chapter some of the necessary mathematical tools used in modern variational image analysis. This includes topics on Tikhonov regularization, maximum a posteriori estimate, convolution, Fourier transform, topologies on Banach spaces, function spaces, calculus of variations and energy minimization, geometric curve evolution, variational level set methods, numerical analysis by finite differences and the gradient descent method, and more. However, this chapter is not a substitute for the much more detailed existing textbooks on these topics. Thus, we strongly encourage the interested reader to consult many of the cited textbooks.

2.1 Tikhonov regularization of ill-posed inverse problems

Tikhonov regularization is the most commonly used method of regularization of ill-posed inverse problems. We refer the reader to the original work by Tikhonov [319], [321], [320], Phillips [262], and Tikhonov and Arsenin [322]. For additional details on Tikhonov regularization, the reader is referred to the more recent publications [323] and [324].

We consider an operator equation of the form

$$Ku = f \tag{2.1}$$

in the unknown u in a Hilbert space X, where K is a mapping from X to Y (another Hilbert space), and $f \in Y$ is the given data (the observation). In image processing, equation (2.1) usually describes the image formation model: f represents the observed image, u is the original true image to be recovered, and K is a transform that links the image u to the observation f. The operator K could represent for instance a blurring operator (convolution with a kernel k), the Radon transform in computerized tomography, or the Fourier transform in magneto-resonance imaging. Often, the observation f may encompass noise inherent in any image acquisition, formation and transmission process.

Equation (2.1) is well-posed if existence, uniqueness, and continuous dependence of the solution on the data hold (if K is injective, and if K^{-1} exists and it is continuous). Otherwise the problem is ill-posed. Thus, if the operator K is not invertible (or even if it would be invertible, its inversion may be ill-conditioned), then the problem becomes ill-posed. In many image processing tasks, $K : X \mapsto Y$ is a linear and continuous (bounded) operator, and the standard approach for solving equation (2.1) is the least squares method that seeks to minimize over $u \in X$,

$$F(u) = \|Ku - f\|_Y^2. \tag{2.2}$$

If u would be a minimizer of (2.2), then it must solve the Euler–Lagrange equation

$$K^*Ku = K^*f,$$

where $K^* : Y \mapsto X$ is the adjoint operator of K. But now K^*K may not be invertible, or its inversion may be ill-conditioned. The simplest form of Tikhonov regularization alleviates these difficulties by solving instead

$$\inf_{u \in X} \left\{ \|Ku - f\|_Y^2 + \lambda \|u\|_X^2 \right\}, \tag{2.3}$$

where $\lambda > 0$ is a regularizing parameter which is often chosen function of the noise level present in the data f. It is possible to show that (2.3) has a unique solution u given by

$$u = (K^*K + \lambda I)^{-1} K^* f,$$

where $I : X \mapsto X$ is the identity operator. The preference is given to solutions with smaller norms. A typical example is obtained when both X and Y coincide with the Hilbert space L^2.

In a more general version of Tikhonov regularization, the term $\lambda\|u\|_X^2$ in (2.3) is substituted by $\lambda\|Lu\|_Z^2$, where Z could be another Hilbert space, and $L : X \mapsto Z$ can be seen as a highpass operator (e.g., a linear differential operator like the gradient or Laplacian; a difference operator). The problem becomes

$$\inf_{u \in X} \left\{ \|Ku - f\|_Y^2 + \lambda \|Lu\|_Z^2 \right\}. \tag{2.4}$$

If u is a solution of (2.4), then $\|Lu\|_Z < \infty$, which is seen as prior information or smoothness constraint on the image u. The regularization $\lambda\|Lu\|_Z^2$ improves the conditioning of the problem, thus enabling a numerical solution. Assuming that $L : X \mapsto Z$ is linear and continuous with adjoint $L^* : Z \mapsto X$, then $u \in X$ is solution of (2.4) if and only if

$$(K^*K + \lambda L^*L)u = K^*f.$$

Throughout the book, we will see many instances of Tikhonov regularization in its various forms for solving inverse problems in image processing.

2.2 Maximum a posteriori (MAP) estimate

Maximum a posteriori (MAP) estimation is a Bayesian statistics technique that can be used to design variational methods in image processing by taking into account the probability distribution function (PDF) of the noise. Let f be a known degraded image (the observation) and u the underlying original image to be restored. We view f and u as random variables with their corresponding (prior) distributions. The MAP estimate of u is the most likely value of u given f, expressed as

$$\arg\max_u \mathbf{P}(u|f),$$

where $\mathbf{P}(u|f)$ is the conditional probability of u given the observation f. Applying Bayes' theorem

$$\mathbf{P}(u|f) = \frac{\mathbf{P}(f|u)\mathbf{P}(u)}{\mathbf{P}(f)},$$

where $\mathbf{P}(f|u)$ is the observation PDF (likelihood of the observation f given u), we obtain by applying $-\log$,

$$\max_u \mathbf{P}(u|f) \Leftrightarrow \max_u \Big\{\mathbf{P}(u)\mathbf{P}(f|u)\Big\} \Leftrightarrow \min_u \Big\{ -\log \mathbf{P}(u) - \log \mathbf{P}(f|u)\Big\}.$$

In the last expression above, the first term $-\log \mathbf{P}(u)$ is called the *prior* on u and it acts as a regularization or assumption on what u is likely to be. The second term $-\log \mathbf{P}(f|u)$ describes the degradation process that produced f from u.

Example 1 *Using the notations from Section 2.1, if* $\mathbf{P}(f|u) = \frac{1}{\sqrt{2\pi}\sigma} e^{-\frac{\|u-f\|_X^2}{2\sigma^2}}$ *and* $\mathbf{P}(u) = \frac{1}{\sqrt{2\pi}\tau} e^{-\frac{\|Lu\|_Z^2}{2\tau^2}}$ *with* $\sigma^2 > 0$ *and* $\tau^2 > 0$ *being the variances of two Gaussian (normal) probability distributions , then we are led to solve*

$$\inf_{u \in X} \Big\{\frac{1}{2\sigma^2}\|u - f\|_X^2 + \frac{1}{2\tau^2}\|Lu\|_Z^2 + constant\Big\},$$

which is equivalent with the general Tikhonov regularization method (2.4) when $K = I : X \mapsto X$ *(factoring out* $\frac{1}{2\sigma^2}$ *and letting* $\lambda = \frac{\sigma^2}{\tau^2}$*).*

2.3 Convolution

The convolution is an important tool in image processing. For example, a blurry image f can be modeled by the convolution of a sharp image u with a kernel k, as $f = k * u$. In image processing, the kernel is called a point spread function; its form can model various types of blurs encountered in reality, such as motion or out-of-focus blurs. We refer the reader to [146] for convolution within the functional spaces framework. In image processing, we refer the reader to [159].

Definition 1 *Let k and u be measurable functions in $\mathbb{R}^N$. The* **convolution** *of k and u is the function $k * u$ defined by*

$$k * u(x) = \int k(x-y)u(y)dy.$$

The elementary properties of convolutions are summarized in the following proposition.

Proposition 1 *Assuming that all integrals in question exist, we have*
(i) $k * u = u * k$.
(ii) $(k * u) * v = k * (u * v)$.
(iii) For $z \in \mathbb{R}^N$, $\tau_z(k*u) = (\tau_z k)*u = k*(\tau_z u)$, where τ_z is the translation by vector z.

The following proposition states a basic fact about convolutions of L^p functions.

Proposition 2 (Young's inequality) *If $k \in L^1$ and $u \in L^p$ $(1 \leq p \leq \infty)$, then $k*u(x)$ exists for almost every x, $k*u \in L^p$, and $\|k*u\|_{L^p} \leq \|k\|_{L^1}\|u\|_{L^p}$.*

2.4 Fourier transform

The Fourier transform is a classical tool in image processing. It expresses a function in the spatial domain as a function of frequency. Many filtering techniques (such as sharpening, smoothing, denoising, deblurring) make use of the Fourier transform. Moreover, based on the convolution theorem, convolution in the spatial domain (an expensive computational operation) is equivalent with pointwise multiplication in the frequency (Fourier) domain. In the discrete case, the transfer from the spatial domain to the frequency domain and vice versa is done in an efficient way using the fast Fourier transform (FFT). The Fourier transform is also used to define the Sobolev spaces $W^{s,p}(\mathbb{R}^N)$ for real exponent $s \in \mathbb{R}$ $(p \geq 1)$.

Definition 2 *The* **Fourier transform** *of $u \in L^1(\mathbb{R}^N)$ is defined by*

$$(\mathcal{F}u)(\xi) = \hat{u}(\xi) = \int_{\mathbb{R}^N} u(x)e^{-2\pi i\xi\cdot x}dx.$$

We observe from the definition that $|\hat{u}(\xi)| \leq \int_{\mathbb{R}^n} |u(x)|dx = \|u\|_{L^1(\mathbb{R}^N)}$ for all $\xi \in \mathbb{R}^N$. We summarize the elementary properties of $\mathcal{F}$ in a theorem [146].

Theorem 1 *Suppose $u, v \in L^1(\mathbb{R}^N)$.*

(i) $\widehat{(\tau_y u)}(\xi) = e^{-2\pi i \xi \cdot y}\hat{u}(\xi)$.

(ii) **(convolution theorem)** $u * v \in L^1(\mathbb{R}^N)$ *and* $\widehat{(u * v)} = \hat{u}\hat{v}$.

(iii) If $x^\alpha u \in L^1(\mathbb{R}^N)$ *for* $|\alpha| \leq k$, *then* $\hat{u} \in C^k$ *and* $\partial^\alpha \hat{u} = [(-2\pi i x)^\alpha u]^\wedge$ *(note that* $\alpha = (\alpha_1, ..., \alpha_N)$ *is a multi-index of natural numbers).*

(iv) If $u \in C^k$, $\partial^\alpha u \in L^1$ *for* $|\alpha| \leq k$, *and* $\partial^\alpha u \in C^0$ *for* $|\alpha| \leq k - 1$, *then* $(\partial^\alpha u)^\wedge(\xi) = (2\pi i \xi)^\alpha \hat{u}(\xi)$.

(v) $\int \hat{u}v = \int u\hat{v}$.

Definition 3 *If $u \in L^1(\mathbb{R}^N)$, we define*

$$\check{u}(x) = \hat{u}(-x) = \int u(\xi)e^{2\pi i \xi \cdot x} d\xi.$$

Theorem 2 (Fourier inversion theorem) *If $u \in L^1$ and $\hat{u} \in L^1$, then u agrees almost everywhere with a continuous function u_0, and $(\hat{u})^\vee = (\check{u})^\wedge = u_0$.*

Theorem 3 (Plancherel theorem) *$\mathcal{F}|_{L^1 \cap L^2}$ extends uniquely to a unitary isomorphism on L^2.*

Plancherel theorem says in particular that the Fourier tranoform is well defined for functions $u \in L^2$; in this case, $\hat{u} \in L^2$ and $\|u\|_{L^2} = \|\hat{u}\|_{L^2}$ (this last equality is often called Parseval's theorem).

2.5 Topologies on Banach spaces

This section is largely inspired by [30].

In the following, we denote by $(V, \|\cdot\|_V)$ a real Banach space, that is, a complete normed vector space endowed with the norm $\|\cdot\|_V$. We denote by V' the topological dual space of V defined as the set of continuous linear forms on V:

$$V' = \left\{ l : V \to \mathbb{R} \text{ linear such that } \|l\|_{V'} = \sup_{u \in V \setminus \{0\}} \frac{|l(u)|}{\|u\|_V} < +\infty \right\}.$$

We now turn to basic elements on both topologies on V and V'.

Definition 4 (topologies on V)

(i) *A sequence (u_n) of V is said to strongly converge to $\bar{u} \in V$, denoted by $u_n \underset{V}{\longrightarrow} \bar{u}$, when $\lim_{n\to+\infty} \|u_n - \bar{u}\|_V = 0$.*

(ii) *A sequence (u_n) of V is said to weakly converge to $\bar{u} \in V$, denoted by $u_n \underset{V}{\rightharpoonup} \bar{u}$, when for every $l \in V'$, $\lim_{n\to+\infty} l(u_n) = l(\bar{u})$.*

Remark 1 *It is easy to prove that strong convergence implies weak convergence. The converse is false in general.*

Before giving the topologies endowing the dual space V', we recall some basic facts about L^p-spaces.

Definition 5

(i) *The Banach space V is said to be reflexive when $(V')' = V$.*

(ii) *The Banach space V is said to be separable if it contains a countable dense subset.*

Example 2 *Let us consider the case $V = L^p(\Omega)$ with $p \in \mathbb{R}$ such that $p \geq 1$.*

(i) *$V = L^p(\Omega)$ is reflexive for $p \in (1, \infty)$ and separable for $p \in [1, \infty)$.*

(ii) *The dual space of $L^p(\Omega)$ for $p \in [1, \infty)$ is $L^q(\Omega)$ with q such that $\frac{1}{p} + \frac{1}{q} = 1$.*

(iii) *The dual space of $L^\infty(\Omega)$ contains $L^1(\Omega)$ but is strictly greater than $L^1(\Omega)$. In particular, it means that there exist continuous linear maps φ on $L^\infty(\Omega)$ that cannot be written in the form $\varphi(f) = \int_\Omega uf \ \forall f \in L^\infty(\Omega)$, with $u \in L^1(\Omega)$.*

Definition 6 (topologies on V')

(i) *A sequence (l_n) of V' is said to strongly converge to $\bar{l}$, denoted by $l_n \underset{V'}{\longrightarrow} \bar{l}$, if $\lim_{n\to+\infty} \|l_n - \bar{l}\|_{V'} = 0$. Coming back to the definition of $\|\cdot\|_{V'}$, it is equivalent to*

$$\lim_{n\to+\infty} \sup_{u\in V\setminus\{0\}} \frac{|l_n(u) - \bar{l}(u)|}{\|u\|_V} = 0.$$

(ii) *Denoting by $(V')'$ the bidual space of V, a sequence (l_n) of V' is said to weakly converge to $\bar{l}$ in V', denoted by $l_n \underset{V'}{\rightharpoonup} \bar{l}$, if for every $v \in (V')'$, $\lim_{n\to+\infty} v(l_n) = v(\bar{l})$.*

(iii) *A sequence (l_n) of V' is said to converge weakly $*$ to $\bar{l}$ in V', denoted by $l_n \underset{V'}{\overset{*}{\rightharpoonup}} \bar{l}$, if for every $u \in V$, $\lim_{n\to+\infty} l_n(u) = \bar{l}(u)$.*

These topology definitions enable us to define weak sequential compactness properties.

Theorem 4 (compactness properties)

(i) Let V be a reflexive Banach space and let $C > 0$ be a positive real constant. Let also (u_n) be a sequence of V such that $\|u_n\|_V \leq C$. Then there exist $\bar{u} \in V$ and a subsequence (u_{n_k}) of (u_n) such that $u_{n_k} \underset{V}{\rightharpoonup} \bar{u}$ when $k \to +\infty$.

(ii) Let V be a separable Banach space and let $C > 0$ be a positive real constant. Let also (l_n) be a sequence of V' such that $\|l_n\|_{V'} \leq C$. Then there exist $\bar{l} \in V'$ and a subsequence (l_{n_k}) of (l_n) such that $l_{n_k} \underset{V'}{\overset{*}{\rightharpoonup}} \bar{l}$ when $k \to +\infty$.

2.6 Sobolev and BV spaces

We review here a few necessary basic definitions, terminology and results for Sobolev and BV spaces, to be used in the later chapters on variational methods for image processing. For the complete theory and proofs, we refer the reader to several books [140, 3, 139, 146, 20, 27, 156].

We first recall the definition of the Hausdorff measure following [140, 120, 20].

Definition 7 *For $K \subset \mathbb{R}^N$ and $n > 0$, set*

$$\mathcal{H}^n(K) = \sup_{\epsilon > 0} \mathcal{H}^n_\epsilon(K),$$

called the n-dimensional Hausdorff measure of the set K, where

$$\mathcal{H}^n_\epsilon(K) = c_n \inf \left\{ \sum_{i=1}^{\infty} (diam\, A_i)^n \right\}$$

and the infimum is taken over all countable families $\{A_i\}_{i=1}^{\infty}$ of closed sets A_i such that

$$K \subset \bigcup_{i=1}^{\infty} A_i \text{ and } diam\ A_i \leq \epsilon \text{ for all } i.$$

Here the constant c_n is chosen so that $\mathcal{H}^n$ coincides with the Lebesgue measure on n-planes. The involved coverings follow the local geometry of K. Also, as $\epsilon \mapsto \mathcal{H}^n_\epsilon(K)$ is monotone decreasing in ϵ,

$$\mathcal{H}^n(K) = \sup_{\epsilon > 0} \mathcal{H}^n_\epsilon(K) = \lim_{\epsilon \to 0} \mathcal{H}^n_\epsilon(K).$$

Remark 2 *When n is an integer and K is contained in a C^1-surface of dimension n, $\mathcal{H}^n(K)$ coincides with its n-dimensional surface measure.*

Let Ω be an open subset of $\mathbb{R}^N$, $N \geq 1$ and let $x = (x_1, \cdots, x_N)$.

2.6.1 Sobolev spaces

Definition 8 *Assume $u \in L^1_{loc}(\Omega)$. We say $v_i \in L^1_{loc}(\Omega)$ is the weak partial derivative of u with respect to x_i in Ω if*

$$\int_\Omega u \frac{\partial \phi}{\partial x_i} dx = - \int_\Omega v_i \phi dx$$

for all $\phi \in C^1_c(\Omega)$.

If the weak partial derivative of u with respect to x_i exists, then it is uniquely defined $\mathcal{L}^N$ a.e. ($\mathcal{L}^N$ denoting the n-dimensional Lebesgue measure). We write $\frac{\partial u}{\partial x_i} = v_i$, for $i = 1, ..., N$, and $\nabla u = \left(\frac{\partial u}{\partial x_1}, ..., \frac{\partial u}{\partial x_N}\right)$.

Definition 9 *Let $1 \leq p \leq \infty$. The function u belongs to the Sobolev space $W^{1,p}(\Omega)$ if $u \in L^p(\Omega)$ and the weak partial derivatives $\frac{\partial u}{\partial x_i}$ exist and belong to $L^p(\Omega)$, $i = 1, ..., N$. The Sobolev space $W^{1,2}(\Omega)$ is often denoted by $H^1(\Omega)$.*

If $u \in W^{1,p}(\Omega)$, define

$$\|u\|_{W^{1,p}(\Omega)} = \left(\int_\Omega (|u|^p + |\nabla u|^p) dx\right)^{1/p} \quad \text{for } 1 \leq p < \infty,$$

and

$$\|u\|_{W^{1,\infty}(\Omega)} = ess \sup_\Omega (|u| + |\nabla u|).$$

We have that $\|u\|_{W^{1,p}(\Omega)}$ is a norm on $W^{1,p}(\Omega)$. Endowed with this norm, $W^{1,p}(\Omega)$ is a Banach space .

Theorem 5 *Assume $1 \leq p < \infty$.*

(i) **(product rule)** *If $u, v \in W^{1,p}(\Omega) \cap L^\infty(\Omega)$, then $uv \in W^{1,p}(\Omega) \cap L^\infty(\Omega)$ and*

$$\frac{\partial (uv)}{\partial x_i} = \frac{\partial u}{\partial x_i} v + u \frac{\partial v}{\partial x_i} \quad \mathcal{L}^N \text{ a.e. } (i = 1, ..., N).$$

(ii) **(chain rule)** *If $u \in W^{1,p}(\Omega)$ and $F \in C^1(\mathbb{R})$, $F' \in L^\infty(\mathbb{R})$, $F(0) = 0$, then $F(u) \in W^{1,p}(\Omega)$ and*

$$\frac{\partial F(u)}{\partial x_i} = F'(u) \frac{\partial u}{\partial x_i} \quad \mathcal{L}^N \text{ a.e. } (i = 1, ..., N).$$

(If $\mathcal{L}^N(\Omega) < \infty$, the condition $F(0) = 0$ is unnecessary).

Theorem 6 (trace theorem) *Assume that Ω is bounded, with Lipschitz boundary $\partial\Omega$, and $1 \leq p < \infty$. There exists a bounded linear operator $T : W^{1,p}(\Omega) \to L^p(\partial\Omega)$ such that $Tu = u$ on $\partial\Omega$ for all $u \in W^{1,p} \cap C(\overline{\Omega})$. Furthermore, for all $\phi \in C^1(\mathbb{R}^N;\mathbb{R}^N)$ and $u \in W^{1,p}(\Omega)$,*

$$\int_\Omega u div\phi dx = -\int_\Omega \nabla u \cdot \phi dx + \int_{\partial\Omega} (\phi \cdot \nu) Tu d\mathcal{H}^{N-1},$$

ν denoting the unit outer normal to $\partial\Omega$.

The function Tu is called the **trace** of u on $\partial\Omega$. We interpret Tu as the "boundary values" of u on $\partial\Omega$.

Theorem 7 (Poincaré–Wirtinger inequality) *Let $\Omega \subset \mathbb{R}^N$ be open, bounded and connected, with Lipschitz boundary $\partial\Omega$, $1 \leq p \leq +\infty$. Then there is a constant $C > 0$ depending only on N, Ω and p such that*

$$\left\| u - \frac{1}{|\Omega|}\int_\Omega u(x)dx \right\|_{L^p(\Omega)} \leq C\|\nabla u\|_{L^p(\Omega)}, \quad \forall u \in W^{1,p}(\Omega).$$

Theorem 8 (Sobolev imbedding theorem, *taken from [114], Theorem 12.11, Chapter 12, p. 511) Let $\Omega \subset \mathbb{R}^N$ be a bounded open set with a Lipschitz boundary. One has*

$$\text{If} \quad 1 \leq p < N, \quad \text{then} \quad W^{1,p}(\Omega) \subset L^q(\Omega) \quad \text{for every } q \in [1, p^*],$$

with p^* the Sobolev exponent defined by $\frac{1}{p^*} = \frac{1}{p} - \frac{1}{N}$. It means that for every $q \in [1, p^*]$, there exists a constant C depending only on p, q and Ω such that $\forall u \in W^{1,p}(\Omega)$

$$\|u\|_{L^q(\Omega)} \leq C\,\|u\|_{W^{1,p}(\Omega)}.$$

$$\text{If} \quad p = N, \quad \text{then} \quad W^{1,p}(\Omega) \subset L^q(\Omega) \quad \text{for every } q \in [1, \infty).$$

It means that for every $q \in [1, \infty)$, there exists a constant C depending only on p, q and Ω such that $\forall u \in W^{1,p}(\Omega)$

$$\|u\|_{L^q(\Omega)} \leq C\,\|u\|_{W^{1,p}(\Omega)}.$$

$$\text{If} \quad p > N, \quad \text{then} \quad W^{1,p}(\Omega) \subset C^{0,\alpha}(\bar{\Omega}) \quad \text{for every } \alpha \in [0, 1 - \tfrac{N}{p}],$$

$C^{0,\alpha}(\bar{\Omega})$ denoting the Hölder space defined as the set of functions $u \in C^0(\bar{\Omega})$ such that $\sup\limits_{\substack{x,y\in\bar{\Omega}\\ x\neq y}} \dfrac{|u(x) - u(y)|}{|x-y|^\alpha} < \infty$.

In particular, there exists a constant C depending only on p and Ω such that $\forall u \in W^{1,p}(\Omega)$

$$\|u\|_{L^\infty(\Omega)} \leq C\,\|u\|_{W^{1,p}(\Omega)}.$$

Theorem 9 (Rellich–Kondrachov theorem, *taken from [114], Theorem 12.12, Chapter 12, p. 512) Let $\Omega \subset \mathbb{R}^N$ be a bounded open set with a Lipschitz boundary. One has*

(i) If $1 \leq p < N$, then the imbedding of $W^{1,p}(\Omega)$ into $L^q(\Omega)$ is compact for every $q \in [1, p^*)$. It means that any bounded set of $W^{1,p}(\Omega)$ is precompact (i.e., its closure is compact) in $L^q(\Omega)$ for every $q \in [1, p^*)$.

(ii) If $p = N$, then the imbedding of $W^{1,p}(\Omega)$ into $L^q(\Omega)$ is compact for every $q \in [1, \infty)$.

(iii) If $p > N$, then the imbedding of $W^{1,p}(\Omega)$ into $C^{0,\alpha}(\bar{\Omega})$ is compact for every $\alpha \in [0, 1 - \frac{N}{p}[$.

Sometimes it is useful to generalize the Sobolev spaces $W^{n,p}(\mathbb{R}^N)$ with integer exponent n for the degree of differentiability (and with $p \geq 1$) to Sobolev spaces with real exponent s, denoted by $W^{s,p}(\mathbb{R}^N)$. This can be done using the Fourier transform introduced in Section 2.4. Here we consider the homogeneous version of these spaces, to be used in some of the later chapters.

Definition 10 *Let $s \in \mathbb{R}$, $1 \leq p \leq \infty$. A function or distribution u belongs to the (homogeneous) Sobolev space $\dot{W}^{s,p}(\mathbb{R}^N)$ if $\|u\|_{\dot{W}^{s,p}(\mathbb{R}^N)} := \|((2\pi|\xi|)^s \hat{u}(\xi))^\vee\|_{L^p} < \infty$.*

Equipped with $\|\cdot\|_{\dot{W}^{s,p}(\mathbb{R}^N)}$, $\dot{W}^{s,p}(\mathbb{R}^N)$ becomes a Banach space.

2.6.2 *BV* space

We now introduce the total variation, an essential tool in variational image processing and analysis.

Definition 11 *A function $u \in L^1(\Omega)$ has* **bounded variation** *in Ω if*

$$\|Du\|(\Omega) = |u|_{BV(\Omega)} := \sup\left\{ \int_\Omega u \; div\, \phi\, dx \middle| \phi \in C_c^1(\Omega; \mathbb{R}^N),\ |\phi| \leq 1 \right\} < \infty.$$

We write $BV(\Omega)$ to denote the space of functions of bounded variation and $\|Du\|(\Omega)$ is called the total variation of u in Ω.

An $\mathcal{L}^N$-measurable subset $E \subset \mathbb{R}^N$ has **finite perimeter** *in Ω if $\chi_E \in BV(\Omega)$; furthermore,*

$$\|D\chi_E\|(\Omega) = \sup\left\{ \int_E div\, \phi\, dx \middle| \phi \in C_c^1(\Omega; \mathbb{R}^N),\ |\phi| \leq 1 \right\}$$

is the perimeter $P_\Omega(E)$ of E in Ω.

For $u \in BV(\Omega)$, define $\|u\|_{BV(\Omega)} = \|u\|_{L^1(\Omega)} + \|Du\|(\Omega)$; $\|u\|_{BV(\Omega)}$ is a norm which makes $BV(\Omega)$ a Banach space.

If $u \in W^{1,1}(\Omega)$, then $\|Du\|(\Omega) = \int_\Omega |\nabla u| dx < \infty$, thus $W^{1,1}(\Omega) \subset BV(\Omega)$. The characteristic function χ_E, with E as above, is an example of an element in $BV(\Omega) \setminus W^{1,1}(\Omega)$. If the boundary ∂E is smooth (contained in a C^1-surface of dimension $N-1$), then $P_\Omega(E)$ coincides with the classical $(N-1)$-dimensional surface measure (counting measure if $N = 1$, length if $N = 2$, and surface area if $N = 3$).

Theorem 10 (lower semicontinuity of total variation) *Suppose $u_n \in BV(\Omega)$ $(n = 1, 2, ...)$ and $u_n \to u$ in $L^1(\Omega)$. Then*

$$\|Du\|(\Omega) \leq \liminf_{n\to\infty} \|Du_n\|(\Omega).$$

Theorem 11 (compactness) *Let $\Omega \subset \mathbb{R}^N$ be open and bounded, with Lipschitz boundary $\partial\Omega$. Assume $\{u_n\}_{n=1}^\infty$ is a sequence in $BV(\Omega)$ satisfying $\sup_n \|u_n\|_{BV(\Omega)} < \infty$. Then there exists a subsequence $\{u_{n_j}\}_{j=1}^\infty$ and a function $u \in BV(\Omega)$ such that $u_{n_j} \to u$ in $L^1(\Omega)$ as $j \to \infty$.*

Next we relate the total variation of u and the perimeter of its superlevel sets. For $u : \Omega \to \mathbb{R}$ and $t \in \mathbb{R}$, define $E_t = \{x \in \Omega | \ u(x) > t\}$.

Theorem 12 (Coarea formula for BV functions) *Let $u \in BV(\Omega)$. Then*

(i) E_t has finite perimeter for $\mathcal{L}^1$ a.e. $t \in \mathbb{R}$.

(ii) $\|Du\|(\Omega) = \int_{-\infty}^{\infty} \|D\chi_{E_t}\|(\Omega)dt$.

(iii) Conversely, if $u \in L^1(\Omega)$ and $\int_{-\infty}^{\infty} \|D\chi_{E_t}\|(\Omega)dt < \infty$, then $u \in BV(\Omega)$.

Theorem 13 (embedding theorem) *Let $\Omega \subset \mathbb{R}^N$ be open and bounded, with Lipschitz boundary $\partial\Omega$. Then the embedding $BV(\Omega) \to L^{\frac{N}{N-1}}(\Omega)$ is continuous and $BV(\Omega) \to L^p(\Omega)$ is compact for all $1 \leq p < \frac{N}{N-1}$.*

Theorem 14 (Poincaré–Wirtinger inequality) *Let $\Omega \subset \mathbb{R}^N$ be open, bounded and connected, with Lipschitz boundary $\partial\Omega$. Then there is a constant $C > 0$ depending only on N and Ω such that*

$$\left\| u - \frac{1}{|\Omega|} \int_\Omega u(x)dx \right\|_{L^p(\Omega)} \leq C\|Du\|(\Omega) \quad \forall u \in BV(\Omega), \ 1 \leq p \leq \frac{N}{N-1}.$$

2.6.2.1 Characterization of BV functions

We want to briefly introduce the characterization of BV functions. The measure gradient Du of $u \in BV(\Omega)$ can be decomposed as

$$Du = D^a u + D^s u = D^a u + D^j u + D^c u,$$

where $D^a u$ is the absolutely continuous part of Du and $D^s u = D^j u + D^c u$ is the singular part composed of the jump part $D^j u$ and the Cantor part $D^c u$. Moreover, we can express

$$Du = \nabla u dx + (u^+ - u^-)\vec{N}_u \mathcal{H}^{N-1}|_{S_u} + D^c u, \tag{2.5}$$

where $\nabla u \in L^1(\Omega)$, S_u is of finite $(N-1)$-dimensional Hausdorff measure, $(u^+ - u^-)\vec{N}_u \chi_{S_u} \in L^1(\Omega, \mathcal{H}^{N-1}|_{S_u}, \mathbb{R}^N)$ with u^+ and u^- the traces of u on each side of the jump part S_u, and $\vec{N}_u$ is the unit normal to S_u; finally, the Cantor part $D^c u$ satisfies $D^c u(B) = 0$ for all B such that $\mathcal{H}^{N-1}(B) < \infty$. We can also express the total variation $\|Du\|(\Omega)$ as

$$\|Du\|(\Omega) = \int_\Omega |\nabla u| dx + \int_{S_u} |u^+ - u^-| d\mathcal{H}^{N-1} + \int_{\Omega \setminus S_u} |D^c u|. \tag{2.6}$$

2.6.2.2 SBV functions

For the image segmentation problem, it is necessary to introduce a (closed) subspace of BV, denoted SBV and called the space of special functions of bounded variation. It was introduced by De Giorgi and Ambrosio [121] to provide a weak formulation for some variational problems with free discontinuities, such as the Mumford and Shah segmentation problem (1.5). We refer the reader to [18], [20], [27].

Definition 12 *We say that $u \in BV(\Omega)$ is a* **special function of bounded variation,** *and we write $u \in SBV(\Omega)$, if the Cantor part $D^c u$ of its derivative Du is zero.*

By the above characterization, we obtain that $u \in BV(\Omega)$ is in $SBV(\Omega)$ if and only if

$$Du = D^a u + D^j u = \nabla u dx + (u^+ - u^-)\vec{N}_u \mathcal{H}^{N-1}|_{S_u}.$$

Moreover, it is possible to show that $W^{1,1}(\Omega) \subset SBV(\Omega) \subset BV(\Omega)$ and that both inclusions are strict. Additionally, $SBV(\Omega)$ is a closed subspace of $BV(\Omega)$. An important result of compactness of SBV due to Ambrosio [18] is given below.

Theorem 15 (compactness of SBV) *Let $u_n \in SBV(\Omega)$ be a sequence of functions such that there exists a constant $C > 0$ with $|u_n(x)| \leq C < \infty$ a.e. $x \in \Omega$ and $\int_\Omega |\nabla u_n|^2 dx + \mathcal{H}^{N-1}(S_{u_n}) \leq C$. Then there exists a subsequence u_{n_k} converging a.e. to a function $u \in SBV(\Omega)$. Moreover, ∇u_{n_k} converges weakly in $L^2(\Omega)^N$ to ∇u, and*

$$\mathcal{H}^{N-1}(S_u) \leq \liminf_{n_k \to \infty} \mathcal{H}^{N-1}(S_{u_{n_k}}).$$

2.6.2.3 *BV* functions of one variable

Following [139], we also define for the illustration the space BV of functions of bounded variation in one dimension, using an equivalent definition of the total variation on the line. This notion can be applied to signal restoration.

Let $-\infty \leq a < b \leq \infty$. Suppose that $u \in L^1(a,b)$.

Definition 13 *The* **essential variation** *of u on the interval (a,b) is*

$$ess\ V_a^b u = \sup \Big\{ \sum_{j=1}^{m} |u(x_{j+1}) - u(x_j)| \Big\},$$

where the supremum is taken over all finite partitions $\{a < x_1 < \ldots < x_{m+1} < b\}$, such that each x_j is a point of approximate continuity of u.

Theorem 16 *Suppose $u \in L^1(a,b)$. Then $\|Du\|(a,b) = ess\ V_a^b u$. Thus $u \in BV(a,b)$ if and only if $ess\ V_a^b u < \infty$.*

2.7 Calculus of variations

The variational methods presented in this book are based on techniques within the calculus of variations framework. We refer the reader to Ekeland and Temam [136], Attouch et al. [27], Ambrosio et al. [20], Aubert and Kornprobst [30], Dacorogna [114], among others.

The scope of this section is to investigate the issue of existence of minimizers of problems defined as follows

$$\inf_{u \in V} I(u),$$

with $I(u) = \displaystyle\int_\Omega f(x, u(x), \nabla u(x))\, dx$ and where:

- $\Omega \subset \mathbb{R}^N$, $N \geq 1$ is a bounded open set.
- $u : \Omega \to \mathbb{R}$ is the unknown function (we thus focus on problems written in integral form, considered to be scalar since u is real-valued). Also $\nabla u = \left(\frac{\partial u}{\partial x_i} \right)_{1 \leq i \leq N}$.
- V is the space of admissible functions.
- $f : \Omega \times \mathbb{R} \times \mathbb{R}^N$, $f = f(x, u, \xi)$, is a given function.

Our intent is to provide the reader with the basic tools concerning optimization of such problems in Banach spaces, and in particular, the direct method of the calculus of variations (at least, to get a general existence theorem since as mentioned in [114, p. 3], the direct method in the calculus of variations encompasses the question of the regularity satisfied by a minimizer).

We thus aim to determine whether there exists $\bar{u} \in V$ such that

$$\inf_{u \in V} I(u) = \min_{u \in V} I(u) = I(\bar{u}) \leq I(u), \; \forall u \in V.$$

The progression of the presentation follows the monograph by Dacorogna [114] titled *Direct Methods in the Calculus of Variations* and the book by Aubert and Kornprobst [30].

For problems involving vector-valued functions u, we refer the reader to part II of the monograph by Dacorogna, dedicated to the notion of quasiconvexity which is a weaker definition of convexity.

Before stating the main results, we review some basic facts on Carathéodory functions, convexity and lower semicontinuity largely inspired by [30] and [114].

2.7.1 Carathéodory functions

Definition 14 *(Taken from [114], Definition 3.5, Chapter 3, p. 75) Let $\Omega \subset \mathbb{R}^N$ be an open set and let $f : \Omega \times \mathbb{R}^N \to \mathbb{R} \cup \{+\infty\}$. Then f is said to be a Carathéodory function if*

(i) $\xi \mapsto f(x, \xi)$ is continuous for almost every $x \in \Omega$.

(ii) $x \mapsto f(x, \xi)$ is measurable for every $\xi \in \mathbb{R}^N$.

Remark 3 *(Taken from [114], Remark 3.6, Chapter 3, p. 75) Let $f : \Omega \times \mathbb{R}^m \times \mathbb{R}^M \to \mathbb{R} \cup \{+\infty\}$, $f = f(x, u, \xi)$. When we speak of Carathéodory functions in this context, we consider the variable ξ as playing the role of (u, ξ) and $\mathbb{R}^N = \mathbb{R}^m \times \mathbb{R}^M$.*

In the following, we denote by $(V, \|\cdot\|_V)$ a real Banach space.

2.7.2 Convex functions

Definition 15 *Let us denote by $\mathrm{dom}\, F = \{x \in V, \, F(x) < +\infty\}$.*

(i) The function $F : V \to \bar{\mathbb{R}}$ is said to be convex if

$$F(\lambda x + (1-\lambda)y) \leq \lambda F(x) + (1-\lambda) F(y)$$

for every $(x, y) \in \mathrm{dom}\, F \times \mathrm{dom}\, F$ and every $\lambda \in [0, 1]$.

(ii) The function $F : V \to \bar{\mathbb{R}}$ is said to be strictly convex if

$$F(\lambda x + (1-\lambda)y) < \lambda F(x) + (1-\lambda) F(y)$$

for every $(x, y) \in \mathrm{dom}\, F \times \mathrm{dom}\, F$ with $x \neq y$ and every $\lambda \in (0, 1)$.

2.7.3 Direct method for integral functional problems formulated on Sobolev space $W^{1,p}(\Omega)$

In this subsection, we are concerned with the case for which the space of admissible functions V is defined by $V = u_0 + W_0^{1,p}(\Omega)$ (with $p > 1$ and $u_0 \in W^{1,p}(\Omega)$ a given function), this notation meaning that $u = u_0$ on $\partial\Omega$ and $u \in W^{1,p}(\Omega)$. The considered minimization problem is thus rephrased as

$$\inf_{u \in V = u_0 + W_0^{1,p}(\Omega)} I(u),$$

with $I(u) = \displaystyle\int_\Omega f(x, u(x), \nabla u(x))\, dx$ and where

- $\Omega \subset \mathbb{R}^N$, $N \geq 1$ is a bounded open set.
- $u : \Omega \to \mathbb{R}$ is the unknown function.
- $f : \Omega \times \mathbb{R} \times \mathbb{R}^N$, $f = f(x, u, \xi)$, is a given Carathéodory function.

The direct method of the calculus of variations relies on three steps:

(i) One constructs a minimizing sequence (u_n) of V (which always exists by the sequential definition of the infimum), that is, a sequence such that $\lim_{n \to +\infty} I(u_n) = \inf_{u \in V} I(u)$.

(ii) One obtains a uniform bound on $\|u_n\|_V$ by deriving a coercivity inequality. Indeed, if I is coercive, meaning that $\lim_{\|u\|_V \to +\infty} I(u) = +\infty$, this uniform bound is straightforwardly extracted. (Arguing by contradiction, let us assume that $\forall C > 0$, $\exists n \in \mathbb{N}$, $\|u_n\|_V > C$. We prove, by construction, that there exists a subsequence (u_{n_k}) of (u_n) such that $\lim_{k \to +\infty} \|u_{n_k}\|_V = +\infty$ and consequently, $\lim_{k \to +\infty} I(u_{n_k}) = +\infty$ owing to the coercivity of I, which contradicts the fact that $\lim_{k \to +\infty} I(u_{n_k}) = \inf_{u \in V} I(u)$). If V is reflexive, from Theorem 4, one can thus find $\bar{u} \in V$ and a subsequence (u_{n_k}) of (u_n) such that $u_{n_k} \underset{V}{\rightharpoonup} \bar{u}$ when $k \to +\infty$.

(iii) One aims to prove that $\bar{u}$ is a minimizer of I. It suffices to have the inequality

$$I(\bar{u}) \leq \liminf_{k \to +\infty} I(u_{n_k}).$$

It turns out that this last property called sequential weak lower semicontinuity, that we investigate hereafter, is related to the convexity of the mapping $\xi \mapsto f(x, u, \xi)$.

Definition 16 (sequential weak lower semicontinuity, *taken from [114], Definition 3.1, Chapter 3, p. 74*) *Let $p \geq 1$ and Ω, u, f be as above. We say that I is sequentially weakly lower semicontinuous in $W^{1,p}(\Omega)$ if for every sequence $u_n \underset{W^{1,p}(\Omega)}{\rightharpoonup} \bar{u}$, then*

$$I(\bar{u}) \leq \liminf_{n \to +\infty} I(u_n).$$

If $p = \infty$, we say that I is sequentially weak $*$ lower semicontinuous in $W^{1,\infty}(\Omega)$ if the same inequality holds for every sequence $u_n \underset{W^{1,\infty}(\Omega)}{\overset{*}{\rightharpoonup}} \bar{u}$.

We now give a necessary and sufficient condition ensuring sequential weak lower semicontinuity of I.

Theorem 17 (necessary condition, *taken from [114], Theorem 3.15, Chapter 3, p. 84*) *Let $\Omega \subset \mathbb{R}^N$ be an open set, $f : \Omega \times \mathbb{R} \times \mathbb{R}^N \to \mathbb{R}$ be a Carathéodory function satisfying, for almost every $x \in \Omega$ and for every $(u, \xi) \in \mathbb{R} \times \mathbb{R}^N$,*

$$|f(x, u, \xi)| \leq a(x) + b(u, \xi),$$

where $a, b \geq 0$, $a \in L^1(\mathbb{R}^N)$ and $b \in C^0(\mathbb{R} \times \mathbb{R}^N)$. Let

$$I(u) = \int_\Omega f(x, u(x), \nabla u(x))\, dx,$$

and assume that there exists $\tilde{u} \in W^{1,\infty}(\Omega)$ such that

$$|I(\tilde{u})| < \infty.$$

If I is weakly $*$ lower semicontinuous in $W^{1,\infty}(\Omega)$, then $\xi \mapsto f(x, u, \xi)$ is convex for almost every $x \in \Omega$ and for every $u \in \mathbb{R}$.

Remark 4 *If Ω is bounded, as any sequence $u_n \underset{W^{1,\infty}(\Omega)}{\overset{*}{\rightharpoonup}} \bar{u}$ is such that $u_n \underset{W^{1,p}(\Omega)}{\rightharpoonup} \bar{u}$ with $p \geq 1$, the convexity of f is also necessary for weak lower semicontinuity of I in $W^{1,p}(\Omega)$.*

Theorem 18 (sufficient condition, *taken from [114], Corollary 3.24, Chapter 3, p. 97*) *Let $p \geq 1$, $\Omega \subset \mathbb{R}^N$ be a bounded open set with Lipschitz boundary and $f : \Omega \times \mathbb{R} \times \mathbb{R}^N \to \mathbb{R} \cup \{+\infty\}$, $f = f(x, u, \xi)$, be a Carathéodory function satisfying*

$$f(x, u, \xi) \geq \langle a(x), \xi \rangle + b(x) + c|u|^r$$

for almost every $x \in \Omega$, for every $(u, \xi) \in \mathbb{R} \times \mathbb{R}^N$, for some $a \in L^{p'}(\Omega, \mathbb{R}^N)$, $\dfrac{1}{p} + \dfrac{1}{p'} = 1$, $b \in L^1(\Omega)$, $c \in \mathbb{R}$, $1 \leq r < \dfrac{Np}{N-p}$ if $p < N$ and $1 \leq r < \infty$ if $p \geq N$, and where $\langle \cdot, \cdot \rangle$ denotes the inner product in $\mathbb{R}^N$. Let

$$I(u) = \int_\Omega f(x, u(x), \nabla u(x))\, dx.$$

Assume that $\xi \mapsto f(x,u,\xi)$ *is convex and that*

$$u_n \underset{W^{1,p}(\Omega)}{\rightharpoonup} \bar{u}.$$

Then

$$I(\bar{u}) \leq \liminf_{n\to+\infty} I(u_n).$$

Remark 5 *Gathering the results of Theorem 17 and Thereom 18 indicates that a necessary and sufficient condition for* I *to be weakly lower semicontinuous in* $W^{1,p}(\Omega)$ *is that* $\xi \mapsto f(x,u,\xi)$ *be convex.*

Remark 6 *Theorem 18 holds if one replaces weak convergence in* $W^{1,p}(\Omega)$ *by weak* $*$ *convergence in* $W^{1,\infty}(\Omega)$.

Equipped with these preliminary results, we are now in position to state the main existence theorem.

Theorem 19 (existence of minimizers, *taken from [114], Theorem 3.30, Chapter 3, p. 106*) *Let* Ω *be a bounded open set of* $\mathbb{R}^N$ *with a Lipschitz boundary. Let* $f : \Omega \times \mathbb{R} \times \mathbb{R}^N \to \mathbb{R} \cup \{+\infty\}$ *be a Carathéodory function satisfying the coercivity condition*

$$f(x,u,\xi) \geq \alpha_1 |\xi|^p + \alpha_2 |u|^q + \alpha_3(x) \tag{2.7}$$

for almost every $x \in \Omega$ *and for every* $(u,\xi) \in \mathbb{R} \times \mathbb{R}^N$, *and for* $\alpha_3 \in L^1(\Omega)$, $\alpha_2 \in \mathbb{R}$, $\alpha_1 > 0$ *and* $p > q \geq 1$. *Assume that* $\xi \mapsto f(x,u,\xi)$ *is convex. Let*

$$I(u) = \int_\Omega f(x,u(x),\nabla u(x))\,dx.$$

Assuming that $I(u_0) < \infty$ *(*I *is said to be proper), then*

$$m = \inf_{u \in u_0 + W_0^{1,p}(\Omega)} I(u)$$

is attained. Furthermore, if $(u,\xi) \mapsto f(x,u,\xi)$ *is strictly convex for almost every* $x \in \Omega$, *then the minimizer is unique.*

Remark 7 *(i) The coercivity inequality combined with the fact that* $I(u_0) < \infty$ *allows to conclude that the infimum is finite.*

(ii) Considering a minimizing sequence (u_n) *of* $u_0 + W_0^{1,p}(\Omega)$, *one obtains a uniform bound on* $\|u_n\|_{W^{1,p}(\Omega)}$ *using the coercivity inequality (2.7), Hölder's inequality and for instance, the following generalized Poincaré inequality.*

Theorem 20 **(generalized Poincaré inequality**, *taken from [125], p. 106*) *Let Ω be a Lipschitz bounded domain in $\mathbb{R}^N$. Let $p \in [1, +\infty)$ and let $\mathcal{N}$ be a continuous seminorm on $W^{1,p}(\Omega)$, that is, a norm on the constant functions. Then there exists a constant $C > 0$ that depends only on Ω, N, p, such that*

$$\|u\|_{W^{1,p}(\Omega)} \leq C \left(\left(\int_\Omega |\nabla u(x)|^p \, dx \right)^{1/p} + \mathcal{N}(u) \right).$$

It suffices to apply this result to $\mathcal{N}(u) = \displaystyle\int_{\partial\Omega} |u(x)| \, d\sigma$, when Ω is a C^1 open set.

(iii) We can thus extract a subsequence of (u_n), still denoted by (u_n), and find $\bar{u} \in W^{1,p}(\Omega)$ such that $u_n \underset{W^{1,p}(\Omega)}{\rightharpoonup} \bar{u}$.

(iv) Appealing to Theorem 18, one gets that $\liminf\limits_{n\to+\infty} I(u_n) \geq I(\bar{u})$.

(v) Due to the continuity of the trace map, $\bar{u}_{|\partial\Omega} = u_0$. Note that the set $\{u \in W^{1,p}(\Omega),\, u = u_0 \text{ on } \partial\Omega\}$ is a strongly closed convex set.

Remark 8 *Of course, the proposed methodology can be enlarged and adapted to more general contexts: when the functional is not necessarily written in an integral form or when it is defined on more complex functional spaces. As an example, let us consider a functional minimization problem phrased in terms of the total variation among others, and well-defined on $BV(\Omega)$ (ensured thanks to the continuous embedding $BV(\Omega) \to L^{\frac{N}{N-1}}(\Omega)$). The first two steps are similar to the proposed methodology. Once a uniform bound is obtained on $\|u_n\|_{BV(\Omega)}$, one can thus extract a subsequence (u_{n_j}) of (u_n) and find $\bar{u} \in BV(\Omega)$ such that $u_{n_j} \underset{L^1(\Omega)}{\longrightarrow} \bar{u}$ when $j \to +\infty$ and $\|Du\|(\Omega) \leq \liminf\limits_{j\to+\infty} \|Du_{n_j}\|(\Omega)$ (owing to the compactness property stated in Theorem 11). Two ingredients might be used to prove that $\bar{u}$ is a minimizer (last step of the direct method):*

(i) If $(x_n)_{n\in\mathbb{N}}$ and $(y_n)_{n\in\mathbb{N}}$ are real-valued sequences, one has

$$\liminf_{n\to+\infty} x_n + \liminf_{n\to+\infty} y_n \leq \liminf_{n\to+\infty} (x_n + y_n).$$

(ii) If $J : V \to \mathbb{R}\cup\{+\infty\}$ (with V a Banach space) is convex, then J is weakly lower semicontinuous if and only if J is strongly lower semicontinuous.

2.7.4 Euler–Lagrange equations

Once the existence of minimizers is achieved, it remains to state the optimality conditions. We follow again the notations of the monograph by Dacorogna [114] and aim to compute the Gâteaux derivative $I'(u)$ of $I(u) =$

$\int_{\Omega} f(x, u(x), \nabla u(x))\, dx$. With $f = f(x, u, \xi)$, we denote by $f_u = \dfrac{\partial f}{\partial u}$ and by $D_\xi f = (f_{\xi_\alpha})_{1 \leq \alpha \leq N} \in \mathbb{R}^N$, where $f_{\xi_\alpha} = \dfrac{\partial f}{\partial \xi_\alpha}$. We are now able to state the main theorem.

Theorem 21 (weak form of Euler–Lagrange equation *partially taken from [114], Theorem 3.37*) *Let $f : \Omega \times \mathbb{R} \times \mathbb{R}^N \to \mathbb{R}$, $f = f(x, u, \xi)$ be a Carathéodory function satisfying, for almost every $x \in \Omega$, for every $(u, \xi) \in \mathbb{R} \times \mathbb{R}^N$*

$$|f(x, u, \xi)| \leq \alpha_1(x) + \beta\left(|u|^p + |\xi|^p\right),$$

where $\alpha_1 \in L^1(\Omega)$ and $\beta \geq 0$. This growth condition on f is complemented by growth conditions on the derivatives of f. The functions $f_u, f_{\xi_\alpha} : \Omega \times \mathbb{R} \times \mathbb{R}^N \to \mathbb{R}$ are assumed to be Carathéodory functions for every $\alpha = 1, \cdots, N$. Moreover, they are assumed to satisfy, for almost every $x \in \Omega$ and for every $(u, \xi) \in \mathbb{R} \times \mathbb{R}^N$,

$$|f_u(x, u, \xi)|, |D_\xi f(x, u, \xi)| \leq \alpha_1(x) + \beta\left(|u|^p + |\xi|^p\right),$$

where $\alpha_1 \in L^1(\Omega)$ and $\beta \geq 0$.

For $\varphi : \Omega \to \mathbb{R}$, let

$$\begin{aligned} L(u, \varphi) &= \int_{\Omega} \left\{ \left[\sum_{\alpha=1}^{N} \frac{\partial f}{\partial \xi_\alpha}(x, u, \nabla u) \frac{\partial \varphi}{\partial x_\alpha} \right] + \frac{\partial f}{\partial u}(x, u, \nabla u)\varphi \right\} dx, \\ &= \int_{\Omega} \left\{ \langle D_\xi f(x, u, \nabla u), \nabla \varphi \rangle + \frac{\partial f}{\partial u}(x, u, \nabla u)\varphi \right\} dx. \end{aligned}$$

Assume that $\bar{u} \in u_0 + W_0^{1,p}(\Omega)$ is a minimizer of (P) defined as

$$\inf_{u \in u_0 + W_0^{1,p}(\Omega)} I(u) = \int_{\Omega} f(x, u(x), \nabla u(x))\, dx. \tag{P}$$

Then

$$L(\bar{u}, \varphi) = 0 \text{ for every } \varphi \in C_0^\infty(\Omega). \tag{2.8}$$

Conversely, if $\bar{u}$ satisfies (2.8) and if $(u, \xi) \mapsto f(x, u, \xi)$ is convex for almost every $x \in \Omega$, then $\bar{u}$ is a minimizer of (P).

Remark 9 *The growth conditions on f and its derivatives can be improved. For further details, we refer the reader to [114, Section 3.4.2, Chapter 3].*

Theorem 22 (Euler–Lagrange equation, *taken from [114], Corollary 3.38, Chapter 3, p. 114*) *Let $f : \Omega \times \mathbb{R} \times \mathbb{R}^N \to \mathbb{R}$ be a C^2-function. Assume that $\bar{u} \in C^2(\bar{\Omega})$ is a minimizer of (P) defined as*

$$\inf_{u \in u_0 + W_0^{1,p}(\Omega)} I(u) = \int_{\Omega} f(x, u(x), \nabla u(x))\, dx. \tag{P}$$

Then $\bar{u}$ satisfies, for every $x \in \Omega$,

$$\sum_{\alpha=1}^{N} \frac{\partial}{\partial x_\alpha}\left[\frac{\partial f}{\partial \xi_\alpha}(x, \bar{u}, \nabla \bar{u})\right] = \frac{\partial f}{\partial u}(x, \bar{u}, \nabla \bar{u}).$$

2.7.5 Lax–Milgram theorem

In this subsection, we recall Lax–Milgram theorem and relate it to the general theorem of existence of minimizers. We assume that V is a Hilbert space, that is, a vector space endowed with an inner product (u, v) and complete for the associated norm $(u, u)^{\frac{1}{2}} = \|u\|_V$.

Theorem 23 (Lax–Milgram theorem) *Let $a : V \times V \to \mathbb{R}$ be a V-elliptic continuous bilinear form. (The property of V-ellipticity means that $\exists \alpha > 0$, such that $\forall v \in V$, $a(v, v) \geq \alpha \|v\|_V^2$). Then for all $L \in V'$, there exists a unique $u \in V$ such that*

$$a(u, v) = L(v) \text{ for every } v \in V.$$

If moreover a is symmetric, then u is characterized by

$$u \in V \text{ and } \frac{1}{2}a(u, u) - L(u) = \min_{v \in V}\left\{\frac{1}{2}a(v, v) - L(v)\right\}.$$

We now relate Lax–Milgram theorem with the direct method of the calculus of variations in the case where a is symmetric. Let us denote

$$I(v) = \frac{1}{2}a(v, v) - L(v).$$

Owing to the coercivity of a and the continuity of L, one gets the following coercivity inequality

$$I(v) \geq \frac{\alpha}{2}\|v\|_V^2 - \|L\|_{V'}\|v\|_V, \tag{2.9}$$

ensuring that the infimum is finite.

Let us now consider a minimizing sequence (v_n) of V. Inequality (2.9) allows us to conclude that, for n large enough,

$$\frac{\alpha}{2}\|v_n\|_V^2 - \|L\|_{V'}\|v_n\|_V \leq m + 1,$$

with $m = \inf\limits_{v \in V} I(v)$. It is then not difficult to see that $\|v_n\|_V \leq \beta$ with β the uniform bound defined by $\beta = \frac{\|L\|_{V'} + \sqrt{\|L\|_{V'}^2 + 2\alpha\,(m+1)}}{\alpha}$. The space V being reflexive, one can thus extract a subsequence of (v_n), still denoted by (v_n) and find $u \in V$ such that $v_n \underset{V}{\rightharpoonup} u$. Functional I is obviously convex, continuous for the strong topology so strongly lower semicontinous. From Remark 8 (ii), it follows that I is weakly lower semicontinous and the existence of minimizers is guaranteed. The uniqueness results from the strict convexity of I. Indeed, if one considers $(u, v) \in V \times V$ and $\lambda \in (0, 1)$, a mere computation gives that

$$\begin{aligned}\lambda I(u) + (1-\lambda) I(v) - I(\lambda u + (1-\lambda)v) &= \frac{\lambda(1-\lambda)}{2} a(u-v, u-v), \\ &\geq \frac{\lambda(1-\lambda)\alpha}{2} \|u-v\|_V^2,\end{aligned}$$

due to the property of V-ellipticity of a. The result is immediate, this last bound vanishing if and only if $u = v$.

Ultimately, $u \in V$ is the unique element such that

$$I(u) = \inf_{v \in V} I(v).$$

By definition, $\forall v \in V$ and $\forall \lambda \in \mathbb{R}$,

$$I(u) \leq I(u + \lambda v). \tag{2.10}$$

The bilinear form a being assumed symmetric, inequality (2.10) amounts to

$$\frac{\lambda^2}{2} a(v, v) + \lambda\,(a(u, v) - L(v)) \geq 0,$$

for every $\lambda \in \mathbb{R}$.

Let us assume in a first step that $\lambda > 0$. Dividing the previous inequality by λ and letting it tend to 0 yields

$$a(u, v) - L(v) \geq 0,$$

for every $v \in V$. Similarly, taking $\lambda < 0$ and proceeding as above yields to

$$a(u, v) - L(v) \leq 0,$$

which, combined with the previous inequality, gives $a(u, v) = L(v)$, $\forall v \in V$. In conclusion, if u is the unique minimizer of $\inf_{v \in V} I(v)$, then it satisfies $a(u, v) = L(v)$, $\forall v \in V$. The converse is straightforward.

2.7.6 Euler–Lagrange equations and energy decrease

The basis of many models for image processing and analysis is an energy minimization problem that can formally be expressed as

$$\inf_{u_1, \ldots, u_M} \int_\Omega f(u_1, \ldots, u_M, \nabla u_1, \ldots, \nabla u_M) dx, \tag{2.11}$$

where $\Omega \subset \mathbb{R}^N$, $x = (x_1, \ldots, x_N) \in \mathbb{R}^N$. The problem has M unknowns belonging to some space of functions. We assume that f is differentiable and that the functions u_i take real values. We use the notations $u = (u_1, ..., u_M)$, $\nabla u = (\nabla u_1, ..., \nabla u_M)$, and $(u_i)_{x_j} = \frac{\partial u_i}{\partial x_j}$.

Using gradient descent and embedding the Euler–Lagrange equations (described in Section 2.7.4 and extended here to the vector-valued case) in a dynamic scheme with an artificial time $t \geq 0$, we associate the following time-dependent coupled PDE's, for $1 \leq i \leq M$:

$$\frac{\partial u_i}{\partial t} = -\frac{\partial f(u, \nabla u)}{\partial u_i} + \sum_{j=1}^{N} \frac{\partial}{\partial x_j}\Big(\frac{\partial f(u, \nabla u)}{\partial((u_i)_{x_j})}\Big), \tag{2.12}$$

with initial conditions $u_i(0, x) = u_{0,i}(x)$ in Ω. Thus now the functions u_i depend on both time $t \geq 0$ and space $x \in \Omega$. On the boundary $\partial\Omega$, we assume free boundary conditions in the form $\sum_{j=1}^{N} \frac{\partial f(u,\nabla u)}{\partial((u_i)_{x_j})} n_j = 0$ for all $t > 0$, where $\vec{n} = (n_1, ..., n_N)$ is the exterior unit normal to $\partial\Omega$.

We formally compute

$$\frac{d}{dt}\int_\Omega f(u_1, ..., u_M, \nabla u_1, ..., \nabla u_M)dx = \frac{d}{dt}\int_\Omega f(u, \nabla u)dx$$

and we show that this quantity is always negative or zero, therefore the energy is decreasing in time:

$$\begin{aligned}
&\frac{d}{dt}\int_\Omega f(u, \nabla u)dx \\
&= \sum_{i=1}^{M}\int_\Omega \Big(\frac{\partial f(u, \nabla u)}{\partial u_i}\Big)\Big(\frac{\partial u_i}{\partial t}\Big)dx + \sum_{i=1}^{M}\int_\Omega \Big[\sum_{j=1}^{N}\Big(\frac{\partial f(u, \nabla u)}{\partial((u_i)_{x_j})}\Big)\Big(\frac{\partial (u_i)_{x_j}}{\partial t}\Big)\Big]dx, \\
&= \sum_{i=1}^{M}\int_\Omega \Big(\frac{\partial f(u, \nabla u)}{\partial u_i}\Big)\Big(\frac{\partial u_i}{\partial t}\Big)dx + \sum_{i=1}^{M}\int_\Omega \Big[\sum_{j=1}^{N}\Big(\frac{\partial f(u, \nabla u)}{\partial((u_i)_{x_j})}\Big)\Big(\frac{\partial (u_i)_{t}}{\partial x_j}\Big)\Big]dx, \\
&= \sum_{i=1}^{M}\int_\Omega \Big(\frac{\partial f(u, \nabla u)}{\partial u_i}\Big)\Big(\frac{\partial u_i}{\partial t}\Big)dx \\
&+ \sum_{i=1}^{M}\int_\Omega \Big\{-\sum_{j=1}^{N}\Big[\frac{\partial}{\partial x_j}\Big(\frac{\partial f(u, \nabla u)}{\partial((u_i)_{x_j})}\Big)\Big]\Big(\frac{\partial u_i}{\partial t}\Big)\Big\}dx \\
&+ \sum_{i=1}^{M}\Big\{\int_{\partial\Omega}\Big(\frac{\partial u_i}{\partial t}\Big)\Big(\sum_{j=1}^{N}\frac{\partial f(u, \nabla u)}{\partial((u_i)_{x_j})}n_j\Big)dS\Big\}, \\
&= \sum_{i=1}^{M}\int_\Omega \Big(\frac{\partial u_i}{\partial t}\Big)\Big[\frac{\partial f(u, \nabla u)}{\partial u_i} - \sum_{j=1}^{N}\frac{\partial}{\partial x_j}\Big(\frac{\partial f(u, \nabla u)}{\partial((u_i)_{x_j})}\Big)\Big]dx, \\
&= -\sum_{i=1}^{M}\int_\Omega \Big(\frac{\partial u_i}{\partial t}\Big)^2 dx \leq 0.
\end{aligned}$$

The time dependent system of PDEs (2.12) will be used in the discrete case to approximate the solution to the minimization problem. In other words, under some assumptions, the solution $u(t, \cdot)$ of (2.12) should approximate a minimizer $\bar{u}(\cdot)$ of (2.11) as t increases.

2.7.7 Γ–convergence

In variational image processing, the notion of Γ–convergence appears for instance when showing that phase-field, elliptic, or other approximations converge to the weak formulation of the Mumford-Shah segmentation functional. We refer the interested reader to Dal Maso [116] for a comprehensive introduction to Γ–convergence.

Definition 17 *Let $X = (X, D)$ be a metric space. We say that a sequence $F_j : X \to [-\infty, +\infty]$ Γ-converges to $F : X \to [-\infty, +\infty]$ (as $j \to \infty$) if for all $u \in X$ we have*

(i) (liminf inequality) for every sequence $(u_j) \subset X$ converging to u,

$$F(u) \leq \liminf_j F_j(u_j); \tag{2.13}$$

(ii) (existence of a recovery sequence) there exists a sequence $(u_j) \subset X$ converging to u such that

$$F(u) \geq \limsup_j F_j(u_j),$$

or, equivalently by (2.13),

$$F(u) = \lim_j F_j(u_j).$$

The function F is called the Γ-limit of (F_j) (with respect to D), and we write $F = \Gamma - \lim_j F_j$.

The following fundamental theorem is essential in the convergence of some of the approximations.

Theorem 24 *(fundamental theorem of Γ–convergence) Let us suppose that $F = \Gamma - \lim_j F_j$, and let a compact set $C \subset X$ exist such that $\inf_X F_j = \inf_C F_j$ for all j. Then there is minimum of F over X such that*

$$\min_X F = \lim_j \inf_X F_j,$$

and if $(u_j) \subset X$ is a converging sequence such that $\lim_j F_j(u_j) = \lim_j \inf_X F_j$, then its limit is a minimum point of F.

2.8 Geometric curve evolution

We review here some notations, terminology and properties of differential geometry of curves and surfaces. These play an important role in image segmentation and boundary detection by curve and surface evolution, as we will see.

2.8.1 Parametrized curves

Let $C = C(q) = (C_1(q), C_2(q))$ (for $q \in I = [a, b]$) be a planar oriented curve in $\mathbb{R}^2$. Assume that C is a regular curve, that is, C_1 and C_2 are differentiable and $|C'(q)| := |(C_1'(q), C_2'(q))| > 0$ at all points $q \in I$.

The tangent vector to the curve at $C(q)$ is $\vec{T}(q) = C'(q) = (C_1'(q), C_2'(q))$. The normal vector to the curve at $C(q)$ is $\vec{N}(q) = (-C_2'(q), C_1'(q))$ (with $(\vec{T}, \vec{N}) = \frac{\pi}{2}$). The arc length of C (or curvilinear abscissa) between a and $q \in (a, b)$ is $s(q) = \int_a^q |C'(r)|dr = \int_a^q \sqrt{C_1'(r)^2 + C_2'(r)^2}dr$.

The regular curve C can be parameterized by arc length s. Then, $\vec{T}(s) = \frac{dC}{ds}(s)$ is a unit vector, $|\vec{T}(s)| = 1$. The curvature tensor is defined by $\frac{d\vec{T}}{ds}(s) = \frac{d^2C}{ds^2}(s)$. It is possible to show that the curvature tensor is collinear to $\vec{N}(s)$, thus $\frac{d\vec{T}}{ds}(s) = k(s)\vec{N}(s)$. For any parameterization we have the curvature

$$k(q) = \frac{C_1'(q)C_2''(q) - C_2'(q)C_1''(q)}{((C_1'(q))^2 + (C_2'(q))^2)^{3/2}}.$$

The following formula will also be useful:

$$\frac{1}{|C'(q)|}\frac{d}{dq}\Big(\frac{C'(q)}{|C'(q)|}\Big) = k(q)\frac{\vec{N}(q)}{|\vec{N}(q)|}.$$

2.8.2 Area and length minimization

Assume that we have a family of regular and smooth curves $C(q,t) = (C_1(q,t), C_2(q,t))$ for $t \geq 0$, $q \in I = [a, b]$ such that

$$C(a,t) = C(b,t), \ \ \frac{\partial C}{\partial q}(a,t) = \frac{\partial C}{\partial q}(b,t).$$

The *length functional* is

$$L(t) = \int_a^b \Big|\frac{\partial C}{\partial q}\Big|dq = \int_a^b \sqrt{\Big(\frac{\partial C_1}{\partial q}\Big)^2 + \Big(\frac{\partial C_2}{\partial q}\Big)^2}\,dq.$$

We compute the time derivative of L,

$$L'(t) = \int_a^b \frac{\langle \frac{\partial C}{\partial q}, \frac{\partial^2 C}{\partial q \partial t} \rangle}{|\frac{\partial C}{\partial q}|} dq.$$

Applying integration by parts yields

$$\begin{aligned}
L'(t) &= -\int_a^b \langle \frac{\partial}{\partial q}\Big(\frac{\frac{\partial C}{\partial q}}{|\frac{\partial C}{\partial q}|}\Big), \frac{\partial C}{\partial t} \rangle dq, \\
&= -\int_a^b \langle \Big(\frac{\partial}{\partial q}\Big(\frac{\frac{\partial C_1}{\partial q}}{|\frac{\partial C}{\partial q}|}\Big), \frac{\partial}{\partial q}\Big(\frac{\frac{\partial C_2}{\partial q}}{|\frac{\partial C}{\partial q}|}\Big)\Big), \frac{\partial C}{\partial t} \rangle dq, \\
&= -\int_a^b |\frac{\partial C}{\partial q}| \langle \frac{1}{|\frac{\partial C}{\partial q}|} \frac{\partial}{\partial q}\Big(\frac{\frac{\partial C}{\partial q}}{|\frac{\partial C}{\partial q}|}\Big), \frac{\partial C}{\partial t} \rangle dq, \\
&= -\int_a^b |\frac{\partial C}{\partial q}| \langle k(q) \frac{\vec{N}(q)}{|\vec{N}(q)|}, \frac{\partial C}{\partial t} \rangle dq.
\end{aligned}$$

This last expression decreases most rapidly if C satisfies

$$\frac{\partial C}{\partial t} = k(q) \frac{\vec{N}(q)}{|\vec{N}(q)|}.$$

If the curve $C = C(s,t)$ is parameterized by arc length s, then the curve shortening flow becomes

$$\frac{\partial C}{\partial t} = k(s) \vec{N}(s),$$

where $k(s)$ is the curvature at $C(s)$ and $\vec{N}(s)$ is the unit normal. In other words, motion of the curve in the normal direction by curvature shortens its length. The reader is referred to [150] for more details on this topic.

The *area functional* is

$$A(t) = -\frac{1}{2} \int_a^b \langle C(q,t), \vec{N}(q) \rangle dq = -\frac{1}{2} \int_a^b \langle C(q,t), \Big(-\frac{\partial C_2}{\partial q}, \frac{\partial C_1}{\partial q} \Big) \rangle dq.$$

Then

$$\begin{aligned}
A'(t) &= -\frac{1}{2}\int_a^b \langle \frac{\partial C}{\partial t}, \vec{N}\rangle dq - \frac{1}{2}\int_a^b \langle C(q,t), \Big(-\frac{\partial^2 C_2}{\partial t \partial q}, \frac{\partial^2 C_1}{\partial t \partial q}\Big)\rangle dq,\\
&= -\frac{1}{2}\int_a^b \langle \frac{\partial C}{\partial t}, \Big(-\frac{\partial C_2}{\partial q}, \frac{\partial C_1}{\partial q}\Big)\rangle dq\\
&\quad + \frac{1}{2}\int_a^b \langle \Big(-\frac{\partial C_1}{\partial q}, \frac{\partial C_2}{\partial q}\Big), \Big(\frac{\partial C_2}{\partial t}, \frac{\partial C_1}{\partial t}\Big)\rangle dq,\\
&= -\int_a^b \langle \frac{\partial C}{\partial t}, \Big(-\frac{\partial C_2}{\partial q}, \frac{\partial C_1}{\partial q}\Big)\rangle dq,\\
&= -\int_a^b \langle \frac{\partial C}{\partial t}, \vec{N}\rangle dq,\\
&= -\int_a^b \langle \frac{\partial C}{\partial t}, \frac{\vec{N}}{|\vec{N}|}\rangle |\vec{N}| dq.
\end{aligned}$$

The last expression decreases most rapidly if C solves

$$\frac{\partial C}{\partial t} = \vec{N},$$

where in the last line $\vec{N}$ is the unit normal. We conclude that the area minimizing flow is the motion in the normal direction by a constant.

2.9 Variational level set methods

The level set approach introduced and developed by Osher and Sethian [254], [285], [286], [256] for curve and surface evolution provides a powerful and elegant tool for image segmentation. Evolving curves and surfaces are represented by the zero level set of a Lipschitz continuous function ϕ, which must solve a time-dependent (geometric) partial differential equation. Change of topology, such as merging or breaking of the curve or surface is automatically handled by the formulation. Numerical discretizations are performed on a fixed rectangular grid. We present here the necessary terminology on variational level sets inspired from the work of Zhao et al. [352] on a variational level set approach for motion of triple junctions in the plane.

Let $\phi : \Omega \to \mathbb{R}$ be a Lipschitz continuous function. We will use the one-dimensional Heaviside function $H : \mathbb{R} \to \mathbb{R}$, defined by

$$H(z) = \begin{cases} 1 \text{ if } z \geq 0 \\ 0 \text{ if } z < 0 \end{cases},$$

and its weak distributional derivative $\delta = H'$. In practice, we may need to work with smooth approximations of the Heaviside and δ functions. Here we will use the following C^∞ approximations as $\epsilon \to 0$ given by [92], [94],

$$H_\epsilon(z) = \frac{1}{2}\Big[1 + \frac{2}{\pi}\arctan\Big(\frac{z}{\epsilon}\Big)\Big], \quad \delta_\epsilon = H'_\epsilon.$$

The area (or the volume) of the region $\{x \in \Omega : \phi(x) > 0\}$ is

$$A\{x \in \Omega : \phi(x) > 0\} = \int_\Omega H(\phi(x))dx,$$

and for a level parameter $l \in \mathbb{R}$, the area (or volume) of the region $\{x \in \Omega : \phi(x) > l\}$ is

$$A\{x \in \Omega : \phi(x) > l\} = \int_\Omega H(\phi(x) - l)dx.$$

The perimeter (or more generally the surface area) of the region $\{x \in \Omega : \phi(x) > 0\}$ is given by

$$L\{x \in \Omega : \phi(x) > 0\} = \int_\Omega |DH(\phi)|,$$

which is the total variation of $H(\phi)$ in Ω, and the perimeter (or surface area) of $\{x \in \Omega : \phi(x) > l\}$ is

$$L\{x \in \Omega : \phi(x) > l\} = \int_\Omega |DH(\phi - l)|.$$

Given any image function $g : \Omega \to \mathbb{R}$, the averages of g over the (nonempty) regions $\{x \in \Omega : \phi(x) > 0\}$ and $\{x \in \Omega : \phi(x) < 0\}$ respectively, are

$$\frac{\int_\Omega g(x)H(\phi(x))dx}{\int_\Omega H(\phi(x))dx} \text{ and } \frac{\int_\Omega g(x)(1 - H(\phi(x)))dx}{\int_\Omega (1 - H(\phi(x)))dx} = \frac{\int_\Omega g(x)H(-\phi(x))dx}{\int_\Omega H(-\phi(x))dx}.$$

More generally, for a given level parameter $l \in \mathbb{R}$, the averages of g over the corresponding (nonempty) regions $\{x \in \Omega : \phi(x) > l\}$ and $\{x \in \Omega : \phi(x) < l\}$ respectively, are

$$\frac{\int_\Omega g(x)H(\phi(x) - l)dx}{\int_\Omega H(\phi(x) - l)dx} \text{ and } \frac{\int_\Omega g(x)H(l - \phi(x))dx}{\int_\Omega H(l - \phi(x))dx}.$$

We prove next that if H and δ are substituted by the above C^∞ approximations H_ϵ, δ_ϵ as $\epsilon \to 0$, we obtain approximations of the area and length (perimeter) measures. We obviously have that $H_\epsilon(z) \to H(z)$ for all $z \in \mathbb{R}$, as $\epsilon \to 0$, and that the approximating area term $A_\epsilon(\phi) = \int_\Omega H_\epsilon(\phi(x))dx$ converges to $A(\phi) = \int_\Omega H(\phi(x))dx$.

Generalizing a result of Samson et al. [280], [279] we can show as in [104] that our approximating functional $L_\epsilon(\phi) = \int_\Omega |\nabla H_\epsilon(\phi)|dx = \int_\Omega \delta_\varepsilon(\phi)|\nabla\phi|dx$ converges to the length $|K|$ of the zero-level line $K = \{x \in \Omega : \phi(x) = 0\}$, under the assumption that $\phi : \Omega \to \mathbb{R}$ is Lipschitz continuous. The same result holds for the case of any l-level curve of ϕ, and not only for the 0-level curve.

Theorem 25 *Let us define*

$$L_\epsilon(\phi) = \int_\Omega |\nabla H_\epsilon(\phi)|dx = \int_\Omega \delta_\epsilon(\phi)|\nabla\phi|dx.$$

Then we have

$$\lim_{\epsilon\to 0} L_\epsilon(\phi) = \int_{\{\phi=0\}} ds = |K|,$$

where $K = \{x \in \Omega : \ \phi(x) = 0\}$.

Proof. Using another form of the coarea formula [140], we have

$$\begin{aligned} L_\epsilon(\phi) &= \int_{\mathbb{R}} \Big[\int_{\phi=\rho} \delta_\epsilon(\phi(x))ds\Big]d\rho, \\ &= \int_{\mathbb{R}} \Big[\delta_\epsilon(\rho)\int_{\phi=\rho} ds\Big]d\rho. \end{aligned}$$

By setting $h(\rho) = \int_{\phi=\rho} ds$, we obtain

$$L_\epsilon(\phi) = \int_{\mathbb{R}} \delta_\epsilon(\rho)h(\rho)d\rho = \int_{\mathbb{R}} \frac{1}{\pi}\frac{\epsilon}{\epsilon^2+\rho^2}h(\rho)d\rho.$$

By the change of variable $\theta = \frac{\rho}{\epsilon}$, we obtain

$$\begin{aligned} \lim_{\epsilon\to 0} L_\epsilon(\phi) &= \lim_{\epsilon\to 0}\int_R \frac{1}{\pi}\frac{\epsilon^2}{\epsilon^2+\epsilon^2\theta^2}h(\theta\epsilon)d\theta, \\ &= \lim_{\epsilon\to 0}\int_{\mathbb{R}} \frac{1}{\pi}\frac{1}{1+\theta^2}h(\theta\epsilon)d\theta, \\ &= h(0)\int_{\mathbb{R}} \frac{1}{\pi}\frac{1}{1+\theta^2}d\theta = h(0)\frac{1}{\pi}\arctan\theta|_{-\infty}^{+\infty} \\ &= h(0) = \int_{\phi=0} ds = |K|, \end{aligned}$$

which concludes the proof. □

In general, this convergence result is valid for any approximations H_ϵ, δ_ϵ, under the assumptions

$$\lim_{\epsilon\to 0} H_\epsilon(z) = H(z) \text{ in } \mathbb{R}\setminus\{0\},$$

$\delta_\epsilon = H'_\epsilon$, $H_\epsilon \in C^1(\mathbb{R})$, $\int_{-\infty}^{+\infty}\delta_1(x)dx = 1$.

2.10 Numerical analysis

When working with minimization problems and partial differential equations, the gradient descent and finite difference methods are the standard approaches for their discretizations. We briefly review here the main terminology, notations and main formulas. The interested reader is referred for instance to [252] and [60] for techniques of numerical optimization, and to [141], [243] for finite difference schemes for differential equations.

2.10.1 Gradient descent method

Assume that we wish to minimize a possibly convex and differentiable function F over a vector space X (such as $\mathbb{R}^N$), with gradient (or Gâteaux derivative) ∇F. The gradient descent method allows us to approximate stationary points of F (those points u for which $\nabla F(u) = 0$) (see also Section 2.7.6). In the convex case, any stationary point of F must be a global minimizer of F. We introduce an artificial time $t \geq 0$. Let $u_0 \in X$ and assume that $u(t) \in X$ for $t \geq 0$ satisfies

$$u(0) = u_0 \text{ and } \frac{du}{dt} = -\nabla F(u(t)) \text{ for } t > 0. \tag{2.14}$$

Then we can easily show that $F(u(t))$ decreases as time t increases. Indeed,

$$\frac{d}{dt}\Big[F(u(t))\Big] = \nabla F(u(t))\frac{du}{dt} = -\|\nabla F(u(t))\|^2 \leq 0 \quad \forall t > 0.$$

In the discrete case, we consider the time step $\triangle t > 0$ and we denote by u^n an approximation to $u(n\triangle t)$, $n = 0, 1, 2, ...$ An explicit forward Euler method applied to (2.14) starts with $u^0 = u_0 \in X$ given and evolves

$$\frac{u^{n+1} - u^n}{\triangle t} = -\nabla F(u^n), \quad n \geq 0,$$

or

$$u^{n+1} = u^n - \triangle t \nabla F(u^n), \quad n \geq 0.$$

It is desirable to ensure that $F(u^{n+1}) = F(u^n - \triangle t \nabla F(u^n)) < F(u^n)$ (unless steady state is reached or $\nabla F(u^n) = 0$). If this inequality is not satisfied, then $\triangle t$ must be decreased. Under additional assumptions on the function F, the sequence u^n converges as $n \to \infty$ to a limit u which is a stationary point or even a minimizer of the function F.

2.10.2 Finite differences

Finite differences are approximation formulas of derivatives of functions. These are derived using Taylor's formula (see Burden, Faires and Burden [74]).

In one dimension, let $u : [a, b] \to \mathbb{R}$ be a function defined on an interval $I = [a, b]$, with $a < b$. Let the space discretization be $\triangle x > 0$. We discretize I by a sequence of points $x_0 = a$, $x_1 = x_0 + \triangle x$, ..., $x_j = x_0 + j\triangle x$, ..., $x_{M+1} = b$. Thus $x_{j+1} - x_j = \triangle x$ for $j = 0, ..., M$ and $\triangle x = \frac{b-a}{M+1}$. We denote by u_j the value $u(x_j)$.

If $u \in C^2[a,b]$, we can approximate its first order derivative at x_j by the one-sided forward finite difference of first order

$$u'(x_j) \approx \frac{u_{j+1} - u_j}{\triangle x}, \; j = 0, ..., M, \tag{2.15}$$

with error $O(\triangle x) = -\frac{\triangle x}{2} u''(\xi)$ for some ξ between x_j and x_{j+1}. The one-sided backward finite difference is

$$u'(x_j) \approx \frac{u_j - u_{j-1}}{\triangle x}, \; j = 1, ..., M+1. \tag{2.16}$$

If $u \in C^3[a,b]$, the first order derivative $u'(x_j)$ can also be approximated by a second order central finite difference, which is

$$u'(x_j) \approx \frac{u_{j+1} - u_{j-1}}{2\triangle x}, \; j = 1, ..., M, \tag{2.17}$$

with error $O(\triangle x^2) = -\frac{\triangle x^2}{6} u'''(\xi)$ for some ξ between x_{j-1} and x_{j+1}.

If $u \in C^4[a,b]$, then its second order derivative at x_j can be approximated by the second order central finite difference

$$u''(x_j) \approx \frac{u_{j+1} - 2u_j + u_{j-1}}{\triangle x^2}, \; j = 1, ..., M, \tag{2.18}$$

with error $O(\triangle x^2) = -\frac{\triangle x^2}{12} u^{(iv)}(\xi)$ for some ξ between x_{j-1} and x_{j+1}.

As we will see throughout the book, these finite difference approximations are easily extended to two or three dimensions in the approximation of partial derivatives.

In two (discrete) spatial dimensions, it will be useful to use the following notations. Let $h > 0$, and let $(x_i, y_j) = (ih, jh)$ for $0 \leq i, j \leq M$ be the $(M+1) \times (M+1)$ discrete points. Let $u_{i,j} \approx u(x_i, y_j)$. Define

$$\begin{aligned}
\triangle_{\pm}^x u_{i,j} &= \pm(u_{i\pm1,j} - u_{i,j})/h, \\
\triangle_{\pm}^y u_{i,j} &= \pm(u_{i,j\pm1} - u_{i,j})/h, \\
\triangle_0^x u_{i,j} &= (u_{i+1,j} - u_{i-1,j})/(2h), \text{ and} \\
\triangle_0^y u_{i,j} &= (u_{i,j+1} - u_{i,j-1})/(2h).
\end{aligned}$$

2.11 Exercises

Exercise 2.1 *Let X and Y be two Hilbert spaces, $K : X \mapsto Y$ a linear and continuous operator with adjoint $K^* : Y \mapsto X$, $f \in Y$, and the least squares functional $F : X \mapsto \mathbb{R}$ given by $F(v) = \|Kv - f\|_Y^2$, for all $v \in X$. Show that $u \in X$ is a minimizer of F if and only if $K^*(Ku) = K^* f$.*

Exercise 2.2 *Let X and Y be two Hilbert spaces, $K : X \mapsto Y$ a linear and continuous operator, $f \in Y$, and $\lambda > 0$. Show that the minimization problem*

$$\inf_{v \in X} \left\{ \|Kv - f\|_Y^2 + \lambda \|v\|_X^2 \right\}$$

has a unique solution $u \in X$, given by $u = (K^*K + \lambda I)^{-1}(K^*f)$, where I is the identity operator on X and $K^* : Y \mapsto X$ is the adjoint of K.

Exercise 2.3 *Let X, Y and Z be Hilbert spaces, $K : X \mapsto Y$ a linear and continuous operator with adjoint $K^* : Y \mapsto X$, $L : X \mapsto Z$ another linear and continuous operator with adjoint $L^* : Z \mapsto X$, $f \in Y$, and $\lambda > 0$. Show that $u \in X$ is solution of the minimization problem*

$$\inf_{v \in X} \left\{ \|Kv - f\|_Y^2 + \lambda \|Lv\|_Z^2 \right\}$$

if and only if u satisfies

$$(K^*K + \lambda L^*L)u = (K^*f).$$

Exercise 2.4 *Prove Proposition 1.*

Exercise 2.5 *Prove Young's inequality from Proposition 2.*

Exercise 2.6 *Prove Theorem 1.*

Exercise 2.7 *Let $f \in L^1$ be a continuous function such that $\hat{f} \in L^1$. Prove that $(\hat{f})^\vee = (\check{f})^\wedge = f$.*

Exercise 2.8 *Take $\Omega = (-1, 1)$ and $u(x) = |x|$.*
(a) Compute the (weak) derivative v of u and show that $u \in W^{1,\infty}(\Omega)$.
(b) Show that v' does not exist in the weak sense.

Exercise 2.9 *Show existence and uniqueness of minimizers of the problem*

$$\inf_{u \in H^1(\Omega)} \left\{ F(u) = \int_\Omega |\nabla u|^2 dx + \lambda \int_\Omega (u - f)^2 dx \right\},$$

where Ω is an open, bounded and connected subset of $\mathbb{R}^2$ with Lipschitz boundary $\partial\Omega$, $f \in L^2(\Omega)$ is a real-valued image function, and $\lambda > 0$ is a given parameter. Also, compute the associated Euler–Lagrange equation of the functional F.

Exercise 2.10 *Let $u : \mathbb{R}^2 \times [0, \infty) \to \mathbb{R}$ be a smooth solution of*

$$\frac{\partial u}{\partial t} = |Du| G(curv(u)),$$

where $curv(u) = div\left(\frac{Du}{|Du|}\right)$ is the curvature operator in two spatial dimensions, $t \geq 0$, and G is a continuous and nondecreasing function such that $kG(k) \geq 0$. Show that the total sum of the perimeters of level lines decreases in time.

Exercise 2.11 *Let $u : \mathbb{R}^2 \to \mathbb{R}$ be a function in $C^2(\mathbb{R}^2)$. Let $x \in \mathbb{R}^2$ be such that $|Du|(x) \neq 0$.*

(a) Verify the formula

$$\triangle u = u_{\vec{N}\vec{N}} + u_{\vec{T}\vec{T}},$$

where $\vec{N}$ and $\vec{T}$ are two unit orthogonal vectors at x, such that $\vec{N} = \frac{Du(x)}{|Du|(x)}$ is normal to the level line of u passing through x (as in Figure 1.3), and $u_{\vec{N}\vec{N}} = \vec{N}^t D^2 u \vec{N}$ and $u_{\vec{T}\vec{T}} = \vec{T}^t D^2 u \vec{T}$ are the second order derivatives of u at x in the $\vec{N}$ and $\vec{T}$ directions, respectively.

(b) Verify the formula at x:

$$div\Big(\frac{Du}{|Du|}\Big) = \frac{1}{|Du|} u_{\vec{T}\vec{T}}.$$

Exercise 2.12 *Using the notations from Section 2.8, show that the length functional is invariant of the parameterization.*

Exercise 2.13 *Consider the one-dimensional length minimization problem*

$$\min_u \Big\{F(u) = \int_0^1 \sqrt{1 + (u'(x))^2} dx\Big\},$$

over functions $u : [0,1] \to \mathbb{R}$ in $C^1([0,1])$ with boundary conditions $u(0) = 0$ and $u(1) = 1$.

(a) Assuming that a minimizer $u \in C^2([0,1])$, compute the associated Euler–Lagrange equation and find the exact solution of the minimization.

(b) Show that the functional $u \mapsto F(u)$ over a convex set, is convex.

(c) Consider a discrete version of the problem: let

$$x_0 = 0 < x_1 < ... < x_n < x_{n+1} = 1$$

be equidistant points, with $x_{i+1} - x_i = h$. For $\vec{u} = (u_1, ..., u_n)$, consider $f(\vec{u}) = h \sum_{i=0}^n \sqrt{1 + \left(\frac{u_{i+1} - u_i}{h}\right)^2}$, with the additional condition that $u_0 = 0$ and $u_{n+1} = 1$. Choose an appropriate discretization integer n. Then numerically and experimentally analyze the behavior of the gradient descent method applied to f. Choose the initial starting point u^0 as a curve joining the points $(0,0)$ and $(1,1)$. Record the number of iterations k and plot the error $u^k - u^*$, where u^* is the exact minimizer.

Exercise 2.14 *Using Taylor's formula, show that each derivative in (2.15), (2.16), (2.17), and (2.18) is approximated by the given formula with the corresponding error.*

Part I

Image Restoration

Chapter 3

Variational Image Restoration Models

We want to address in this chapter one of the most important problems in image processing: the image restoration one. Given a noisy and/or blurry image f (the observation), we wish to restore its clean sharp version u. Noise and blur are inherent degradations during the acquisition process. Noise can be of random nature (described as a random variable following a probability distribution) or of deterministic nature (as a periodic pattern due to physical interference within the imaging device). In simpler cases, the noise term is additive or multiplicative. In more complicated cases, there is a more nonlinear, intricate relation between noise, the observed image and the underlying clean image. Noise is usually removed by averaging pixel values (in other words, smoothing). In contrast, blur is produced by averaging or smoothing. Therefore, in order to remove noise (process called denoising), we want to smooth the observed noisy image f, in such a way that edges, texture and other features are preserved. In order to remove blur (process called deblurring), we want to sharpen the observed image f (deblurring). These two opposite processes (denoising and deblurring) render the restoration problem very difficult. Due to these and other difficulties, the problem of restoring u from the observed image f (even knowing something about the noise and blur) is an ill-posed inverse problem. Inspired from Tikhonov regularization, described in Chapter 2 in a general setting, this can be addressed via regularization. This method introduces a priori assumptions on the unknown image u and the problem becomes well-posed. Moreover, interesting mathematical and numerical problems of variational nature arise.

We assume that Ω is an open and bounded domain in $\mathbb{R}^N$, $N \geq 1$ integer. In one dimension, for signals, Ω is an interval in $\mathbb{R}$. In two dimensions, we deal with planar images and Ω is a rectangle in $\mathbb{R}^2$. In three dimensions, we deal with volumetric images and Ω is a rectangular parallelepiped in $\mathbb{R}^3$. Let $f : \Omega \to \mathbb{R}$ be a given image function, possibly degraded by noise, blur, and other types of degradations. As we have already mentioned, one of the most important problems in image processing is the inverse problem called restoration, that is, recovering a clean and sharp image $u : \Omega \to \mathbb{R}$ from its degraded version f.

A general degradation model is of the form

$$f = S(Ku, n),$$

where K is a degradation operator (such as blur), and n represents noise.

3.1 Linear degradation model with Gaussian noise and total variation regularization

One of the most studied and used methods in variational image restoration is the total variation regularization introduced in image processing by Rudin, Osher and Fatemi [277, 276]. Also, one-dimensional discrete cases have been proposed by Alliney [11, 12, 13]. Related work is also presented and analyzed in Acar–Vogel [2] and Chambolle–Lions [86], with a generalization by Vese in [336]. We also refer the reader to the texts [23], [30], [90] and [236].

We assume that the degraded image data (the observation) is a function $f \in L^2(\Omega)$, and the linear degradation model linking f to u (the image to be recovered) is

$$f = Ku + n,$$

where $K : L^p(\Omega) \mapsto L^2(\Omega)$ is a linear and continuous operator responsible for blur, n is additive Gaussian noise of zero mean and standard deviation σ^2, $p = \frac{N}{N-1}$ if $N \geq 2$ and $p = 2$ if $N = 1$.

Following Rudin, Osher and Fatemi [277, 276], the *total variation* restoration model is formulated in two (related) ways.

First, assuming that the variance of the noise $\sigma^2 > 0$ is known or can be estimated, we have the constrained minimization

$$\inf_{u \in BV(\Omega)} \left\{ \int_\Omega |Du|, \text{ subject to } \int_\Omega (Ku - f)^2 dx = \sigma^2 \right\}. \tag{3.1}$$

Second, using a Lagrange multiplier $\lambda > 0$ for the constraint, we have the unconstrained minimization

$$\inf_{u \in BV(\Omega)} \left\{ \int_\Omega |Du| + \lambda \int_\Omega (Ku - f)^2 dx \right\}. \tag{3.2}$$

We focus in this chapter on the unconstrained restoration model (3.2). Theoretical results of existence and uniqueness of minimizers, optimality conditions and characterization of minimizers, numerical approximations, and experimental results obtained using this model or some of its variants are presented.

3.1.1 Signal restoration

In the case of one-dimensional signals $f : \Omega \to \mathbb{R}$, where $\Omega = (a, b) \subset \mathbb{R}$, the model can be expressed as

$$\inf_{u \in BV(a,b)} F(u) = \lambda \int_a^b (Ku - f)^2 dx + \int_a^b |u'(x)|. \tag{3.3}$$

We can formally compute the Euler–Lagrange equation associated with the minimization (3.3), thus imposing that $F'(u) = 0$, as in Chapter 2. Indeed, by the standard calculus, we obtain that, if u is a minimizer of (3.3), then it must formally satisfy (for instance in the distributional sense), at points where $|u'| \neq 0$,

$$2\lambda K^*(Ku - f) = \left(\frac{u'}{|u'|}\right)' \text{ for } x \in (a, b), \tag{3.4}$$

with boundary conditions $\frac{u'(x)}{|u'(x)|} = 0$ for $x = a$ and $x = b$.

We give a numerical discretization of the one-dimensional functional. Let $x_0 = a$, $x_j = a + jh$, $1 \leq j \leq n$, and $x_{n+1} = b$ be the discrete points of the signal, with $h = \frac{b-a}{n+1}$ the space step, and n a positive integer. The discrete energy can be expressed as,

$$F(u) = \lambda h \sum_{j=0}^{n} \left((Ku)_j - f_j\right)^2 + \sum_{j=0}^{n} |u_{j+1} - u_j|,$$

which could be minimized in practice in several ways, for instance using a numerical optimization technique from [252].

3.1.2 Existence of minimizers

We prove here existence of minimizers in $BV(\Omega)$ for a slightly more general version of the unconstrained minimization problem for image restoration. We follow Acar-Vogel [2] and Chambolle-Lions [86] when the regularization is the total variation, and [336] for a more general regularization. We also refer the reader to [23], [30], and [90].

Let Ω be an open, bounded and connected subset of $\mathbb{R}^N$, with Lipschitz boundary $\Gamma = \partial\Omega$. Let $p \geq 1$, $1 \leq q < \infty$, $\lambda > 0$. We make the following hypotheses:

- (H1) $\phi : \mathbb{R} \mapsto \mathbb{R}^+$ is even and convex, nondecreasing in $\mathbb{R}^+$, such that:

  (i) $\phi(0) = 0$ (without loss of generality).

  (ii) There exist constants $c > 0$ and $b \geq 0$ such that

$$cz - b \leq \phi(z) \leq cz + b, \quad \forall z \in \mathbb{R}^+.$$

- (H2) $K : L^1(\Omega) \mapsto L^p(\Omega)$ is a linear and continuous operator.
- (H3) $\|K\chi_\Omega\|_{L^1(\Omega)} > 0$.
- (H4) K is injective or ϕ is strictly convex.

Examples of such (convex) functions ϕ are:
- $\phi(t) = |t|$, leading to the total variation regularization;
- $\phi(t) = \sqrt{1 + t^2} - 1$ (function of minimal surfaces);

- $\phi(t) = \begin{cases} \frac{1}{2\epsilon}t^2 & \text{if} \quad |t| \leq \epsilon \\ |t| - \frac{\epsilon}{2} & \text{if} \quad |t| > \epsilon \end{cases}$ (Huber function).
- $\phi(t) = \log\cosh t$ (logistic function).

We want to study the problem

$$\inf_{u\in BV(\Omega)} \Big\{ F(u) = \int_\Omega \phi(|Du|) + \lambda \|f - Ku\|^q_{L^p(\Omega)} \Big\}. \tag{3.5}$$

Remark 10 *Thanks to (H1)(ii), the functional $u \mapsto \int_\Omega \phi(|Du|)dx$ is well defined and finite on the Sobolev space $W^{1,1}(\Omega)$. However, $W^{1,1}(\Omega)$ is not a reflexive Banach space and the minimization problem (3.5) may not have a solution in this space because minimizing sequences u_n may not converge to a limit u in $W^{1,1}(\Omega)$. To overcome this difficulty, we must work with functions u of bounded variation. We will use the notion of convex functions of measures ([126, 158]) to properly define $\int_\Omega \phi(|Du|)$ for $u \in BV(\Omega)$; additionally, the functional $\phi(|\cdot|)(\Omega)$ is weakly* lower semicontinuous on $\mathcal{M}(\Omega)$, space of Radon measures ([126, 158]). Moreover, $BV(\Omega)$ is the appropriate space of functions for many basic image processing tasks, since it allows for discontinuities along curves (edges in images), while $W^{1,p}(\Omega)$, $p \geq 1$, cannot. Finally, it is easy to verify that, under the above assumptions, the functional F is convex on $BV(\Omega)$.*

Theorem 26 *Let $f \in L^p(\Omega)$. If (H1)-(H3) hold, then problem (3.5) has a minimizer $u \in BV(\Omega)$. Moreover, the minimizer u is unique if (H4) also holds.*

Proof. For the sake of brevity, the L^p-norm is sometimes denoted $\|\cdot\|_p$. Infimum of F is finite since $F(u) \geq 0$ on $BV(\Omega)$, and $F(0) = \lambda\|f\|_p^q < \infty$. Let $u_n \in BV(\Omega)$ be a minimizing sequence, thus $\inf_v F(v) = \lim_{n\to\infty} F(u_n)$. Then $F(u_n) \leq C < \infty$, $\forall n \geq 1$. Note that C may change from line to line. Poincaré–Wirtinger inequality from Theorem 14 in Chapter 2 implies that there is a constant $C' = C'(N, \Omega) > 0$ such that for all $n \geq 1$, we have $\|u_n - u_{n,\Omega}\|_1 \leq C'|u_n|_{BV}$, where $u_{n,\Omega}$ is the mean of u_n over Ω and where $|\cdot|_{BV}$ denotes the semi-norm on $BV(\Omega)$. Let $v_n = u_n - u_{n,\Omega}$, thus $v_{n,\Omega} = 0$ and $Dv_n = Du_n$. Similarly, we have $\|v_n\|_1 \leq C'|v_n|_{BV}$.

Since Ω is bounded, we have for some constant $C_1 > 0$,

$$\begin{aligned}
(C/\lambda)^{2/q} \geq \|f - Ku_n\|_p^2 \geq C_1\|f - Ku_n\|_1^2 &= C_1\|(Kv_n - f) + Ku_{n,\Omega}\|_1^2, \\
&\geq C_1\Big|\|Kv_n - f\|_1 - \|Ku_{n,\Omega}\|_1\Big|^2, \\
&\geq C_1\|Ku_{n,\Omega}\|_1(\|Ku_{n,\Omega}\|_1 - 2\|Kv_n - f\|_1), \\
&\geq C_1\|Ku_{n,\Omega}\|_1\Big(\|Ku_{n,\Omega}\|_1 - 2(\|K\|\|v_n\|_1 + \|f\|_1)\Big).
\end{aligned}$$

Let $x_n = \|Ku_{n,\Omega}\|_1$ and $a_n = (\|K\|\|v_n\|_1 + \|f\|_1)$. Then

$$0 \leq a_n \leq \|K\|\|v_n\|_1 + \|f\|_1 \leq C'\|K\|\|v_n\|_{BV} + \|f\|_1 \leq C'\|K\|\frac{b|\Omega| + C}{c} + \|f\|_1$$

and $x_n(x_n - 2a_n) \leq \frac{(C/\lambda)^{2/q}}{C_1} = \bar{c}$. Thus we obtain $0 \leq x_n \leq a_n + \sqrt{a_n^2 + \bar{c}} \leq C_2$ for some constant $C_2 > 0$, which implies

$$\|Ku_{n,\Omega}\|_1 = \frac{|\int_\Omega u_n dx|}{|\Omega|}\|K\chi_\Omega\|_1 \leq C_2.$$

Thanks to assumptions on K, we deduce that the sequence $|u_{n,\Omega}|$ is uniformly bounded. By Poincaré–Wirtinger inequality we obtain $\|u_n\|_1 \leq C$. Thus, $|u_n|_{BV(\Omega)} + \|u_n\|_{L^1(\Omega)}$ is uniformly bounded. Following [140], we deduce that there is a subsequence $\{u_{n_j}\}$ of $\{u_n\}$, and $u \in BV(\Omega)$ such that u_{n_j} converges to u strongly in $L^1(\Omega)$ and Du_{n_j} converges to Du weakly* in $\mathcal{M}(\Omega)$. Then we also have

$$\int_\Omega \phi(|Du|) \leq \liminf_{n_j \to \infty} \int_\Omega \phi(|Du_{n_j}|).$$

Since $(u_{n_j} - f) \to (u - f)$ in $L^1(\Omega)$, and K is continuous from $L^1(\Omega)$ to $L^p(\Omega)$, we deduce that $\|K(u_{n_j} - f)\|_p \to \|K(u - f)\|_p$ as $n_j \to \infty$. We conclude that

$$F(u) \leq \liminf_{n_j \to \infty} F(u_{n_j}) = \inf_{v \in BV(\Omega)} F(v);$$

thus u is a minimizer of F.

Let $u, v \in BV(\Omega)$ be two minimizers of F. We first show that $Ku = Kv$: if, by contradiction, $Ku \neq Kv$, then

$$F(\frac{1}{2}u + \frac{1}{2}v) < \frac{1}{2}F(u) + \frac{1}{2}F(v) = \min F$$

(because F is convex, and it is the sum of two convex functions in the independent variables Du and Ku, the second one being strictly convex), which is impossible and thus $Ku = Kv$. If K is injective, we obtain that $u = v$. Otherwise, if K is not injective, but if ϕ is strictly convex, then $Du = Dv$, which implies that $u = v + C$ with C a constant. From $Ku = Kv + CK\chi_\Omega$ and $Ku = Kv$, we obtain $C = 0$ (assumption (H3)), thus $u = v$. □

3.1.3 Dual problem and optimality conditions

We apply in this section the general duality techniques of Ekeland–Temam [136] and in particular of Demengel–Temam [126] to restoration functionals with total variation regularization, following [336], in order to obtain the optimality conditions in a rigorous way.

Let $f \in L^p(\Omega)$ be the given data, with $p \geq 1$ and $\Omega \subset \mathbb{R}^N$ open, bounded, and connected, with Lipschitz boundary $\partial\Omega$. Let k be a smoothing convolution kernel, such as the Gaussian kernel or the Poisson kernel, and define $Ku = k * u$, with $K : L^1(\Omega) \to L^p(\Omega)$ linear and continuous, and adjoint K^*. We can consider the restoration functionals

$$\inf_{u \in BV(\Omega)} \Big\{E(u) = |u|_{BV(\Omega)} + \lambda\|k * u - f\|^q_{L^p(\Omega)}\Big\}, \tag{3.6}$$

where $q \geq 1$. We assume that both p and q are finite. For $p = q = 2$ we recover (3.2).

For the purpose of illustration, we show the derivation of the optimality conditions when $p = q = 1$. We will also mention the differences for the other values of p and q.

The minimization problem for $p = 1$, $q = 1$ is

$$(P_1) \qquad \inf_{u \in BV(\Omega)} E(u) = |u|_{BV(\Omega)} + \lambda \|Ku - f\|_{L^1(\Omega)},$$

using the notation $|u|_{BV(\Omega)} = \int_\Omega |Du|$ for the semi-norm of u in $BV(\Omega)$. For $u \in L^1(\Omega)$, we will also use the operator notation $Ku = k * u$ to be the corresponding linear and continuous operator from $L^1(\Omega)$ to $L^1(\Omega)$, with adjoint K^* (with radially symmetric kernel k, the operator K is self-adjoint). We wish to characterize a solution u of (P_1) using duality techniques.

We have

$$\inf_{u \in BV(\Omega)} E(u) = \inf_{u \in W^{1,1}(\Omega)} E(u),$$

since for any $u \in BV(\Omega)$, we can find $u_n \in W^{1,1}(\Omega)$ such that $u_n \to u$ strongly in $L^1(\Omega)$ and $|u_n|_{BV(\Omega)} \to |u|_{BV(\Omega)}$. Thus, we first consider the simpler problem

$$(P_2) \qquad \inf_{u \in W^{1,1}(\Omega)} E(u) = \int_\Omega |\nabla u| dx + \lambda \|Ku - f\|_{L^1(\Omega)}.$$

We now write (P_2^*), the dual of (P_2), in the sense of Ekeland-Temam [136].

We first recall the definition of the Legendre transform (or polar) of a function: let V and V^* be two normed vector spaces in duality by a bilinear pairing denoted $\langle \cdot, \cdot \rangle$. Let $\varphi : V \to \bar{\mathbb{R}}$ be a function. Then the Legendre transform $\varphi^* : V^* \to \bar{\mathbb{R}}$ is defined by

$$\varphi^*(u^*) = \sup_{u \in V} \left\{ \langle u^*, u \rangle - \varphi(u) \right\}.$$

We let $G_1(w_0) = \lambda \int_\Omega |w_0 - f| dx$ and $G_2(\bar{w}) = \int_\Omega |\bar{w}| dx$, with $G_1 : L^1(\Omega) \to \mathbb{R}$, $G_2 : L^1(\Omega)^N \to \mathbb{R}$, and using $w = (w_0, w_1, w_2, ..., w_N) = (w_0, \bar{w}) \in L^1(\Omega)^{N+1}$, we define $G(w) = G_1(w_0) + G_2(\bar{w})$.

Let $\Lambda : u \mapsto (Ku, \nabla u), W^{1,1}(\Omega) \to L^1(\Omega)^{N+1}$ and let Λ^* be its adjoint. Then $E(u) = F(u) + G(\Lambda u)$, with $F(u) \equiv 0$.

Then (P_2^*) is ([136], Chapter III, Section 4):

$$(P_2^*) \qquad \sup_{p^* \in L^\infty(\Omega)^{N+1}} -F^*(\Lambda^* p^*) - G^*(-p^*).$$

We have $F^*(\Lambda^* p^*) = 0$ if $\Lambda^* p^* = 0$, and $F^*(\Lambda^* p^*) = +\infty$ otherwise. It is easy to see that

$$G^*(p^*) = G_1^*(p_0^*) + G_2^*(\bar{p}^*), \text{ for } p^* = (p_0^*, \bar{p}^*).$$

We have

$$G_1^*(p_0^*) = \begin{vmatrix} \int_\Omega p_0^*(f)dx & \text{if } |p_0^*| \le \lambda \text{ a.e.,} \\ +\infty & \text{otherwise,} \end{vmatrix}$$

and

$$G_2^*(\bar{p}^*) = \begin{vmatrix} 0 & \text{if } |\bar{p}^*| \le 1 \text{ a.e.,} \\ +\infty & \text{otherwise.} \end{vmatrix}$$

Thus we have

$$(P_2^*) \qquad \sup_{p^* \in X} -\int_\Omega (-p_0^*)(f)dx,$$

where

$X = \{(p_0^*, p_1^*, ..., p_N^*) = (p_0^*, \bar{p}^*) \in L^\infty(\Omega)^{N+1},\ |p_0^*| \le \lambda,\ |\bar{p}^*| \le 1,\ \Lambda^* p^* = 0\}$.

Under the satisfied assumptions, $\inf(P_1) = \inf(P_2) = \sup(P_2)^*$ and (P_2^*) has at least one solution p^*.

Using the definition of Λ, we can show that [336]

$$X = \{(p_0^*, p_1^*, ..., p_N^*) = (p_0^*, \bar{p}^*) \in L^\infty(\Omega)^{N+1},\ |p_0^*| \le \lambda,\ |\bar{p}^*| \le 1,$$
$$K^* p_0^* - \text{div}\bar{p}^* = 0,\ \bar{p}^* \cdot \nu = 0 \text{ on } \partial\Omega\}.$$

Now let $u \in BV(\Omega)$ be a solution of (P_1) and $p = (p_0, \bar{p}) \in X$ be a solution of (P_2^*). We must have the extremality relation

$$|u|_{BV(\Omega)} + \lambda \|Ku - f\|_{L^1(\Omega)} = \int_\Omega p_0(f)dx.$$

We have that $Du \cdot \bar{p}$ is an unsigned measure, satisfying a generalized Green's formula

$$\int_\Omega Du \cdot \bar{p} = -\int_\Omega u\text{div}\bar{p}\, dx + \int_{\partial\Omega} u(\bar{p} \cdot \nu)ds.$$

Since $\bar{p} \cdot \nu = 0$ $\partial\Omega$ a.e., we have

$$\int_\Omega |Du| + \lambda \int_\Omega |Ku - f|dx + \int_Q p_0 Kudx + \int_\Omega Du \cdot \bar{p} - \int_\Omega p_0(f)dx = 0,$$

or using the decomposition $Du = \nabla u dx + D_s u = \nabla u dx + C_u + J_u = \nabla u dx + C_u + (u^+ - u^-)\nu \mathcal{H}^{N-1}|_{S_u}$ [140], with S_u the support of the jump measure J_u, we get

$$\int_\Omega |\nabla u| dx + \int_{\Omega \setminus S_u} |C_u| + \int_{S_u} |u^+ - u^-| d\mathcal{H}^{N-1} + \int_\Omega \nabla u \cdot \bar{p} dx + \int_{\Omega \setminus S_u} \bar{p} \cdot C_u$$

$$+ \int_{S_u} (u^+ - u^-)\bar{p} \cdot \nu d\mathcal{H}^{N-1} + \lambda \int_\Omega |Ku - f| dx + \int_\Omega p_0 Kudx - \int_\Omega p_0(f)dx = 0.$$

Since for any function φ and its polar φ^* we must have $\varphi^*(u^*) - \langle u^*, u \rangle + \varphi(u) \ge 0$ for any $u \in V$ and $u^* \in V^*$, we obtain:

(i) $\lambda\,|Ku - f| - (-p_0)(Ku) + (-p_0)(f) \geq 0$ for dx a.e. in Ω.

(ii) $|\nabla u| - \nabla u \cdot (-\bar{p}) + 0 \geq 0$ for dx a.e. in Ω where $\nabla u(x)$ is defined.

(iii) $0-(-\bar{p})\cdot C_u+|C_u| = (1+\bar{p}\cdot h)|C_u| \geq 0$, since $|\bar{p}| \leq 1$ (letting $C_u = h\cdot|C_u|$, $h \in L^1(|C_u|)^N$, $|h| = 1$).

(iv) $0 - (-\bar{p}\cdot\nu)(u^+ - u^-) + |u^+ - u^-| = (u^+ - u^-)(1 + \bar{p}\cdot\nu) \geq 0$ for $d\mathcal{H}^{N-1}$ a.e. in S_u (again since $|\bar{p}| \leq 1$ and by convention of notation, $u^+ > u^-$).

Therefore, each expression in (i)-(iv) must be exactly 0 and we obtain a characterization of extremals u.

Theorem 27 *u is a minimizer of (P_1) if and only if there is $(p_0, p_1, ..., p_N) = (p_0, \bar{p}) \in (L^\infty(\Omega))^{N+1}$ such that*

$$
\begin{aligned}
|p_0| \leq \lambda, \quad |\bar{p}| \leq 1,\\
\bar{p}\cdot\nu = 0 \text{ on } \partial\Omega,\\
K^* p_0 - div\bar{p} = 0,\\
\lambda\,|Ku - f| + p_0(Ku - f) = 0,\\
|\nabla u| + \nabla u \cdot \bar{p} = 0,\\
1 + \bar{p}\cdot\nu = 0 \text{ on } S_u \text{ and } |\bar{p}| = 1 \text{ on } S_u,
\end{aligned}
\tag{3.7}
$$

and

$$supp|C_u| \subset \{x \in \Omega\backslash S_u, 1+\bar{p}(x)\cdot h(x) = 0, h \in L^1(|C_u|)^N, |h| = 1, C_u = h|C_u|\}.$$

A similar statement as in Theorem 27 can be shown for the general case $1 \leq p, q < \infty$. The main change is in the definition of G_1, which becomes $G_1(w_0) = \lambda\|w_0 - f\|_p^q = \lambda\Big(\int_\Omega |w_0 - f|^p dx\Big)^{q/p}$ for $w_0 \in L^p(\Omega)$. For example, if $1 < q < \infty$, then $G_1^*(p_0^*) = \lambda q\Big[\frac{1}{q'}\|\frac{p_0^*}{\lambda q}\|_{p'}^{q'} + \int_\Omega (f)\frac{p_0^*}{\lambda q}dx\Big]$ and (3.7) changes accordingly, $\frac{1}{p} + \frac{1}{p'} = 1$ and $\frac{1}{q} + \frac{1}{q'} = 1$ (similarly in the case $1 \leq p < \infty$, $q = 1$).

For more details we refer the reader to [336] where the special case $p = q = 2$ was considered with a more general regularizing term $\int_\Omega \phi(|Du|)$.

3.1.4 Characterization of minimizers via dual norm

We present in this section a characterization of minimizers inspired from the work of Meyer [236] and of Andreu-Vaillo et al. [23], in two dimensions, for problem (3.2).

We assume that $\Omega \subset \mathbb{R}^2$ is open, bounded and connected, and with Lipschitz boundary $\partial\Omega$.

Due to Poincaré–Wirtinger inequality (Theorem 14 from Chapter 2), we have the embedding $BV(\Omega) \subset L^2(\Omega)$. For any $u \in BV(\Omega)$, we will use the notation $|u|_{BV(\Omega)} = \int_\Omega |Du|$ for the total variation of u. Also, $\langle \cdot, \cdot \rangle$ denotes here the $L^2(\Omega)$ inner product: $\langle v, w \rangle = \int_\Omega v(x)w(x)dx$, for all $v, w \in L^2(\Omega)$, and we recall the Cauchy–Schwarz inequality, $\langle v, w \rangle \leq \|v\|_{L^2(\Omega)}\|w\|_{L^2(\Omega)}$. We will also use the notation for the average of an arbitrary function $v \in L^2(\Omega)$: $v_\Omega = \frac{\int_\Omega v(x)dx}{|\Omega|}$.

Definition 18 *For any $v \in L^2(\Omega)$, we define*

$$\|v\|_* = \sup_{u \in BV(\Omega), |u|_{BV(\Omega)} \neq 0} \frac{\langle u, v \rangle}{|u|_{BV(\Omega)}} \in [0, \infty].$$

Lemma 1 *Let $v \in L^2(\Omega)$. Then $\|v\|_* < \infty$ if and only if $\int_\Omega v(x)dx = 0$.*

Proof. If $\int_\Omega v(x)dx = 0$, then for any $u \in BV(\Omega)$, with $|u|_{BV(\Omega)} \neq 0$, we have (using Cauchy–Schwarz and Poincaré–Wirtinger inequalities):

$$\begin{aligned} \frac{\langle u, v \rangle}{|u|_{BV(\Omega)}} &= \frac{\langle v, u - u_\Omega \rangle}{|u|_{BV(\Omega)}} \leq \frac{\|v\|_{L^2(\Omega)}\|u - u_\Omega\|_{L^2(\Omega)}}{|u|_{BV(\Omega)}} \\ &\leq \frac{\|v\|_{L^2(\Omega)} C |u|_{BV(\Omega)}}{|u|_{BV(\Omega)}} = C\|v\|_{L^2(\Omega)} < \infty, \end{aligned}$$

where $C > 0$ is a constant depending only on Ω. Thus

$$\|v\|_* = \sup_{u \in BV(\Omega), |u|_{BV(\Omega)} \neq 0} \frac{\langle u, v \rangle}{|u|_{BV(\Omega)}} \leq C\|v\|_{L^2(\Omega)} < \infty.$$

The converse is also true. Assume that $\|v\|_* < \infty$, $v \in L^2(\Omega)$. Let $u \in BV(\Omega)$ be fixed, with $|u|_{BV(\Omega)} \neq 0$, and c a constant. Then $u + c \in BV(\Omega)$, $|u|_{BV(\Omega)} = |u + c|_{BV(\Omega)} \neq 0$. Then

$$\frac{\langle (u + c), v \rangle}{|u + c|_{BV(\Omega)}} \leq \|v\|_* < \infty,$$

and

$$c \int_\Omega v(x)dx \leq \|v\|_* |u|_{BV(\Omega)} - \langle u, v \rangle \in \mathbb{R}.$$

Since this must hold for any constant c, we must have $\int_\Omega v(x)dx = 0$. □

We recall that $\lambda > 0$ and for any $u \in BV(\Omega)$ we define

$$F(u) = \lambda\|Ku - f\|^2_{L^2(\Omega)} + |u|_{BV(\Omega)},$$

where $f \in L^2(\Omega)$ and $K : L^2(\Omega) \to L^2(\Omega)$ is linear and continuous with adjoint K^*, such that $K\chi_\Omega = \chi_\Omega$. With the above notations, we are ready to state and prove a characterization of minimizers of F using the above "dual norm" $\| \cdot \|_*$.

Theorem 28

(i) $u \equiv f_\Omega$ *is a minimizer of* F *if and only if* $\|K^*(f - f_\Omega)\|_* \leq \frac{1}{2\lambda}$.

(ii) Assume $\|K^*(f - f_\Omega)\|_* > \frac{1}{2\lambda}$. *Then* $u \in BV(\Omega)$ *is a minimizer of* F *if and only if*

$$\|K^*(f - Ku)\|_* = \frac{1}{2\lambda} \text{ and } \langle u, K^*(f - Ku)\rangle = \frac{1}{2\lambda}|u|_{BV(\Omega)}. \tag{3.8}$$

Proof. (i) If $\|K^*(f - f_\Omega)\|_* \leq \frac{1}{2\lambda}$, then for all $h \in BV(\Omega)$,

$$0 \leq |h|_{BV(\Omega)} - 2\lambda\langle K^*(f - f_\Omega), h\rangle.$$

We deduce that

$$\lambda\|f - f_\Omega\|^2_{L^2(\Omega)} + \lambda\|Kh\|^2_{L^2(\Omega)} \leq |h|_{BV(\Omega)} + \lambda\|f - f_\Omega - Kh\|^2_{L^2(\Omega)},$$

thus

$$\lambda\|f - f_\Omega\|^2_{L^2(\Omega)} \leq |h|_{BV(\Omega)} + \lambda\|f - f_\Omega - Kh\|^2_{L^2(\Omega)},$$

or $F(f_\Omega) = \lambda\|f - Kf_\Omega\|^2_{L^2(\Omega)} \leq F(h + f_\Omega)$ for all $h \in BV(\Omega)$, using $K\chi_\Omega = \chi_\Omega$. Therefore, $u \equiv f_\Omega$ is a minimizer of F.

For the converse, if $u \equiv f_\Omega$ is a minimizer of F, then for any other $h \in BV(\Omega)$ we have

$$F(f_\Omega) = \lambda\|f - Kf_\Omega\|^2_{L^2(\Omega)} \leq \lambda\|K(h + f_\Omega) - f\|^2_{L^2(\Omega)} + |h + f_\Omega|_{BV(\Omega)}.$$

Changing h into ϵh in the last inequality, $\epsilon \in \mathbb{R}$, and after cancellations, we obtain

$$0 \leq \lambda\epsilon^2\|Kh\|^2_{L^2(\Omega)} - 2\lambda\epsilon\langle Kh, f - f_\Omega\rangle + |\epsilon||h|_{BV(\Omega)}.$$

First dividing by $\epsilon > 0$ and then taking the limit as $\epsilon \searrow 0$, we finally obtain $2\lambda\langle Kh, f - f_\Omega\rangle \leq |h|_{BV(\Omega)}$, for all $h \in BV(\Omega)$, or that $\|K^*(f - f_\Omega)\|_* \leq \frac{1}{2\lambda}$.

(ii) For the direct implication in (ii), if u is a minimizer of F, then for any $h \in BV(\Omega)$ and $\epsilon \in \mathbb{R}$, we have $F(u + \epsilon h) \geq F(u)$, or

$$\lambda\|K(u + \epsilon h) - f\|^2_{L^2(\Omega)} + |u + \epsilon h|_{BV(\Omega)} \geq \lambda\|Ku - f\|^2_{L^2(\Omega)} + |u|_{BV(\Omega)},$$

or after expanding the quadratic term and cancellation,

$$\lambda\epsilon^2\|Kh\|^2_{L^2(\Omega)} - 2\lambda\epsilon\langle f - Ku, Kh\rangle + |u + \epsilon h|_{BV(\Omega)} \geq |u|_{BV(\Omega)}. \tag{3.9}$$

Using the triangle inequality for the $BV(\Omega)$ semi-norm, we obtain

$$\lambda\epsilon^2\|Kh\|^2_{L^2(\Omega)} - 2\lambda\epsilon\langle f - Ku, Kh\rangle + |u|_{BV(\Omega)} + |\epsilon|\,|h|_{BV(\Omega)} \geq |u|_{BV(\Omega)},$$

or

$$\lambda\epsilon^2\|Kh\|^2_{L^2(\Omega)} + |\epsilon||h|_{BV(\Omega)} \geq 2\lambda\epsilon\langle f - Ku, Kh\rangle.$$

First dividing by $\epsilon > 0$ and then taking the limit as $\epsilon \searrow 0$, we obtain for any $h \in BV(\Omega)$ that

$$|h|_{BV(\Omega)} \geq 2\lambda\langle K^*(f - Ku), h\rangle, \tag{3.10}$$

which implies that $\|K^*(f-Ku)\|_* \leq \frac{1}{2\lambda}$. Taking now $h = u$ in (3.9), we obtain

$$\lambda\epsilon^2\|Ku\|^2_{L^2(\Omega)} - 2\lambda\epsilon\langle f - Ku, Ku\rangle + |1+\epsilon||u|_{BV(\Omega)} \geq |u|_{BV(\Omega)}.$$

For $\epsilon \in (-1, 0)$, $|1+\epsilon| = 1+\epsilon$ and the above inequality simplifies to

$$\lambda\epsilon^2\|Ku\|^2_{L^2(\Omega)} - 2\lambda\epsilon\langle f - Ku, Ku\rangle + \epsilon|u|_{BV(\Omega)} \geq 0.$$

We first divide by $\epsilon < 0$,

$$\lambda\epsilon\|Ku\|^2_{L^2(\Omega)} + |u|_{BV(\Omega)} \leq 2\lambda\langle f - Ku, Ku\rangle$$

then take $\epsilon \nearrow 0$ to obtain

$$|u|_{BV(\Omega)} \leq 2\lambda\langle f - Ku, Ku\rangle.$$

Together with (3.10), we deduce that $|u|_{BV(\Omega)} = 2\lambda\langle K^*(f-Ku), u\rangle$ and that we must have $\|K^*(f-Ku)\|_{L^2(\Omega)} = \frac{1}{2\lambda}$ (the minimizer u cannot be a constant in the second case (ii)).

For the converse implication in (ii), we have to show that, if $u \in BV(\Omega)$ satisfies (3.8), then u is a minimizer of F. Take $h \in BV(\Omega)$ arbitrary, and let $w = h - u$, or $h = u + w$. We have, using the first condition in (3.8), the first inequality below,

$$\begin{aligned}
&|u+w|_{BV(\Omega)} + \lambda\|K(u+w) - f\|^2_{L^2(\Omega)} \\
&\geq 2\lambda\langle u+w, K^*(f-Ku)\rangle + \lambda\|Ku - f\|^2_{L^2(\Omega)} \\
&+ \lambda\|Kw\|^2_{L^2(\Omega)} + 2\lambda\langle Kw, Ku - f\rangle \\
&= 2\lambda\langle u, K^*(f-Ku)\rangle + 2\lambda\langle w, K^*(f-Ku)\rangle \\
&+ \lambda\|Ku-f\|^2_{L^2(\Omega)} + \lambda\|Kw\|^2_{L^2(\Omega)} + 2\lambda\langle w, K^*(Ku-f)\rangle \\
&= 2\lambda\langle u, K^*(f-Ku)\rangle + \lambda\|Ku-f\|^2_{L^2(\Omega)} + \lambda\|Kw\|^2_{L^2(\Omega)} \\
&= |u|_{BV(\Omega)} + \lambda\|Ku-f\|^2_{L^2(\Omega)} + \lambda\|Kw\|^2_{L^2(\Omega)} \\
&\geq |u|_{BV(\Omega)} + \lambda\|Ku-f\|^2_{L^2(\Omega)},
\end{aligned}$$

while the second condition in (3.8) is used in the last equality above. This implies that for any $h \in BV(\Omega)$, we have $F(h) \geq F(u)$, thus u satisfying (3.8) is a minimizer of F. □

Remark 11 *Assume that K is the identity operator (denoising only). Then the relation $\|f - u\|_* = \frac{1}{2\lambda}$ from the previous theorem gives us a size on the residual, $f - u$, function of the weight λ.*

3.1.5 Euler–Lagrange equation of regularized problem using Gâteaux differentiability

Let

$$F(u) = \lambda\int_\Omega |f - Ku|^2 dx + \int_\Omega |Du|.$$

The calculations and results from Section 3.1.3 show us how to rigorously derive the optimality conditions associated with the minimization $\inf_{u\in BV(\Omega)} F(u)$. In those results we take into account that Du is a measure and not only the absolutely continous part $\nabla u dx$ of the gradient of u, and we directly work with the total variation which is not differentiable.

Here we want to formally compute the Gâteaux derivative of the functional using the standard calculus presented in Chapter 2, which is easier, general, and sufficient for the numerical discretizations. First, we remove the singularity when $|Du| = 0$, approximating $F(u)$ by $F_\epsilon(u)$, where

$$F_\epsilon(u) = \lambda \int_\Omega |f - Ku|^2 dx + \int_\Omega \sqrt{\epsilon^2 + |Du|^2} dx,$$

with $\epsilon > 0$ a small parameter.

For any test function $v \in C^1(\overline{\Omega})$, we define $g(\varepsilon) = F_\epsilon(u + \varepsilon v)$, $\varepsilon \in \mathbb{R}$. In order to have $F_\epsilon(u) \leq F_\epsilon(u+\varepsilon v)$ for all such v and all ε, necessarily $g(0) \leq g(\varepsilon)$ for all ε. Since g is now differentiable, we compute $g'(\varepsilon)$ and we must have $g'(0) = 0$. One has

$$g(\varepsilon) = \lambda \int_\Omega |f - K(u + \varepsilon v)|^2 dx + \int_\Omega \sqrt{\epsilon^2 + |D(u + \varepsilon v)|^2} dx,$$

and

$$g'(\varepsilon) = 2\lambda \int_\Omega [f - K(u + \varepsilon v)](-Kv) dx + \int_\Omega \frac{(Du + \varepsilon Dv) \cdot Dv}{\sqrt{\epsilon^2 + |D(u + \varepsilon v)|^2}} dx.$$

We thus have

$$g'(0) = 2\lambda \int_\Omega (f - Ku)(-Kv) dx + \int_\Omega \frac{Du \cdot Dv}{\sqrt{\epsilon^2 + |Du|^2}} dx,$$

and using the adjoint K^* of K for the first term, and integration by parts in the second term, we obtain

$$g'(0) = -2\lambda \int_\Omega K^*(f - Ku) v\, dx - \int_\Omega \operatorname{div}\Big(\frac{Du}{\sqrt{\epsilon^2 + |Du|^2}}\Big) v dx$$
$$+ \int_{\partial\Omega} \frac{Du}{\sqrt{\epsilon^2 + |Du|^2}} \cdot \vec{n} v ds = 0,$$

with $\vec{n}$ the unit outward normal to $\partial\Omega$. Taking now $v \in C_0^1(\Omega)$, we deduce that we must have

$$-2\lambda K^*(f - Ku) - \operatorname{div}\Big(\frac{Du}{\sqrt{\epsilon^2 + |Du|^2}}\Big) = 0$$

in Ω. Now, using the last two relations and taking $v \in C^1(\overline{\Omega})$ arbitrary, we obtain the implicit boundary condition

$$\frac{Du}{\sqrt{\epsilon^2 + |Du|^2}} \cdot \vec{n} = 0$$

on $\partial\Omega$.

Summarizing, the Euler–Lagrange equation associated with $F_\epsilon(u)$ becomes:

$$0 = -2\lambda K^*(f - Ku) - \operatorname{div}\Big(\frac{Du}{\sqrt{\epsilon^2 + |Du|^2}}\Big) \text{ in } \Omega, \tag{3.11}$$

together with the boundary condition $\frac{\partial u}{\partial \vec{n}} = 0$ on $\partial\Omega$.

As presented in Chapter 2, we can also associate a time-dependent gradient descent, introducing a time $t \geq 0$ and assuming that $u = u(t, \cdot)$ and $u(0, \cdot) = f(\cdot)$ in Ω:

$$\frac{\partial u}{\partial t} = 2\lambda K^*(f - Ku) + \operatorname{div}\Big(\frac{Du}{\sqrt{\epsilon^2 + |Du|^2}}\Big) \text{ in } (0, \infty) \times \Omega, \tag{3.12}$$

$$\frac{\partial u}{\partial \vec{n}} = 0 \text{ on } (0, \infty) \times (\partial\Omega). \tag{3.13}$$

If u satisfies the above time-dependent problem, then $\frac{d}{dt}F_\epsilon(u(t, \cdot)) \leq 0$, thus $F_\epsilon(u(t, \cdot))$ is decreasing in time.

3.1.6 Numerical discretization of Euler–Lagrange equation

We discretize in several ways the Euler–Lagrange equation associated with the minimization of the total variation model of Rudin–Osher–Fatemi in two dimensions $(x, y) \in \Omega$.

Assume $h > 0$ and let $(x_i, y_j) = (ih, jh)$ for $0 \leq i, j \leq M$ be the $(M+1) \times (M+1)$ discrete points. Let $u_{i,j} \approx u(x_i, y_j)$ and $f_{i,j} = f(x_i, y_j)$. We will use the notations from Section 2.10.

A discrete (nonlinear) form of the Euler–Lagrange equation (3.11) is:

$$\begin{aligned}(K^*Ku)_{i,j} &= (K^*f)_{i,j} + \frac{1}{2\lambda}\triangle_-^x\Big[\frac{1}{\sqrt{\epsilon^2 + (\triangle_+^x u_{i,j})^2 + (\triangle_+^y u_{i,j})^2}}\triangle_+^x u_{i,j}\Big] \\ &\quad + \frac{1}{2\lambda}\triangle_-^y\Big[\frac{1}{\sqrt{\epsilon^2 + (\triangle_+^x u_{i,j})^2 + (\triangle_+^y u_{i,j})^2}}\triangle_+^y u_{i,j}\Big],\end{aligned}$$

or

$$\begin{aligned}(K^*Ku)_{i,j} &= (K^*f)_{i,j} + \frac{1}{2\lambda h^2}\frac{u_{i+1,j} - u_{i,j}}{\sqrt{\epsilon^2 + (\frac{u_{i+1,j} - u_{i,j}}{h})^2 + (\frac{u_{i,j+1} - u_{i,j}}{h})^2}} \\ &\quad - \frac{1}{2\lambda h^2}\frac{u_{i,j} - u_{i-1,j}}{\sqrt{\epsilon^2 + (\frac{u_{i,j} - u_{i-1,j}}{h})^2 + (\frac{u_{i-1,j+1} - u_{i-1,j}}{h})^2}} \\ &\quad + \frac{1}{2\lambda h^2}\frac{u_{i,j+1} - u_{i,j}}{\sqrt{\epsilon^2 + (\frac{u_{i+1,j} - u_{i,j}}{h})^2 + (\frac{u_{i,j+1} - u_{i,j}}{h})^2}} \\ &\quad - \frac{1}{2\lambda h^2}\frac{u_{i,j} - u_{i,j-1}}{\sqrt{\epsilon^2 + (\frac{u_{i+1,j-1} - u_{i,j-1}}{h})^2 + (\frac{u_{i,j} - u_{i,j-1}}{h})^2}}.\end{aligned}$$

The above nonlinear discrete equation is obtained from imposing that the gradient of the following discrete energy is zero:

$$\sum_{(i,j)\in\Omega} h^2\left[\lambda(f_{i,j}-(Ku)_{i,j})^2+\sqrt{\epsilon^2+\left(\frac{u_{i+1,j}-u_{i,j}}{h}\right)^2+\left(\frac{u_{i,j+1}-u_{i,j}}{h}\right)^2}\right].$$

Introducing the notations

$$\begin{aligned} c_1(u) &= \frac{1}{\sqrt{\epsilon^2+(\frac{u_{i+1,j}-u_{i,j}}{h})^2+(\frac{u_{i,j+1}-u_{i,j}}{h})^2}}, \\ c_2(u) &= \frac{1}{\sqrt{\epsilon^2+(\frac{u_{i,j}-u_{i-1,j}}{h})^2+(\frac{u_{i-1,j+1}-u_{i-1,j}}{h})^2}}, \\ c_3(u) &= \frac{1}{\sqrt{\epsilon^2+(\frac{u_{i+1,j}-u_{i,j}}{h})^2+(\frac{u_{i,j+1}-u_{i,j}}{h})^2}}, \\ c_4(u) &= \frac{1}{\sqrt{\epsilon^2+(\frac{u_{i+1,j-1}-u_{i,j-1}}{h})^2+(\frac{u_{i,j}-u_{i,j-1}}{h})^2}}, \end{aligned}$$

(note that $c_1 = c_3$), then the discrete nonlinear finite difference scheme inside discrete Ω becomes

$$\begin{aligned} (K^*Ku)_{i,j} &= (K^*f)_{i,j} \\ &+ \frac{1}{2\lambda h^2}\Big[c_1(u)u_{i+1,j}+c_2(u)u_{i-1,j}+c_3(u)u_{i,j+1}+c_4(u)u_{i,j-1} \\ &- \Big(c_1(u)+c_2(u)+c_3(u)+c_4(u)\Big)u_{i,j}\Big]. \end{aligned}$$

3.1.6.1 Denoising, $K = I$

In the case of denoising only ($K = I$), we can use fixed point iteration in the above equation (letting $u^0_{i,j} = f_{i,j}$) so we introduce linearized equations in u^{n+1} for $n \geq 0$.

Stationary explicit scheme In this case, the explicit scheme for the stationary Euler–Lagrange equation is given by

$$\begin{aligned} u^{n+1}_{i,j} &= f_{i,j} \\ &+ \frac{1}{2\lambda h^2}\Big[c_1(u^n)u^n_{i+1,j}+c_2(u^n)u^n_{i-1,j}+c_3(u^n)u^n_{i,j+1}+c_4(u^n)u^n_{i,j-1} \\ &- \Big(c_1(u^n)+c_2(u^n)+c_3(u^n)+c_4(u^n)\Big)u^n_{i,j}\Big]. \end{aligned}$$

Stationary semi-implicit scheme In this case, a semi-implicit scheme for the stationary Euler–Lagrange equation is given by

$$\begin{aligned} u_{i,j}^{n+1} &= f_{i,j} \\ &+ \frac{1}{2\lambda h^2}\Big[c_1(u^n)u_{i+1,j}^n + c_2(u^n)u_{i-1,j}^n + c_3(u^n)u_{i,j+1}^n + c_4(u^n)u_{i,j-1}^n \\ &- \Big(c_1(u^n) + c_2(u^n) + c_3(u^n) + c_4(u^n)\Big)u_{i,j}^{n+1}\Big], \end{aligned}$$

and solving for $u_{i,j}^{n+1}$, we obtain

$$\begin{aligned} u_{i,j}^{n+1} &= \left(\frac{1}{1 + \frac{1}{2\lambda h^2}(c_1(u^n) + c_2(u^n) + c_3(u^n) + c_4(u^n))}\right) \\ &\cdot \Big[f_{i,j} + \frac{1}{2\lambda h^2}(c_1(u^n)u_{i+1,j}^n + c_2(u^n)u_{i-1,j}^n + c_3(u^n)u_{i,j+1}^n \\ &+ c_4(u^n)u_{i,j-1}^n)\Big]. \end{aligned}$$

Evolutionary explicit scheme with $\triangle t > 0$ the time step, is given by

$$\begin{aligned} \frac{u_{i,j}^{n+1} - u_{i,j}^n}{\triangle t} &+ u_{i,j}^n \\ &= f_{i,j} + \frac{1}{2\lambda h^2}\Big[c_1(u^n)u_{i+1,j}^n + c_2(u^n)u_{i-1,j}^n + c_3(u^n)u_{i,j+1}^n \\ &+ c_4(u^n)u_{i,j-1}^n - \Big(c_1(u^n) + c_2(u^n) + c_3(u^n) + c_4(u^n)\Big)u_{i,j}^n\Big], \end{aligned}$$

from which we obtain

$$\begin{aligned} u_{i,j}^{n+1} &- u_{i,j}^n + \triangle t(f_{i,j} - u_{i,j}^n) \\ &+ \frac{\triangle t}{2\lambda h^2}\Big[c_1(u^n)u_{i+1,j}^n + c_2(u^n)u_{i-1,j}^n + c_3(u^n)u_{i,j+1}^n + c_4(u^n)u_{i,j-1}^n \\ &- \Big(c_1(u^n) + c_2(u^n) + c_3(u^n) + c_4(u^n)\Big)u_{i,j}^n\Big]. \end{aligned}$$

Evolutionary semi-implicit scheme is obtained from

$$\begin{aligned} \frac{u_{i,j}^{n+1} - u_{i,j}^n}{\triangle t} &+ u_{i,j}^{n+1} \\ &= f_{i,j} + \frac{1}{2\lambda h^2}\Big[c_1(u^n)u_{i+1,j}^n + c_2(u^n)u_{i-1,j}^n + c_3(u^n)u_{i,j+1}^n \\ &+ c_4(u^n)u_{i,j-1}^n - \Big(c_1(u^n) + c_2(u^n) + c_3(u^n) + c_4(u^n)\Big)u_{i,j}^{n+1}\Big], \end{aligned}$$

and solving for $u_{i,j}^{n+1}$, we obtain

$$\begin{aligned} u_{i,j}^{n+1} &= \left(\frac{1}{1 + \triangle t + \frac{\triangle t}{2\lambda h^2}(c_1(u^n) + c_2(u^n) + c_3(u^n) + c_4(u^n))}\right) \\ &\cdot \Big[u_{i,j}^n + \triangle t f_{i,j} + \frac{\triangle t}{2\lambda h^2}(c_1(u^n)u_{i+1,j}^n + c_2(u^n)u_{i-1,j}^n + c_3(u^n)u_{i,j+1}^n \\ &+ c_4(u^n)u_{i,j-1}^n)\Big]. \end{aligned}$$

We use any of the above equations for $u_{i,j}^{n+1}$ for all interior points (x_i, y_j) such that $1 \leq i, j \leq M-1$.

Boundary conditions can be implemented in the following way: if $u_{i,j}^n$ has been computed using one of the above numerical schemes for $1 \leq i, j \leq M-1$, then we let

$$u_{0,j}^n = u_{1,j}^n,\ u_{M,j}^n = u_{M-1,j}^n,\ u_{i,0}^n = u_{i,1}^n,\ u_{i,M}^n = u_{i,M-1}^n,$$

and

$$u_{0,0}^n = u_{1,1}^n,\ u_{0,M}^n = u_{1,M-1}^n,\ u_{M,0}^n = u_{M-1,1}^n,\ u_{M,M}^n = u_{M-1,M-1}^n.$$

Remark 12 *For the semi-implicit schemes, using $u_{i,j}^0 = f_{i,j}$, we note that if $m_1 \leq f_{i,j} \leq m_2$ for any $0 \leq i, j \leq M$, then we have $m_1 \leq u_{i,j}^n \leq m_2$ for any $n \geq 0$ (thus such schemes are unconditionally stable).*

3.1.6.2 Denoising and deblurring

When K is a blur operator (such as a convolution), similarly we present two numerical discretizations in the time-dependent case.

Evolutionary explicit scheme is given by

$$\begin{aligned}
&\frac{u_{i,j}^{n+1} - u_{i,j}^n}{\triangle t} + (K^*Ku^n)_{i,j} = (K^*f)_{i,j} \\
&+ \frac{1}{2\lambda h^2}\Big[c_1(u^n)u_{i+1,j}^n + c_2(u^n)u_{i-1,j}^n + c_3(u^n)u_{i,j+1}^n \\
&+ c_4(u^n)u_{i,j-1}^n - \Big(c_1(u^n) + c_2(u^n) + c_3(u^n) + c_4(u^n)\Big)u_{i,j}^n\Big],
\end{aligned}$$

from which we obtain

$$\begin{aligned}
u_{i,j}^{n+1} &= u_{i,j}^n + \triangle t((K^*f)_{i,j} - (K^*Ku^n)_{i,j}) \\
&+ \frac{\triangle t}{2\lambda h^2}\Big[c_1(u^n)u_{i+1,j}^n + c_2(u^n)u_{i-1,j}^n + c_3(u^n)u_{i,j+1}^n + c_4(u^n)u_{i,j-1}^n \\
&- \Big(c_1(u^n) + c_2(u^n) + c_3(u^n) + c_4(u^n)\Big)u_{i,j}^n\Big].
\end{aligned}$$

Evolutionary semi-implicit scheme is obtained from

$$\begin{aligned}
&\frac{u_{i,j}^{n+1} - u_{i,j}^n}{\triangle t} + (K^*Ku^n)_{i,j} = (K^*f)_{i,j} \\
&+ \frac{1}{2\lambda h^2}\Big[c_1(u^n)u_{i+1,j}^n + c_2(u^n)u_{i-1,j}^n + c_3(u^n)u_{i,j+1}^n \\
&+ c_4(u^n)u_{i,j-1}^n - \Big(c_1(u^n) + c_2(u^n) + c_3(u^n) + c_4(u^n)\Big)u_{i,j}^{n+1}\Big],
\end{aligned}$$

and solving for $u_{i,j}^{n+1}$, we obtain

$$\begin{aligned} u_{i,j}^{n+1} &= \left(\frac{1}{1+\frac{\triangle t}{2\lambda h^2}(c_1(u^n)+c_2(u^n)+c_3(u^n)+c_4(u^n))}\right) \\ &\cdot \Big[u_{i,j}^n + \triangle t(K^*f)_{i,j} - \triangle t(K^*Ku^n)_{i,j} \\ &+ \frac{\triangle t}{2\lambda h^2}(c_1(u^n)u_{i+1,j}^n + c_2(u^n)u_{i-1,j}^n \\ &+ c_3(u^n)u_{i,j+1}^n + c_4(u^n)u_{i,j-1}^n)\Big]. \end{aligned}$$

Note that the above schemes may introduce some asymmetry that is not visible in general. Other schemes can be proposed, for instance alternating at each iteration the discretization of the divergence operator, with all four (schematic) choices given below (as proposed in [23]):

$$\begin{array}{l} \triangle_-^x(\triangle_+^x),\ \triangle_-^y(\triangle_+^y) \text{ (the above scheme)} \\ \triangle_+^x(\triangle_-^x),\ \triangle_+^y(\triangle_-^y) \\ \triangle_+^x(\triangle_-^x),\ \triangle_-^y(\triangle_+^y) \\ \triangle_-^x(\triangle_+^x),\ \triangle_+^y(\triangle_-^y). \end{array}$$

3.2 Numerical results for image restoration

Several numerical results for total variation-based image restoration are presented in this section on synthetic and real images. The restoration experiments (taken from Jung and Vese [183, 179]) are applied to noisy and blurry images. The blur Ku is modeled by a convolution $Ku = k * u$ and the point spread function k is mentioned in each experiment. The time-dependent gradient descent approach combined with the proposed semi-implicit scheme presented in this chapter has been used. These results can be compared with others obtained by more recent methods and presented in later chapters.

Input images f shown in Figures 3.1 through 3.3 were degraded by blur and additive Gaussian noise of zero mean. These were restored using the (BV, L^2) Rudin–Osher–Fatemi model (3.2) with L^2 data fidelity term, appropriate for image restoration in the presence of Gaussian noise.

Sometimes, the input image f is degraded by blur and additive Laplace noise, or random-valued impulse noise (such as salt-and-pepper noise which might be caused by bit errors in transmissions). In such cases, the restoration works better using the (BV, L^1) model,

$$\inf_{u\in BV(\Omega)} \Big\{\lambda\int_\Omega |f-Ku|dx + \int_\Omega |Du|\Big\}.$$

When $K = I$, this model was developed and analyzed by Alliney [11], [12], [13]

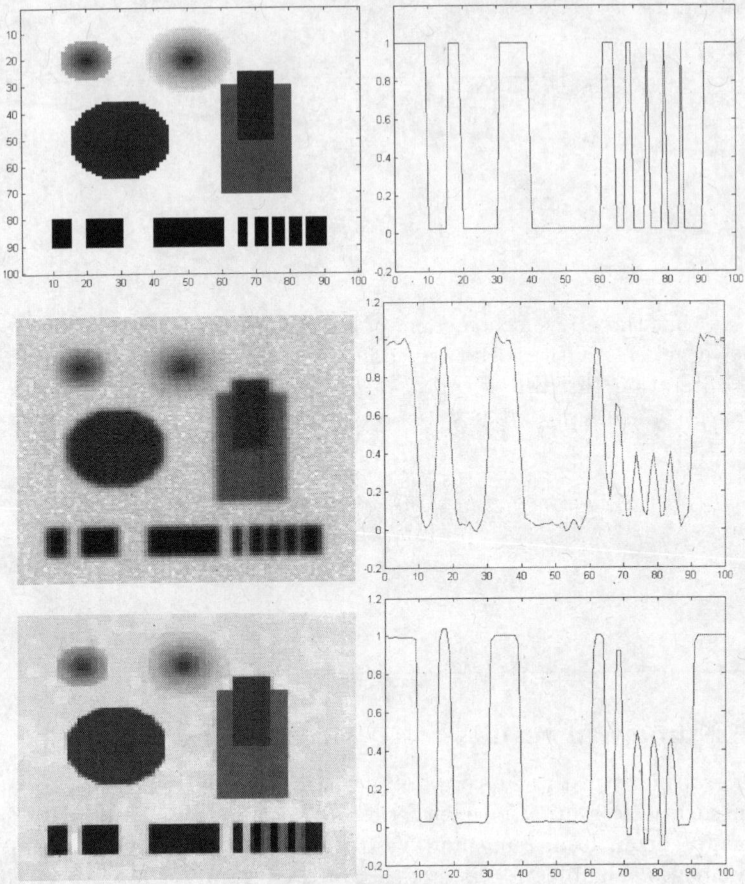

FIGURE 3.1: Image recovery: Gaussian blur kernel with $\sigma_b = 1$ and Gaussian noise with $\sigma_n = 5$. Top: original image and its cross section. Middle: noisy blurry version f and its cross section. Bottom: recovered image u and its cross section (signal-to-noise ratio (SNR) = 32.9485).

in the one-dimensional discrete case, and by Chan and Esedoḡlu [87] in the continuous two-dimensional case. For fine theoretical results on the (BV, L^1) model without blur, we also refer the reader to Allard [8], [9], [10]. The L^1 data fidelity term is more appropriate for removing random-valued impulse noise than the L^2 data fidelity due to its ability to remove outliers without much affecting the image quality. The L^1 data fidelity term $\int_\Omega |f - Ku| dx$ can be made differentiable by approximating it with $\int_\Omega \sqrt{\epsilon^2 + (f - Ku)^2} dx$, with $\epsilon > 0$ a very small parameter.

FIGURE 3.2: Left to right: original image; noisy blurry version f with Gaussian kernel with $\sigma_b = 1$ contaminated by additive Gaussian noise with $\sigma_n = 5$; recovered image u (SNR = 14.4240).

FIGURE 3.3: Top: original image and noisy blurry version f using the pillbox kernel of radius 2 contaminated by additive Gaussian noise with $\sigma_n = 5$. Bottom: recovered image u (SNR = 25.0230) and residual $f - Ku$.

3.3 Compressive sensing for computerized tomography reconstruction

The classical assumption in signal and image processing is that a signal or image must be sampled at a rate at least twice its highest frequency in order to be represented without error. However, in practice, we often compress the data soon after sensing, trading off signal representation complexity (bits) for some error (consider, for example, JPEG compression of images in digital cameras). Over the past few years, a new theory of "compressive sensing" has begun to emerge, in which the signal or image is sampled (and simultaneously compressed) at a greatly reduced rate.

For original work on compressive sensing theory and applications, we refer the reader to seminal work of Candes, Romberg, Tao [78, 79] and Donoho [129]. We also refer the reader to the online *Compressive Sensing Resources* http://dsp.rice.edu/cs site maintained by Rice University.

One of the main applications of compressive sensing is in medical image acquisition. For example, in computerized tomography (CT), the aim is to reduce the radiation dose by reducing the number of measurements (samples or projections) without compromising the image quality.

This section presents an application of the total variation regularization to image reconstruction in computerized tomography, from a reduced number of measurements in the Radon (sinogram) domain. This section follows the work of Yan et al. [347, 226].

3.3.1 Expectation maximization and total variation-based model for computed tomography reconstruction from undersampled data

We focus here on computerized tomography. We recall the continuous Radon transform in two dimensions, applied to an image function $u(x, y)$. Assuming that a straight line in the plane can be represented by the equation depending on two parameters (θ, t),

$$x \cos\theta + y \sin\theta = t,$$

the Radon transform $u \mapsto f$ (with $(\theta, t) \mapsto f(\theta, t)$ the image in the sinogram domain (θ, t) of the function $(x, y) \mapsto u(x, y)$) is defined by the projections along lines of u,

$$f(\theta, t) = \int_{-\infty}^{+\infty} \int_{-\infty}^{+\infty} u(x, y)\delta(x \cos\theta + y \sin\theta - t)dxdy,$$

where δ is the one-dimensional Dirac delta function, with support on the line of equation $x \cos\theta + y \sin\theta = t$. In the two-dimensional discrete case, u represents a discrete image matrix and f (or a noisy version of f) is known only at a finite number of samples (θ_i, t_i). In computed tomography, we must recover u from its projections f. Although the Radon transform is an invertible operator, in the real discrete case only a small number of projections (θ_i, t_i) are given, which may be also noisy. Thus, in the real case, the recovery of u from projections f is a difficult inverse problem. Given discrete data f in the sinogram domain obtained from an image u in the spatial domain, we denote by K

the discrete Radon transform, thus $f = Ku$ (possibly with noise). Let us assume that the discrete spatial domain has $j = 1, ..., N$ points, that the discrete samples f are given at $i = 1, ..., M$ points, and $K = (K_{i,j})_{i=1,...,M,j=1,...,N}$ is the discrete Radon matrix.

For the reconstruction of two-dimensional and three-dimensional images from the projections of the object, the filtered back projection (FBP) method [292] is the most commonly used algorithm in practice by manufacturers. However, the filtered back projection method is sensitive to noise and it is not flexible (the projections must be distributed unformly in angle and must be complete for high accuracy; the data cannot be collected over any set of lines).

One example of reconstruction algorithm is expectation maximization (EM) [127, 293]. This is based on the assumption that the noise in f is Poisson noise. If u is given, the probability of obtaining f is

$$P(f|Ku) = \Pi_{i=1}^{M} \frac{e^{-(Ku)_i}((Ku)_i)^{f_i}}{f_i!},$$

the product being taken over the discrete sinogram domain composed of points (θ_i, t_i).

Therefore, given f and K, the objective is to find u such that the above probability is maximized. However, instead of maximizing the probability, we can minimize $-\log P(f|Ku) = \sum_{i=1}^{M}((Ku)_i - f_i \log((Ku)_i)) + C$, with C being a constant, and the sum being taken over the discrete sinogram domain. Then the EM iteration is as follows: starting with an initial guess u^0, compute for $n \geq 0$ at each discrete spatial domain,

$$u_j^{n+1} = \frac{\sum_{i=1}^{M}(K_{ij}(\frac{f_i}{(Ku^n)_i}))}{\sum\limits_{i=1}^{M} K_{ij}} u_j^n. \tag{3.14}$$

Following Yan et al. [347, 226], we present here a combination of the expectation maximization (EM) method using Poisson noise with the total variation TV regularization, when K is the discrete Radon transform and the data f represents a small number of projections in computerized tomography. The assumption is that the reconstructed image cannot have too large a total variation (thus noise and reconstruction artifacts are removed). We will see that the presented EM+TV method gives superior results to those obtained by filtered back projection or by expectation maximization only.

For relevant work, we refer to the Compressive Sensing Resources site of Rice University [1]. Additionally, we refer the reader to Jia et al. [176], Jung et al. [182], Setzer et al. [287], Le et al. [203], Brune et al. [70, 71], Jafarpour et al. [175], Harmany et al. [169], Willet et al. [344], among other work.

3.3.1.1 EM+TV reconstruction method

The objective is to reconstruct an image with both minimal total variation and maximal probability, given fewer noisy projections in the sinogram domain. So we can consider finding a Pareto optimal point by solving a scalarization of these two objective functions and the problem is to solve

$$\begin{cases} \min\limits_{u} & \int_\Omega |\nabla u| + \alpha \sum\limits_{i=1}^{M} \left((Ku)_i - f_i \log(Ku)_i\right), \\ \text{subject to} & u_j \geq 0, \quad j = 1, \cdots, N, \end{cases} \tag{3.15}$$

with $\alpha > 0$ a tuning parameter and Ω the spatial domain of u. This is a convex constrained optimization problem and we can find the optimal solution by solving the Karush–Kuhn–Tucker (KKT) conditions [200, 186]:

$$\begin{aligned} -\operatorname{div}\Big(\frac{\nabla u}{|\nabla u|}\Big)_j + \alpha \sum_{i=1}^{M} \Big(K_{ij}(1 - \frac{f_i}{(Ku)_i})\Big) - v_j = 0, \qquad & j = 1, \cdots, N, \\ v_j \geq 0, \quad u_j \geq 0, \qquad & j = 1, \cdots, N, \\ v^T u = 0. & \end{aligned}$$

By positivity of $\{u_j\}$, $\{v_j\}$ and the complementary slackness condition $v^T u = 0$, we have $u_j v_j = 0$ for every $j = 1, \cdots, N$. Thus, if we multiply the partial differential equation by u_j at pixel j, we obtain

$$-u_j \operatorname{div}\Big(\frac{\nabla u}{|\nabla u|}\Big)_j + \alpha \sum_{i=1}^{M} \Big(K_{ij}(1 - \frac{f_i}{(Ku)_i})\Big) u_j = 0, \qquad j = 1, \cdots, N,$$

or equivalently,

$$-\frac{u_j}{\sum\limits_{i=1}^{M} K_{ij}} \operatorname{div}\Big(\frac{\nabla u}{|\nabla u|}\Big)_j + \alpha u_j - \alpha \frac{\sum\limits_{i=1}^{M} \Big(K_{ij}(\frac{f_i}{(Ku)_i})\Big)}{\sum\limits_{i=1}^{M} K_{ij}} u_j = 0, \qquad j = 1, \cdots, N.$$

After plugging the EM step from (3.14)

$$u_j^{EM} = \frac{\sum\limits_{i=1}^{M} (K_{ij}(\frac{f_i}{(Ku)_i}))}{\sum\limits_{i=1}^{M} K_{ij}} u_j \tag{3.16}$$

into the last KKT condition, we obtain

$$-\frac{u_j}{\sum\limits_{i=1}^{M} K_{ij}} \operatorname{div}\Big(\frac{\nabla u}{|\nabla u|}\Big)_j + \alpha u_j - \alpha u_j^{EM} = 0, \qquad j = 1, \cdots, N. \tag{3.17}$$

To solve the above EM+TV problem in the form of (3.17), we adapt the iterative semi-implicit finite-difference scheme described in Section 3.1.6. Each iteration is called a TV step. Given u^{EM}, already computed from the EM step, we compute the new u discretizing (3.17). Finally, the two steps (EM and TV) are solved in an alternating fashion. Usually, for each main iteration, we apply two or three EM steps, followed by five to eight TV steps. For the TV steps, the initial guess can be defined as the result from the previous EM update or from the last TV update.

3.3.1.2 Numerical results

We present in this subsection reconstruction results in two dimensions using the fan-beam computed tomography geometry, and in three dimensions using the cone-beam computed tomography geometry, on artificially constructed experiments. These were obtained by Yan [347, 226].

In two dimensions, we compare the reconstruction results obtained by the described EM+TV method with those obtained by filtered back projection (FBP). For the numerical experiments, we choose the two dimensional Shepp–Logan phantom of dimension 256 $\times$ 256. The projections are obtained using Siddon's algorithm [294, 351]. We consider both the noise-free and noise cases. With the FBP method, we present results using 36 views (every 10 degrees), 180 views, and 360 views; for each view there are 301 measurements. In order to show that we can reduce the number of views by using EM+TV, we only use 36 views for the described method. We show in Figure 3.4 the sinogram data without noise corresponding to 360, 180 and 36 views, respectively (the missing projection values are substituted by zero or black). The reconstruction results are shown in Figure 3.5. We notice the much improved results obtained with EM+TV using only 36 views (both visually and according to the root mean square error (RMSE) between the original and reconstructed images, scaled between 0 and 255), by comparison with FBP using 36, 180 or even 360 views. Using the proposed EM+TV method, with only few samples, we obtained sharp results and without artifacts.

We have seen that, in two dimensions, the presented EM+TV method gives superior results over the standard filtered back projection. In three dimensions, we compare the reconstruction results obtained by the described EM+TV method with those obtained by EM only, visually and according to the root mean square error between original and reconstructed images after rescaling to [0,255]. For the numerical experiments, we similarly choose the three dimensional Shepp–Logan phantom of dimension 128 $\times$ 128 $\times$ 128. The projections are obtained using Siddon's algorithm [294, 351]. We consider only the noise-free case here. For both methods, we present results using 36 views, and for each view, there are 301 $\times$ 257 measurements. The reconstruction results are shown in Figure 3.6. We notice again the much improved results obtained using EM+TV, by comparison with EM only. The results obtained

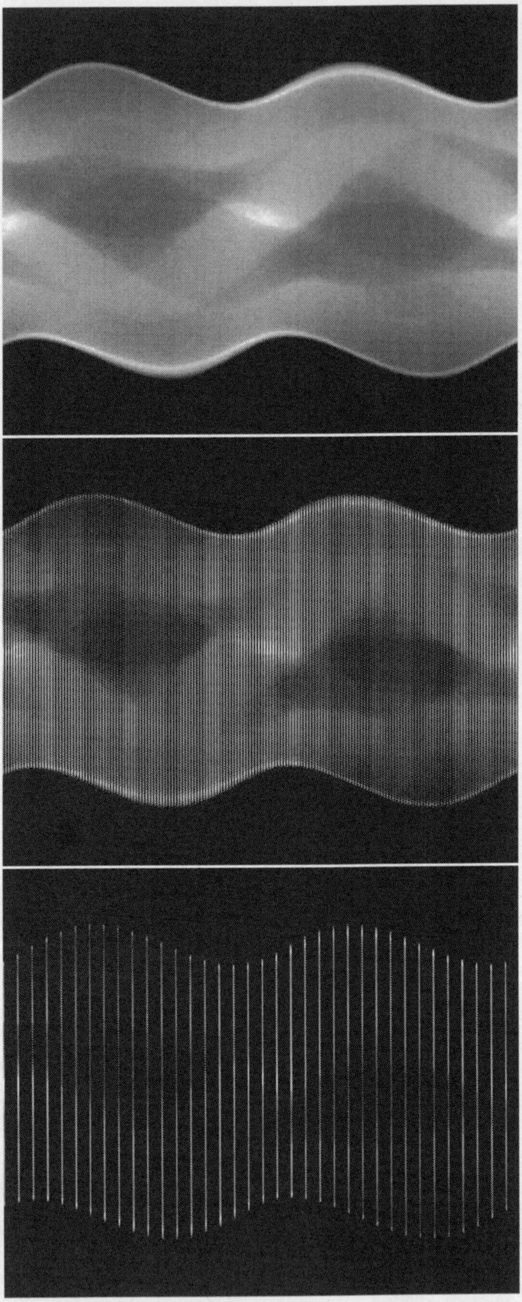

FIGURE 3.4: The sinogram data without noise in two dimensions, generated from the two-dimensional Shepp–Logan phantom. Top to bottom: 360 views, 180 views, 36 views. The x-scale is [0,360] in angle, and the y-scale is [1,301] for the second line parameter.

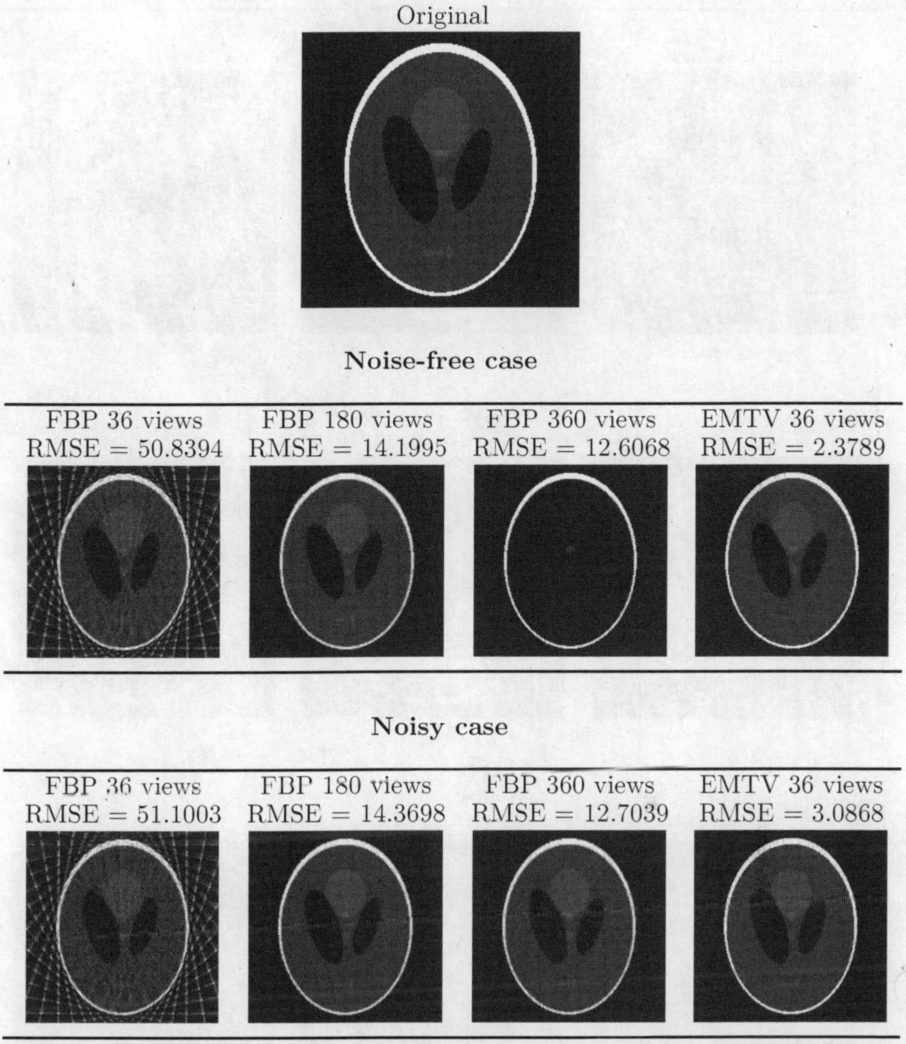

FIGURE 3.5: Reconstruction results in two dimensions. Top: original image (the Shepp–Logan phantom). Middle from left to right: reconstruction results in the noise-free case using FBP with 36, 180 and 360 views, and result using EM+TV with 36 views. Bottom from left to right: reconstruction results in the noisy case using FBP with 36, 180 and 360 views, and result using EM+TV with 36 views. The root mean square errors are also given.

using EM only (without the total variation regularization) still have many very clear artifacts and needed many more EM steps.

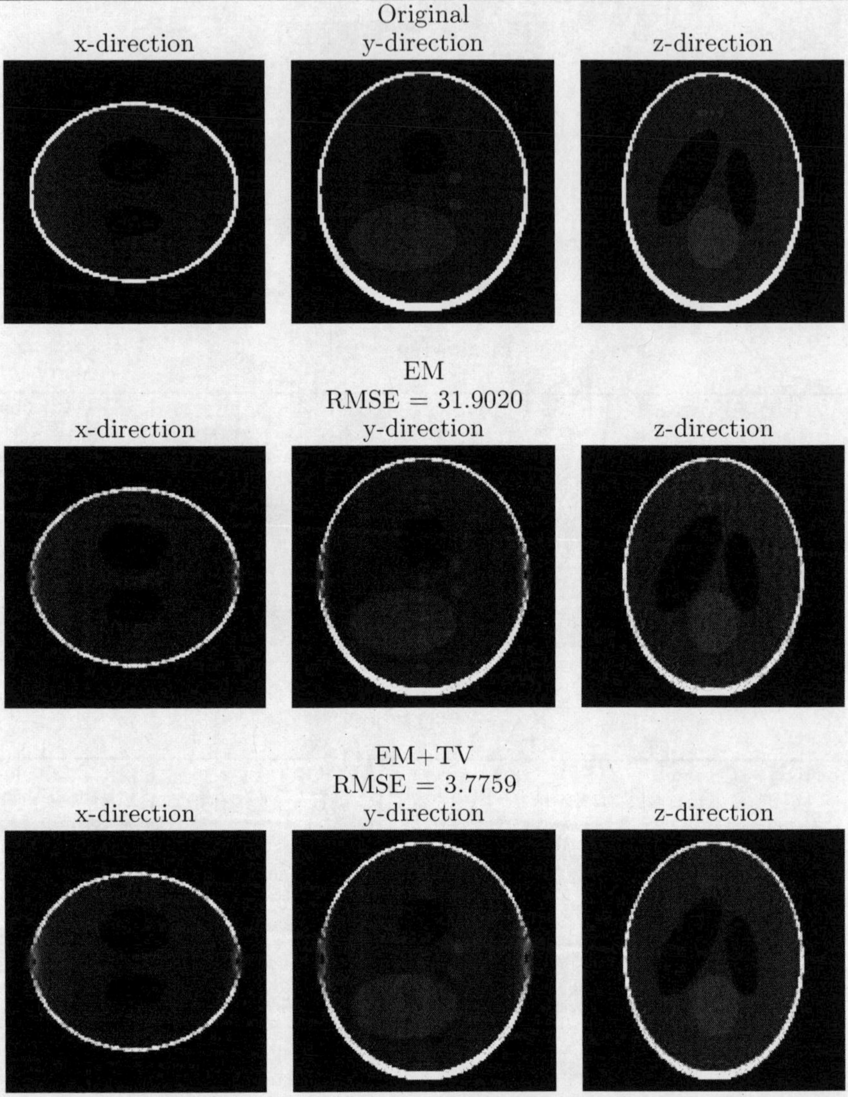

FIGURE 3.6: Reconstruction results in three dimensions in the noise-free case. Top row: two-dimensional views of the original three-dimensional Shepp–Logan phantom. Middle row: two-dimensional views of reconstruction results obtained using the EM method. Bottom row: two-dimensional views of reconstruction results obtained using the described EM+TV method. The root mean square errors are also given.

3.4 Exercises

Exercise 3.1 *Consider the general image restoration functional*

$$F(u) = \int_\Omega \phi(|Du|) + \lambda \int_\Omega (f - Ku)^2 dx, \tag{3.18}$$

with $f, u : \Omega \to \mathbb{R}$, Ω *an open and bounded subset of* $\mathbb{R}^2$, $f \in L^2(\Omega)$, $K : L^2(\Omega) \to L^2(\Omega)$ *a linear and continuous operator, and* $\phi : [0, \infty) \mapsto [0, \infty)$ *continuous and increasing, with* $\phi(|t|) \leq a|t| + b$ *for all* t, a, b *constants and* $a > 0$. *If* ϕ *is convex, show that the functional* F *is convex on* $BV(\Omega)$. *When is* F *strictly convex?*

Exercise 3.2 *Assume that* ϕ *satisfies the conditions given in Exercise 3.1, that it is differentiable, and that all quantities are well defined.*

(a) Formally derive the following Euler-Lagrange equation associated with the minimization of F *from (3.18):*

$$2\lambda K^*(Ku - f) = div\Big(\frac{\phi'(|Du|)}{|Du|} Du\Big), \quad \frac{\phi'(|Du|)}{|Du|} Du \cdot \vec{n} = 0_{|\partial\Omega}.$$

(b) When ϕ *is twice differentiable, show that the divergence operator thus obtained can be expressed as*

$$div\Big(\frac{\phi'(|Du|)}{|Du|} Du\Big) = \frac{\phi'(|Du|)}{|Du|} u_{\vec{T}\vec{T}} + \phi''(|Du|) u_{\vec{N}\vec{N}}$$

at points $x \in \Omega$ *with* $|Du|(x) \neq 0$, *where* $\vec{N} = \frac{Du}{|Du|}$, $\vec{T}$ *is orthogonal to* $\vec{N}$ *and of unit length, and* $u_{\vec{T}\vec{T}} = \vec{T}^t D^2 u \vec{T}$, $u_{\vec{N}\vec{N}} = \vec{N}^t D^2 u \vec{N}$ *are the second order (directional) derivatives of* u *in the* $\vec{T}$ *and* $\vec{N}$ *directions. Moreover, verify that* $u_{\vec{T}\vec{T}} + u_{\vec{N}\vec{N}} = \triangle u$.

(c) Give a geometric interpretation of the obtained diffusion equation, function of the convexity of ϕ.

(d) Under what conditions on ϕ *does the quasilinear second order operator* $div\Big(\frac{\phi'(|Du|)}{|Du|} Du\Big) = \sum_{i,j=1}^2 a_{i,j}(Du) u_{x_i,x_j}$ *satisfy the degenerate ellipticity property*

$$\sum_{i,j=1}^2 a_{i,j}(Du)\xi_i\xi_j \geq 0 \text{ for all } \xi_1, \xi_2 \in \mathbb{R} \text{ ?}$$

Exercise 3.3 *Consider in two dimensions* $f \in L^2(\Omega)$, $\lambda > 0$, *and* $u(\cdot, \lambda) \in BV(\Omega)$ *the (unique) minimizer of the ROF denoising functional*

$$F(u) = \lambda \int_\Omega |Du| + \frac{1}{2} \int_\Omega (f - u)^2 dx.$$

(a) Show that the L^2*-norm of* $u(\cdot,\lambda)$ *given by* $\sqrt{\int_\Omega u(x,\lambda)^2 dx}$ *is bounded from above by a constant independent of* λ.

(b) Show that $\int_\Omega u(x,\lambda)dx = \int_\Omega f(x)dx$*, for any* $\lambda > 0$.

(c) Show that $u(\cdot,\lambda)$ *converges in the* $L^1(\Omega)$ *norm to the average of the data* f*, as* $\lambda \to \infty$*. In other words, prove*

$$\lim_{\lambda\to\infty} \int_\Omega \left| u(x,\lambda) - \frac{\int_\Omega f(x)dx}{|\Omega|} \right| dx = 0.$$

Exercise 3.4 *Consider in two dimensions* $f \in L^2(\mathbb{R}^2)$, $\lambda > 0$*, and the ROF denoising functional*

$$F(u) = \lambda \int |Du| + \frac{1}{2}\int (f-u)^2 dx.$$

Show the following invariance properties:

(a) For any constant c*, if* u *is a minimizer of* F *with data* f*, then* $u + c$ *is a minimizer of* F *with data* $f + c$ *(gray-level invariance).*

(b) For any translation τ_h *defined by* $(\tau_h)(f)(x) = f(x+h)$, $h \in \mathbb{R}^2$*, if* u *is a minimizer of* F *with data* f*, then* $\tau_h(u)$ *is a minimizer of* F *with data* $\tau_h(f)$ *(translation invariance).*

(c) Let R *be an isometry of* $\mathbb{R}^2$ *and define* $(Rf)(x) = f(Rx)$*. If* u *is a minimizer of* F *with data* f*, then* Ru *is a minimizer with data* Rf *(isometry invariance).*

Exercise 3.5 *Let* $\alpha > 0$, $\lambda > 0$ *and* $f : \mathbb{R}^2 \mapsto \mathbb{R}$ *be defined by* $f(x) = \alpha\chi_D$*, where* $D = B(0,R)$ *is the open disk centered at the origin of radius* $R > 0$*. Consider*

$$F(u) = \lambda \int_{\mathbb{R}^2} (f-u)^2 dx + \int_{\mathbb{R}^2} |Du|, \forall u \in BV(\mathbb{R}^2).$$

Show in an elementary way that f *cannot be a minimizer of* F *over* $BV(\mathbb{R}^2)$.

Exercise 3.6 *Let* Ω *be an open, bounded and connected subset of* $\mathbb{R}^2$, $\lambda > 0$, $f \in L^2(\Omega)$*, and* u_λ *be a solution of*

$$\inf_{v\in BV(\Omega)} \int_\Omega |Dv| + \lambda \int_\Omega (Kv - f)^2 dx$$

where $K : L^2(\Omega) \to L^2(\Omega)$ *is linear and continuous. Show that the function* $\sigma(\lambda) = \|Ku_\lambda - f\|_{L^2(\Omega)}$ *is well defined and decreasing.*

Exercise 3.7 *Let* $\epsilon > 0$ *and* $\Omega \subset \mathbb{R}^N$ *be an open and bounded subset of* $\mathbb{R}^N$.

(a) Show that $\sqrt{\epsilon^2 + |x|^2} = \sup\{x \cdot y + \sqrt{\epsilon^2(1-|y|^2)} :\ y \in \mathbb{R}^N,\ |y| \le 1\}$.

(b) For $u \in L^1(\Omega)$*, define*

$$J_0(u) = \sup\left\{ \int_\Omega u \, div\, \phi \, dx \middle| \phi \in C_c^1(\Omega;\mathbb{R}^N),\ \ |\phi| \le 1 \right\},$$

and

$$J_\epsilon(u) = \sup\left\{ \int_\Omega \left(u\ div\phi + \sqrt{\epsilon^2(1-|\phi(x)|^2)}\right) dx \;\middle|\; \phi \in C_c^1(\Omega;\mathbb{R}^N),\quad |\phi| \le 1 \right\}$$

($J_\epsilon(u) = \int_\Omega \sqrt{\epsilon^2 + |Du|^2}$). For any $\epsilon > 0$ and $u \in L^1(\Omega)$, show that $J_0(u) < \infty$ if and only if $J_\epsilon(u) < \infty$. Moreover, for any $u \in BV(\Omega)$,

$$\lim_{\epsilon \to 0} J_\epsilon(u) = J_0(u).$$

(c) For any $\epsilon \ge 0$, J_ϵ is (weakly) lower semicontinuous with respect to the weak L^p topology for $1 \le p < \infty$.

(d) For any $\epsilon \ge 0$, J_ϵ is convex.

(e) For any $\epsilon \ge 0$, J_ϵ is not a strictly convex function.

Note that, if $u \in BV(\Omega)$, then $J_\epsilon(u)$ coincides with $\int_\Omega \sqrt{\epsilon^2 + |Du|^2}$ which is well defined also based on the notion of convex functions of measure [126].

Exercise 3.8 *In the purely denoising case ($K = I$), show that the stationary semi-implicit scheme presented in Section 3.1.6 is unconditionally stable in the L^∞-norm.*

Exercise 3.9 *In the purely denoising case ($K = I$), show that the evolutionary semi-implicit scheme presented in Section 3.1.6 is unconditionally stable in the L^∞-norm.*

Chapter 4

Nonlocal Variational Methods in Image Restoration

The traditional regularization terms previously discussed (depending on the local image derivatives) are based on local image operators (like the image gradient), which denoise and preserve edges very well, but may induce loss of fine structures like texture during the restoration process. More recently, Buades et al. [72] introduced the nonlocal means filter or NL–means (as a type of neighborhood filter), producing excellent denoising results. Gilboa and Osher [154, 155] formulated the variational framework of NL–means by proposing nonlocal regularizing functionals using nonlocal operators such as the nonlocal gradient and nonlocal divergence. These have been applied very successfully to image denoising, inpainting, and restoration [154, 155, 261, 223, 183].

4.1 Introduction to neighborhood filters and NL–means

Nonlocal methods in image processing have been explored in many papers because they are well adapted to texture denoising while the standard denoising models working with local image information seem to consider texture as noise, which results in losing texture. Nonlocal methods are generalized from the neighborhood filters and patch-based methods. The idea of neighborhood filters is to restore a pixel by averaging the values of neighboring pixels with a similar grey level value.

Buades et al. [72] generalized the idea of using patch-based methods, and proposed a neighborhood filter called nonlocal means (or NL–means). Let $f : \Omega \to \mathbb{R}$ be an input noisy image, $\Omega \subset \mathbb{R}^2$. Define for $x \in \Omega$

$$u(x) = NLf(x) = \frac{1}{C(x)} \int_\Omega e^{-\frac{d_a(f(x),f(y))}{h^2}} f(y)dy$$

where d_a is the patch distance

$$d_a(f(x), f(y)) = \int_{\mathbb{R}^2} G_a(t)|f(x+t) - f(y+t)|^2 dt,$$

G_a is the Gaussian kernel with standard deviation a determining the patch size, $C(x) = \int_\Omega e^{-\frac{d_a(f(x),f(y))}{h^2}} dy$ is the normalization factor, and h is the filtering parameter which corresponds to the noise level (usually the standard deviation of the noise). The NL means not only compares the grey level at a single point but the geometrical configuration in a whole neighborhood (patch). Thus, to denoise a pixel, it is better to average the nearby pixels with similar structures rather than just with similar intensities.

In practice, to reduce the computational cost, a search window $\Omega_r = \{y \in \Omega : |y - x| \leq r\}$ is used instead of the entire domain Ω (the method in this case is called semi-local).

The weight function at $(x, y) \in \Omega \times \Omega$ depending on $f : \Omega \to \mathbb{R}$, is defined by

$$w(x,y) = exp\Big(-\frac{d_a(f(x), f(y))}{h^2}\Big).$$

The weight function $w(x, y)$ gives the similarity of image features between two pixels x and y, which is normally computed using the blurry noisy image f, or an improved version of it.

4.2 Variational nonlocal regularization for image restoration

Based on the gradient and divergence definitions on graphs in the context of machine learning, Gilboa and Osher [154, 155] derived the nonlocal operators. Let $u : \Omega \to \mathbb{R}$ be a function, and $w : \Omega \times \Omega \to \mathbb{R}$ be a weight function assumed to be nonnegative and symmetric. For $x \in \Omega$, the nonlocal gradient $\nabla_w u(x)$ at x is defined as the vector of all discrete derivatives $(\nabla_w u)(x, y) := (u(y) - u(x))\sqrt{w(x, y)}$, for $y \in \Omega$. Hence, the norm of the nonlocal gradient of u at $x \in \Omega$ is defined as

$$|\nabla_w u|(x) := \sqrt{\int_\Omega (u(y) - u(x))^2 w(x,y)dy}.$$

The nonlocal divergence $div_w \overrightarrow{v}(x) : \Omega \times \Omega \to \Omega$ is defined as the adjoint of the nonlocal gradient

$$(div_w \overrightarrow{v})(x) := \int_\Omega (v(x,y) - v(y,x))\sqrt{w(x,y)}dy.$$

The nonlocal Laplacian can now be defined by

$$\triangle_w u := \frac{1}{2}\text{div}_w(\nabla_w u(x)) = \int_\Omega (u(y) - u(x))w(x,y)dy.$$

Based on these nonlocal operators, Gilboa and Osher introduced *gradient-based* nonlocal regularizing functionals of the general form

$$\Psi(u) = \int_\Omega \phi(|\nabla_w u|^2)dx$$

where $\phi(s)$ is a positive function, convex in $\sqrt{s}$ with $\phi(0) = 0$.

By taking $\phi(s) = s$, they proposed the nonlocal H^1 regularizer (NL/H^1)

$$\Psi^{NL/H^1}(u) := \int_\Omega |\nabla_w u|^2 dx = \int_\Omega \int_\Omega (u(y) - u(x))^2 w(x,y)dydx.$$

By taking $\phi(s) = \sqrt{s}$, they proposed the nonlocal total variation regularizer (NL/TV)

$$\Psi^{NL/TV}(u) := \int_\Omega |\nabla_w u| dx = \int_\Omega \sqrt{\int_\Omega (u(y) - u(x))^2 w(x,y)dydx},$$

which extends the original notion of (local) total variation $\int_\Omega |\nabla u| dx$ to the nonlocal case.

Lou et al. [223] applied these last two nonlocal regularizers to image restoration (denoising and deblurring) and to reconstruction in computerized tomography. Peyré et al. [261] proposed a fast algorithm that adapts the nonlocal total variation penalization to the geometry of the underlying graph with applications to image inpainting, super-resolution, and compressive sampling. The results given in the remaining part of this chapter have been presented in the work of Jung [179] and Jung and Vese [183]. First, it is shown that $\Psi^{NL/TV}$ is a semi-norm, a useful result for proving a characterization of minimizers based on dual norm as in Chapter 3. Then the derivation of the corresponding Euler-Lagrange equations is given, together with numerical results for image denoising and deblurring.

4.2.1 Nonlocal total variation as semi-norm

We show that the following nonlocal total variation is a semi-norm, a necessary step in the proof of characterization of minimizers. We let $|u|_{NL/TV} = \int_\Omega \sqrt{\int_\Omega (u(y) - u(x))^2 w(x,y)dydx}$, with $u : \Omega \to \mathbb{R}$, $v : \Omega \to \mathbb{R}$ and $w : \Omega \times \Omega \to \mathbb{R}$ is nonnegative and symmetric. We only need to show that this functional satisfies the triangle inequality. We want to show that $|u + v|_{NL/TV} \leq |u|_{NL/TV} + |v|_{NL/TV}$ for $u, v : \Omega \to \mathbb{R}$. First, we have the equality

$$\begin{aligned}
&\int_\Omega ((u+v)(y)-(u+v)(x))^2 w(x,y)dy\\
&= \int_\Omega (u(y)-u(x))^2 w(x,y)dy\\
&+ \int_\Omega (v(y)-v(x))^2 w(x,y)dy\\
&+2\int_\Omega (u(y)-u(x))(v(y)-v(x))w(x,y)dy.
\end{aligned}$$

Using Cauchy–Schwarz inequality, we have

$$\begin{aligned}
&\int_\Omega (u(y)-u(x))(v(y)-v(x))w(x,y)dy\\
&\le \left(\int_\Omega \left((u(y)-u(x))\sqrt{w(x,y)}\right)^2 dy\right)^{1/2}\\
&\cdot \left(\int_\Omega \left((v(y)-v(x))\sqrt{w(x,y)}\right)^2 dy\right)^{1/2}.
\end{aligned}$$

Denoting

$$\begin{aligned}
a &= \int_\Omega (u(y)-u(x))^2 w(x,y)dy,\\
b &= \int_\Omega (v(y)-v(x))^2 w(x,y)dy,
\end{aligned}$$

we obtain

$$\begin{aligned}
&\int_\Omega ((u+v)(y)-(u+v)(x))^2 w(x,y)dy\\
&\le (a+b+2\sqrt{a}\sqrt{b}),\\
&= \left(\sqrt{a}+\sqrt{b}\right)^2,
\end{aligned}$$

which finally leads to

$$\begin{aligned}
&\sqrt{\int_\Omega ((u+v)(y)-(u+v)(x))^2 w(x,y)dy}\\
&\le \sqrt{\int_\Omega (u(y)-u(x))^2 w(x,y)dy}\\
&+\sqrt{\int_\Omega (v(y)-v(x))^2 w(x,y)dy}.
\end{aligned}$$

Integrating both sides with respect to x, we obtain

$$\begin{aligned}\int_\Omega \sqrt{\int_\Omega ((u+v)(y)-(u+v)(x))^2 w(x,y)dy}dx \\ \leq \int_\Omega \sqrt{\int_\Omega (u(y)-u(x))^2 w(x,y)dy}dx \\ + \int_\Omega \sqrt{\int_\Omega (v(y)-v(x))^2 w(x,y)dy}dx.\end{aligned}$$

Thus $|u|_{NL/TV}$ satisfies the triangle inequality and we conclude that $|u|_{NL/TV}$ is a semi-norm.

4.2.2 Characterization of minimizers

In this section we characterize the minimizers of the restoration functional with NL/TV regularization using [236, 309, 180]. Assuming that a functional $\|\cdot\|$ on a subspace of $L^2(\Omega)$ is a semi-norm, we can define the dual norm (where $\langle\cdot,\cdot\rangle$ denotes the $L^2(\Omega)$ inner product) of $f \in L^2(\Omega)$ as $\|f\|_* := \sup_{\|\varphi\|\neq 0} \frac{\langle f,\varphi\rangle}{\|\varphi\|} \leq +\infty$, so that the usual duality $\langle f,\varphi\rangle \leq \|\varphi\|\|f\|_*$ holds for $\|\varphi\| \neq 0$.

We define the functional (here $Ku := k * u$),

$$F(u) = \lambda \int_\Omega |f - Ku|^2 dx + |u|_{NL/TV}, \tag{4.1}$$

where $\lambda > 0$ and

$$|u|_{NL/TV} = \int_\Omega |\nabla_w u|(x)dx.$$

We have shown that the regularizer $|u|_{NL/TV}$ is a semi-norm. The following characterization of minimizers allows us to set conditions on the existence of minimizers (including the case of trivial minimizers), and allows us to associate a dual "texture" norm on the residual and to quantify its size, as in Chapter 3.

Proposition 3 *Let $K : L^2(\Omega) \to L^2(\Omega)$ be a linear and continuous blurring operator with adjoint K^* and let F be the associated functional in (4.1). Then*

(i) $\|K^* f\|_* \leq \frac{1}{2\lambda}$ if and only if $u \equiv 0$ is a minimizer of F.

(ii) Assume that $\frac{1}{2\lambda} < \|K^* f\|_* < \infty$. Then u is a minimizer of F if and only if $\|K^*(f - Ku)\|_* = \frac{1}{2\lambda}$ and $\langle u, K^*(f - Ku)\rangle = \frac{1}{2\lambda}|u|_{NL/TV}$,

where $\|\cdot\|_*$ is the corresponding dual norm of $|\cdot|_{NL/TV}$.

We omit the proof of Proposition 3, since it is similar with the proof given in Chapter 3.

4.2.3 Euler–Lagrange equations

We first consider the functional E defined by

$$
\begin{aligned}
E(u) &= \lambda \int_\Omega |f - Ku|^2\, dx + |u|_{NL/H^1}, \\
&= \lambda \int_\Omega |f - Ku|^2\, dx + \int_\Omega \int_\Omega (u(y) - u(x))^2\, w(x,y)\, dy\, dx,
\end{aligned}
$$

where $Ku := k * u$ and $\lambda > 0$.

Let us assume that u is a minimizer of E and let us define $J(\varepsilon) = E(u + \varepsilon h)$ for $\varepsilon \in \mathbb{R}$ and h, a test function. As u minimizes E, $J'(0) = 0$. By differentiating J with respect to ε, we obtain

$$
\begin{aligned}
J'(\varepsilon) = &- 2\lambda \int_\Omega (f - Ku - \varepsilon K h)\, Kh\, dx \\
&+ 2 \int_\Omega \int_\Omega (u(y) + \varepsilon h(y) - u(x) - \varepsilon h(x))(h(y) - h(x)) w(x,y)\, dy\, dx.
\end{aligned}
$$

Then

$$
\begin{aligned}
J'(0) = &- 2\lambda \int_\Omega (f - Ku)\, Kh\, dx + 2 \int_\Omega \int_\Omega (u(y) - u(x)) h(y) w(x,y)\, dy\, dx \\
&- 2 \int_\Omega \int_\Omega (u(y) - u(x)) h(x) w(x,y)\, dy\, dx.
\end{aligned}
$$

Switching x and y in the second integral and denoting by K^* the adjoint operator of K ($K^* u = k(-\cdot) * u$) yields

$$
\begin{aligned}
J'(0) = &- 2\lambda \int_\Omega K^*(f - Ku)\, h\, dx + 2 \int_\Omega \int_\Omega (u(x) - u(y)) h(x) w(y,x)\, dx\, dy \\
&- 2 \int_\Omega \int_\Omega (u(y) - u(x)) h(x) w(x,y)\, dy\, dx.
\end{aligned}
$$

Assuming that we can interchange the order of integration and due to the symmetry of w,

$$
J'(0) = -2\lambda \int_\Omega K^*(f - Ku)\, h\, dx + 4 \int_\Omega \left[\int_\Omega (u(x) - u(y)) w(x,y)\, dy \right] h(x)\, dx.
$$

We deduce that we must have

$$
-\lambda K^*(f - Ku) + 2 \left[\int_\Omega (u(x) - u(y)) w(x,y)\, dy \right] = 0
$$

on Ω, Euler–Lagrange equation associated with E. This equation can also be expressed as

$$
\lambda K^*(Ku - f) = 2 \triangle_w u.
$$

We now consider the functional F defined by

$$
\begin{aligned}
F(u) &= \lambda \int_\Omega |f - Ku|^2\,dx + |u|_{NL/TV},\\
&= \lambda \int_\Omega |f - Ku|^2\,dx + \int_\Omega \sqrt{\int_\Omega (u(y)-u(x))^2\, w(x,y)\,dy}\,dx,
\end{aligned}
$$

where $Ku := k * u$ and $\lambda > 0$.

Let us assume that u is a minimizer of F and let us define $J(\varepsilon) = F(u + \varepsilon h)$ for $\varepsilon \in \mathbb{R}$ and h, a test function. As u minimizes F, $J'(0) = 0$. By differentiating J with respect to ε, we obtain

$$
\begin{aligned}
J'(\varepsilon) =& -2\lambda \int_\Omega (f - Ku - \varepsilon Kh)\, Kh\,dx\\
&+ \int_\Omega \frac{\int_\Omega (u(y)+\varepsilon h(y) - u(x) - \varepsilon h(x))(h(y)-h(x)) w(x,y)\,dy}{\sqrt{\int_\Omega (u(y)+\varepsilon h(y)-u(x)-\varepsilon h(x))^2\, w(x,y)\,dy}}\,dx.
\end{aligned}
$$

Then

$$
\begin{aligned}
J'(0) =& -2\lambda \int_\Omega (f - Ku)\, Kh\,dx\\
&+ \int_\Omega\int_\Omega \frac{(u(y)-u(x))(h(y)-h(x))w(x,y)}{\sqrt{\int_\Omega (u(z)-u(x))^2\, w(x,z)\,dz}}\,dy\,dx,\\
=& -2\lambda \int_\Omega (f - Ku)\, Kh\,dx\\
&+ \int_\Omega\int_\Omega \frac{(u(y)-u(x))h(y)w(x,y)}{\sqrt{\int_\Omega (u(z)-u(x))^2\, w(x,z)\,dz}}\,dy\,dx\\
&- \int_\Omega\int_\Omega \frac{(u(y)-u(x))h(x)w(x,y)}{\sqrt{\int_\Omega (u(z)-u(x))^2\, w(x,z)\,dz}}\,dy\,dx.
\end{aligned}
$$

Switching x and y in the second integral and denoting by K^* the adjoint operator of K ($K^*u = k(-\cdot) * u$) yields

$$\begin{aligned} J'(0) = &-2\lambda \int_\Omega K^*(f - Ku)\, h\, dx \\ &+ \int_\Omega\int_\Omega \frac{(u(x) - u(y))h(x)w(y,x)}{\sqrt{\int_\Omega (u(z) - u(y))^2\, w(y,z)\, dz}}\, dx\, dy \\ &- \int_\Omega\int_\Omega \frac{(u(y) - u(x))h(x)w(x,y)}{\sqrt{\int_\Omega (u(z) - u(x))^2\, w(x,z)\, dz}}\, dy\, dx. \end{aligned}$$

Assuming that we can interchange the order of integration and due to the symmetry of w,

$$\begin{aligned} J'(0) = &-2\lambda \int_\Omega K^*(f - Ku)\, h\, dx \\ &+ \int_\Omega \Bigg[\int_\Omega (u(x) - u(y))w(x,y) \Bigg\{ \frac{1}{\sqrt{\int_\Omega (u(z) - u(y))^2\, w(y,z)\, dz}} \\ &+ \frac{1}{\sqrt{\int_\Omega (u(z) - u(x))^2\, w(x,z)\, dz}} \Bigg\} dy \Bigg] h(x)\, dx. \end{aligned}$$

We deduce that we must have

$$\begin{aligned} &-2\lambda\, K^*(f - Ku) \\ &+ \int_\Omega (u(x) - u(y))w(x,y) \Bigg\{ \frac{1}{\sqrt{\int_\Omega (u(z) - u(y))^2\, w(y,z)\, dz}} \\ &+ \frac{1}{\sqrt{\int_\Omega (u(z) - u(x))^2\, w(x,z)\, dz}} \Bigg\} dy = 0 \end{aligned}$$

on Ω, the Euler–Lagrange equation associated with F. This equation can also be expressed as

$$2\lambda K^*(Ku - f) = \nabla_w \cdot \Big(\frac{\nabla_w u(x)}{|\nabla_w u|(x)} \Big),$$

or

$$2\lambda K^*(Ku - f) = \int_\Omega (u(y) - u(x))w(x,y) \Big\{ \frac{1}{|\nabla_w u|(y)} + \frac{1}{|\nabla_w u|(x)} \Big\} dy.$$

4.3 Numerical results for image restoration

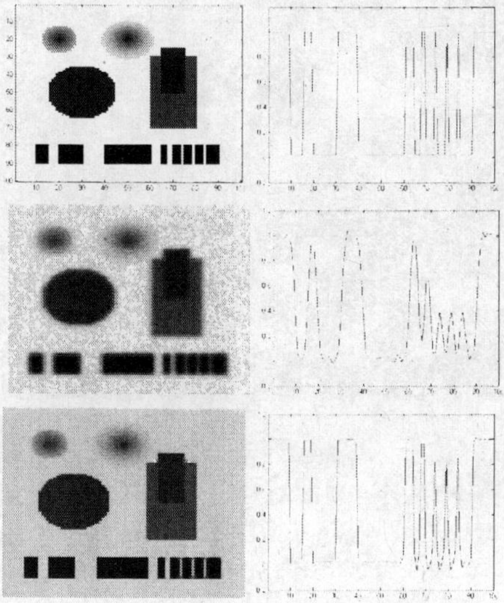

FIGURE 4.1: Image recovery: Gaussian blur kernel with $\sigma_b = 1$ and Gaussian noise with $\sigma_n = 5$. Top: original image and its cross section. Middle: noisy blurry version and its cross section. Bottom row: recovered image and recovered cross section using NL/TV (SNR = 45.1943).

FIGURE 4.2: Left to right: original image; noisy blurry version with Gaussian kernel with $\sigma_b = 1$ contaminated by Gaussian noise with $\sigma_n = 5$; and recovered image with NL/TV (SNR = 17.4165).

The minimization of the nonlocal total variation functional (4.1) presented in this chapter was tested on several restoration experiments with different blur kernels and additive Gaussian noise. We can compare them with their local versions presented in Chapter 3. We test the Gaussian noise model in Figures 4.1 through 4.3. Specifically, in Figure 4.1, we use a simple image and also show its one-dimensional cross section. In this example, we use for the

FIGURE 4.3: Recovery of noisy blurry boat image. Top: original image, noisy blurry version with the pill-box kernel of radius 2, then contaminated by Gaussian noise with $\sigma_n = 5$. Bottom: recovered image u using NL/TV (SNR = 26.4554) and residual.

computation a 31×31 search window size and 5×5 patch. In Figure 4.2, the NL/TV regularizer produces clear edges leading to high SNR. Moreover, as we can see, the residual shown in Figure 4.3 contains much less geometry from the image, by comparison to the residual shown in Figure 3.3.

In conclusion, it is interesting to compare the numerical results presented in this section with those presented in Chapter 3. We can clearly see that the nonlocal total variation method gives much improved restoration results over its local version. On the other hand, (local) total variation requires smaller computational cost.

4.4 Exercises

Exercise 4.1 *Prove Proposition 3.*

Exercise 4.2 *Show the adjoint relation*

$$\langle \nabla_w u, \vec{v} \rangle = \langle u, -div_w \vec{v} \rangle.$$

Exercise 4.3 *Show (as a version of the classical divergence theorem):*

$$\int_\Omega div_w \vec{v} dx = 0.$$

Exercise 4.4 *Show that the nonlocal Laplacian $\triangle_w$ is self adjoint:*

$$\langle \triangle_w u, u \rangle = \langle u, \triangle_w u \rangle.$$

Exercise 4.5 *Show that the nonlocal Laplacian $\triangle_w$ is negative semidefinite:*

$$\langle \triangle_w u, u \rangle = -\langle \nabla_w u, \nabla_w u \rangle.$$

Exercise 4.6 *Using the notations from this chapter, consider the difference-based regularizing functional*

$$\Phi(u) = \int_{\Omega\times\Omega} \phi\left((u(y) - u(x))^2 w(x,y)\right) dydx.$$

Assuming that ϕ is differentiable, compute the Euler–Lagrange equation associated with the minimization with respect to u of

$$\Phi(u) + \lambda \|f - u\|^2_{L^2(\Omega)}.$$

Exercise 4.7 *In two dimensions, let f be a given noisy image degraded by additive Gaussian noise of zero mean. Apply the following minimization in u with NL/H^1 regularization for denoising the image f:*

$$\inf_u \left\{ \lambda \|f - u\|^2_{L^2(\Omega)} + \int_\Omega |\nabla_w u|^2 dx \right\}.$$

Compute the associated Euler–Lagrange equation and give its discretization and numerical algorithm using gradient descent. Repeat the same exercise using NL/TV regularization instead and compare the results.

Chapter 5

Image Decomposition into Cartoon and Texture

In this chapter we present several variational models for the decomposition of an image f into a cartoon component u and a texture component v, inspired by ideas proposed by Meyer [236]. This topic has been of much interest in recent years. The models discussed here can be seen as extensions and generalizations of those presented in Chapter 3, now applied to the separation of cartoon from texture. The cartoon component u should be a simplified version of the image f and can be seen as a piecewise-smooth image made of homogeneous regions with sharp boundaries. Such cartoon component u can be modeled by a function in the space BV of functions of bounded variation, or in its closed subspace, SBV. The texture component v should contain the small oscillatory features of the image f and can be represented by a function (or distribution) belonging to a larger or rougher space, such as L^p or a dual space of distributions. However, when the oscillatory component v belongs to a dual space of distributions, difficulties arise in the numerical computation of the decomposition. This chapter presents an overview of existing cartoon+texture models in variational form and a simpler one, together with experimental results.

5.1 Modeling

A variational model for decomposing a given function f into two components $u + v$ can be given by

$$\inf_{(u,v)\in X_1\times X_2} \Big\{F_1(u) + \lambda F_2(v) : f = u + v\Big\},$$

where $F_1, F_2 \geq 0$ are functionals and X_1, X_2 are function spaces such that $F_1(u) < \infty$ and $F_2(v) < \infty$ if and only if $(u, v) \in X_1 \times X_2$. The constant $\lambda > 0$ is a tuning (scale) parameter. A good model is given by a choice of X_1 and X_2 so that with the given desired properties of u and v, we have $F_1(u) << F_1(v)$ and $F_2(u) >> F_2(v)$. The above decomposition model is equivalent with:

$$\inf_{u\in X_1} \Big\{F_1(u) + \lambda F_2(f - u)\Big\}.$$

When we wish to decompose f into cartoon and texture, it is appropriate to choose as X_1 one of the spaces BV or SBV, while X_2 should be a rougher space of functions or distributions whose norm is small for oscillatory functions. In this chapter, the main choice for X_1 will be BV, the space of functions of bounded variation, presented in Chapter 2 and used in Chapter 3 for image restoration, with $F_1(u) = |u|_{BV} = \int |Du|$ the total variation of u, which is a semi-norm on BV. The discussion will focus on the choice of X_2 and of $F_2 = \|\cdot\|_{X_2}$.

Assume $f \in L^2(\mathbb{R}^N)$. We present several variational models that can be used for image decomposition into cartoon and texture.

Rudin–Osher–Fatemi (ROF) [277] proposed for image denoising (Chapter 3) the minimization

$$\inf_{u \in BV \cap L^2} \left\{ |u|_{BV} + \lambda \int |f-u|^2 dx \right\}. \tag{5.1}$$

This model can also be seen as a decomposition of f into a cartoon component $u \in BV$ and an oscillatory component $v = f - u \in L^2$ (that should only represent the additive Gaussian noise of zero mean).

However, there are several limitations of this model. On one side, many images such as natural images with fine details may not be well represented by BV functions. In this direction, we refer the reader to the work of Alvarez et al. [201], Gousseau and Morel [160], and also of Mumford and Gidas [244].

Another limitation of this model is illustrated in the following example [236, 300].

Example 3 *In two dimensions $N = 2$, let $f = \alpha\chi_D$ in (5.1) with $D = B(0,R)$ the disk centered at the origin and of radius R, and $\alpha \neq 0$.*

If $\lambda R \geq \frac{1}{\alpha}$, then $u = (\alpha - \frac{1}{\lambda R})\chi_D$ and $v = f - u = \frac{1}{\lambda R}\chi_D$.

If $\lambda R \leq \frac{1}{\alpha}$, then $u = 0$ and $v = f$.

Thus, for any $\lambda > 0$, although $f = \alpha\chi_D$ is already in BV and without any noise, we do not obtain $u = f$. The model (5.1) is of the form

$$\inf_{u \in BV \cap L^p} \left\{ |u|_{BV} + \lambda \|f-u\|^q_{L^p} \right\},$$

with $p \geq 1$, $q \geq 1$ and the loss of intensity is always present when $q > 1$. If we consider for instance the model with $p = q = 1$ analyzed by Chan and Esedoḡlu [87] (see also Alliney [12] for the one-dimensional discrete case),

$$\inf_{u \in BV} \left\{ |u|_{BV} + \lambda \int |f-u| dx \right\}, \tag{5.2}$$

then we obtain the following.

Example 4 *In two dimensions, let $f = \chi_D$ in (5.2), again with $D = B(0,R)$ and $\alpha = 1$.*

If $R > \frac{2}{\lambda}$ then $u = f$ and $v = 0$.

If $R < \frac{2}{\lambda}$ *then* $u = 0$ *and* $v = f$.

Thus, by models like (5.2) with $q = 1$, for simple BV functions f, it is possible to obtain $u = f$ and $v = 0$.

Y. Meyer [236] in his book titled *Oscillatory Patterns in Image Processing* analyzed further the ROF minimization [277] and refined it by proposing three other cartoon + texture minimization models of the form

$$\inf_{u \in BV} \left\{ |u|_{BV} + \lambda \|u - f\|_{X_2} \right\},$$

where X_2 is one of the spaces of functions on $\mathbb{R}^N$,

$$X_2 = \left\{ \text{div}\vec{g} : \ \vec{g} \in L^\infty \right\} = G, \quad X_2 = \left\{ \text{div}\vec{g} : \ \vec{g} \in BMO \right\} = F,$$

or

$$X_2 = \left\{ \triangle g : \ g \text{ Zygmund function} \right\} = E.$$

For the above definitions, we recall that:

(i) A function $g \in L^1_{loc}$ belongs to BMO if there is a finite constant c such that $\frac{1}{|Q|} \int_Q |g - g_Q| dx \leq c$ for all squares $Q \subset \mathbb{R}^N$, where g_Q denotes the mean of g over Q.

(ii) A function g belongs to the Zygmund class if there is a finite constant c such that $|g(x+y) + g(x-y) - 2g(x)| \leq c|y|$ for all $x, y \in \mathbb{R}^N$.

The Banach space G is endowed with the norm, for any $v \in G$,

$$\|v\|_G = \inf_{v = div\vec{g}, \vec{g} \in L^\infty} \|\vec{g}\|_{L^\infty},$$

while the norm in the larger space $F = \dot{BMO}^{-1}$ is defined by replacing L^∞ above with BMO. The even larger space E is the generalized homogeneous Besov space $\dot{B}^{-1}_{\infty,\infty} = \triangle \dot{B}^1_{\infty,\infty}$.

We say that a function g belongs to $\dot{B}^\alpha_{p,\infty}$ if there is a finite constant c such that $\|g(\cdot + y) - 2g(\cdot) + g(\cdot - y)\|_{L^p} \leq c|y|^\alpha$ for all $y \in \mathbb{R}^N$.

In two dimensions, the case of planar images, we have the following embeddings:

$$\dot{B}^1_{1,1} \subset BV \subset L^2 \subset G \subset F \subset E,$$

where $\dot{B}^1_{1,1}$ is the predual of $\dot{B}^{-1}_{\infty,\infty}$ and $G = \dot{W}^{-1,1}$ can be seen as a good approximation of the dual of BV.

From the next example (illustrated for the space G) we can see why spaces like G, F or E are better candidates for the space X_2 to model oscillatory components v such as texture.

Example 5 *Let* $a > 0$, $n > 0$ *be fixed, and let* φ *be a smooth function defined on* $\mathbb{R}$ *such that*

$$\varphi(x) = \begin{cases} a & \text{if } |x| < n, \\ 0 & \text{if } |x| > n + 1, \end{cases}$$

and φ is increasing on $(-\infty, 0]$ and decreasing on $[0, \infty)$. Let $m > 0$, $f(x) = \frac{1}{m}\varphi'(x)\sin(mx) + \varphi(x)\cos(mx)$ and $g(x) = \frac{\varphi(x)}{m}\sin(mx) + c$. Then $f = g'$.

- *We have $\|f\|_G = \frac{a}{m}$.*
- $\|f\|_{L^2}^2 \geq 2a^2 \int_0^n |\cos(mx)|^2 \, dx = a^2(n + \frac{1}{2m}\sin(2mn)) \to a^2 n > 0$ *as* $m \to \infty$.
- *Let $M = \lfloor \frac{mn}{2\pi} \rfloor$ be the number of complete periods of $\cos(mx)$ in the interval $[0, n]$. We may assume $M \geq 1$. Then $\|f\|_{L^1} \geq 2a\int_0^n |\cos(mx)| \, dx \geq 8aM \int_0^{\frac{\pi}{2m}} \cos(mx) \, dx = \frac{8aM}{m} \approx 4an/\pi$.*

Therefore, an oscillatory function has small G (F or E) norm which does not depend on the domain $\Omega = (-n, n)$ and approaches 0 as the frequency of oscillations increases, but with important, not so small, L^2 and L^1 norms.

Since the variational image decomposition models proposed by Meyer in [236] deal with spaces of generalized functions, dual norms and the L^∞ norm, it is not straightforward to analyze them and discretize them in practice. Inspired by Meyer's proposals, a rich literature of variational models has been developed and analyzed theoretically and computationally. We list the most relevant ones.

Osher and Vese [338] proposed to minimize

$$\inf_{u,\vec{g}} \left\{ |u|_{BV} + \mu\|f - (u + \text{div}\vec{g})\|_2^2 + \lambda\|\vec{g}\|_p \right\}, \ \mu, \ p \to \infty \tag{5.3}$$

to approximate the (BV, G) Meyer's model and make it computationally amenable. By minimizing (5.3) with respect to u and g_i, $i = 1, ..., N$, f will be decomposed as $f \approx (u + \text{div}\vec{g})$ into cartoon u and texture $v = \text{div}\vec{g}$, with $g_i \in L^p$, and approximating the L^∞–norm of $|\vec{g}|$ (in the definition of $\|v\|_G$) by its L^p–norm. The first term in (5.3) insures that $u \in BV$, the second term guarantees that f is close to $u + \text{div}\,\vec{g}$ in the L^2-norm, while the last term penalizes the L^p-norm of $|\vec{g}|$ (again, this component is related to the definition of the G-norm). As we can see, at least formally, as $\mu \to \infty$ and $p \to \infty$, we obtain the (BV, G) decomposition model. The Euler-Lagrange equations associated with the minimization of (5.3) can now directly be computed and discretized.

Osher et al. [255] proposed an exact decomposition using the minimization

$$\inf_u \left\{ |u|_{BV} + \lambda\|f - u\|_{\dot{H}^{-1}}^2 \right\}. \tag{5.4}$$

We explain how their model has been constructed in two dimensions, over a domain Ω being an open subset of $\mathbb{R}^2$ with Lipschitz boundary. Using the Hodge decomposition of $\vec{g} \in \left(L^2(\Omega)\right)^2$ into $\vec{g} = \nabla P + \vec{Q}$ with $P \in H^1(\Omega)$ a single-valued function and $\vec{Q}$ a divergence-free vector field, yielding to $v = \text{div}\,\vec{g} = f - u = \Delta P$ (expressed further by $P = \Delta^{-1}(f - u)$) and then neglecting $\vec{Q}$, Osher et al. [255] propose the following convex minimization problem for decomposition and restoration:

$$\inf_u E(u) = \int_\Omega |Du| + \lambda \int_\Omega |\nabla(\Delta^{-1})(f-u)|^2 \, dx. \tag{5.5}$$

Compared to Meyer's (BV, G) model, the penalization of the L^∞-norm of $|\vec{g}|$ is replaced by a penalization of the L^2-norm: $\int_\Omega |\vec{g}|^2 \, dx = \int_\Omega |\nabla P|^2 \, dx = \int_\Omega |\nabla(\Delta^{-1})(f-u)|^2 \, dx$. The precise assumptions are the following ones: $u \in BV(\Omega)$ and $v \in L^2(\Omega)$ with $\int_\Omega v(x) \, dx = 0$ (so that $||v||_G < +\infty$). The continuous embedding of $BV(\Omega)$ into $L^2(\Omega)$ thus requires $f \in L^2(\Omega)$.

Later Lieu and Vese [218] generalized (5.4) to

$$\inf_u \left\{ |u|_{BV} + \lambda \|f-u\|^2_{H^{-s}} \right\}, \quad s > 0.$$

Similarly, Le and Vese [206] approximated (BV, F) Meyer's model by

$$\inf_{u,\vec{g}} \left\{ |u|_{BV} + \mu \|f - (u + \mathrm{div}\vec{g})\|_2^2 + \lambda \|\vec{g}\|_{BMO} \right\}.$$

Aujol et al. [28, 32] addressed the original (BV, G) Meyer's problem and proposed an alternate method to minimize

$$\inf_u \left\{ |u|_{BV} + \lambda ||f - (u+v)||_2^2 \right\},$$

subject to the constraint $||v||_G \leq \mu$.

Garnett et al. [152] proposed reformulations and generalizations of Meyer's (BV, E) model, where $E = \dot{B}^{-1}_{\infty,\infty}$ (see also Aujol and Chambolle [33]), given by

$$\inf_{u,\vec{g}} \left\{ |u|_{BV} + \mu \|f - (u + \triangle\vec{g})\|_2^2 + \lambda \|\vec{g}\|_{\dot{B}^\alpha_{p,\infty}} \right\}$$

where $p \geq 1$, $0 < \alpha < 2$, and exact decompositions given by

$$\inf_u \left\{ |u|_{BV} + \lambda \|f-u\|_{\dot{B}^{\alpha-2}_{p,q}} \right\}.$$

In a subsequent work, Garnett et al. [151] proposed different formulations,

$$\inf_{u,\vec{g}} \left\{ |u|_{BV} + \mu \|f - (u + \triangle\vec{g})\|_2^2 + \lambda \|\vec{g}\|_{B\dot{M}O^\alpha} \right\},$$

with $B\dot{M}O^\alpha = I_\alpha(BMO)$, $\|v\|_{B\dot{M}O^\alpha} = \|I_\alpha v\|_{BMO}$, and

$$\inf_{u,\vec{g}} \left\{ |u|_{BV} + \mu \|f - (u + \triangle\vec{g})\|_2^2 + \lambda \|\vec{g}\|_{\dot{W}^{\alpha,p}} \right\},$$

with $\|v\|_{\dot{W}^{\alpha,p}} = \|I_\alpha v\|_p$, $0 < \alpha < 2$, where we recall that the Riesz potential is defined (using the Fourier transform) by

$$I_\alpha = (-\triangle)^{\alpha/2}. \tag{5.6}$$

5.2 A simple cartoon + texture decomposition model

Generalizing (BV, H^{-s}), $(BV, \dot{B}^{\alpha}_{p,\infty})$, and the $TV - Hilbert$ model [34], a simpler cartoon + texture decomposition model (previously introduced in [152]) can be defined using a smoothing convolution kernel K. Thus we assume that K is a positive, even, bounded kernel on $\mathbb{R}^N$ such that $\int K dx = 1$ and such that $K * u$ determines u (i.e. the map $L^p \ni u \to K * u$ is injective). For example, we may take K to be a Gaussian or a Poisson kernel. We fix $\lambda > 0$, $1 \leq p < \infty$ and $1 \leq q < \infty$.

For $f(x) \in L^p$ we consider the extremal problem

$$m_{p,q,\lambda} = \inf\{|u|_{BV} + \mathcal{F}_{p,q,\lambda}(f - u) : u \in BV\} \tag{5.7}$$

where

$$\mathcal{F}_{p,q,\lambda}(h) = \lambda||K * h||^q_{L^p}. \tag{5.8}$$

Thus the simplified image decomposition model is

$$\inf_{u \in BV} \Big\{ |u|_{BV} + \lambda \|K * (f - u)\|^q_{L^p} \Big\}. \tag{5.9}$$

This can be seen as a simplified version of all the previous models, with the appropriate choice of kernel K.

5.2.1 Numerical results for cartoon + texture decomposition

We show in this subsection a few numerical results for decomposing the Barbara image presented in Figure 5.1 into "cartoon" u and "texture" $v = f - u$, with different (symmetric) kernels K, minimizing in u,

$$\mathcal{E}(u) = \int_\Omega |\nabla u| + \lambda \|K * (f - u)\|_{L^1}, \tag{5.10}$$

thus $p = q = 1$ and we have denoted by Ω the planar image domain. These are obtained by discretizing, using finite differences, the Euler–Lagrange equation, which can formally be written as

$$\lambda K * \frac{K * (u - f)}{|K * (u - f)|} = \operatorname{div}\Big(\frac{\nabla u}{|\nabla u|}\Big).$$

Averaging convolution kernel K

Let B be a set containing 0 and $K_B(x) = \frac{1}{|B|}\chi_B(x)$ be the averaging kernel. We have

$$K_B * f(x) = \int_\Omega K_B(x - y) f(y)\, dy = \frac{1}{|B|} \int_B f(x - y)\, dy.$$

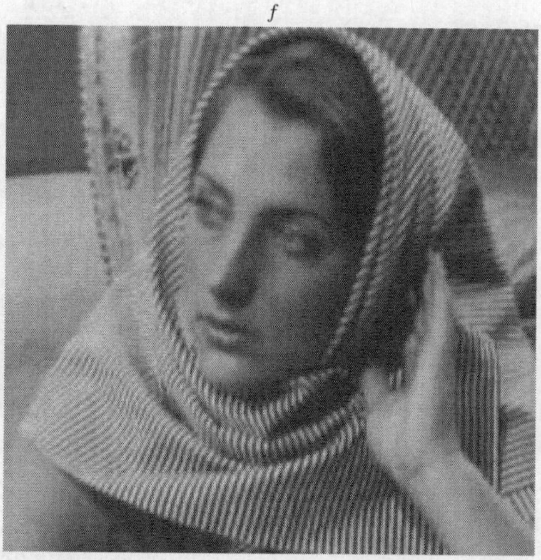

FIGURE 5.1: Test image to be decomposed.

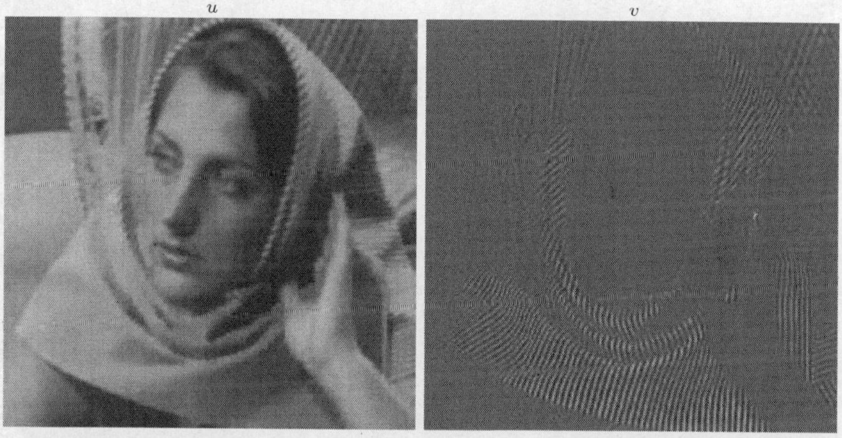

FIGURE 5.2: A decomposition of f from Figure 5.1 using the model (5.10) with the kernel being the characteristic function of a square centered at 0 having 3-pixel length for the sides and $\lambda = 1.5$.

Figures 5.2 through 5.3 [152] show decompositions of f from Figure 5.1 using model (5.10) with a non-smooth averaging kernel K, where B is a square centered at 0 with sides parallel to the axes. Both decompositions use $p = 1$, and $\lambda = 1.5$. However, the decomposition in Figure 5.2 uses the square with 3-pixel length for the sides, while the other uses the square with 5-pixel length.

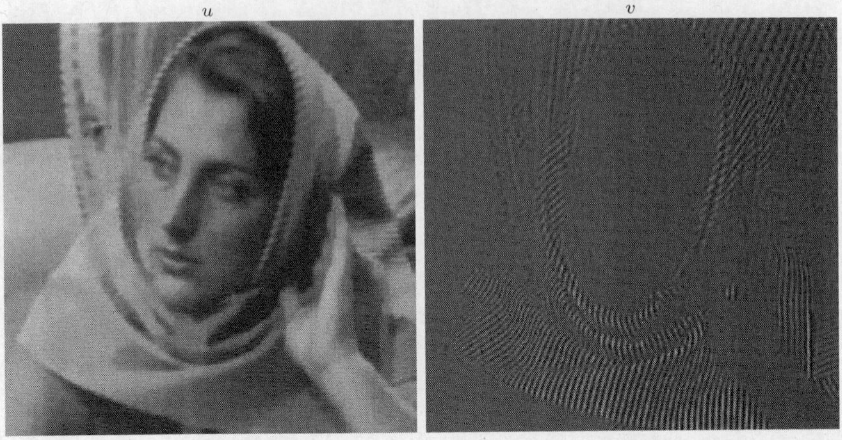

FIGURE 5.3: A decomposition of f from Figure 5.1 using the model (5.10) with the kernel being the characteristic function of a square centered at 0 having 5-pixel length for the sides and $\lambda = 1.5$.

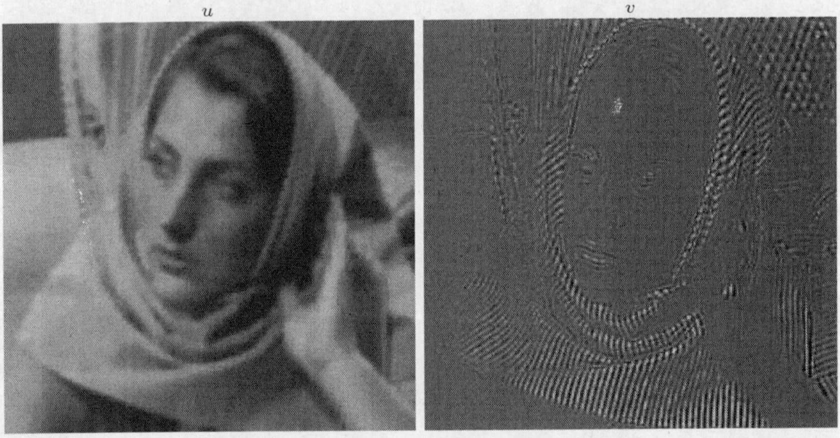

FIGURE 5.4: A decomposition of f from Figure 5.1 using the model (5.10) with the kernel being the Gaussian kernel of standard deviation $\sigma = 1$, $p = 1$, and $\lambda = 1$.

Gaussian convolution kernel K

In Figure 5.4, we show a decomposition result using a smooth kernel K, given by the Gaussian kernel of standard deviation σ.

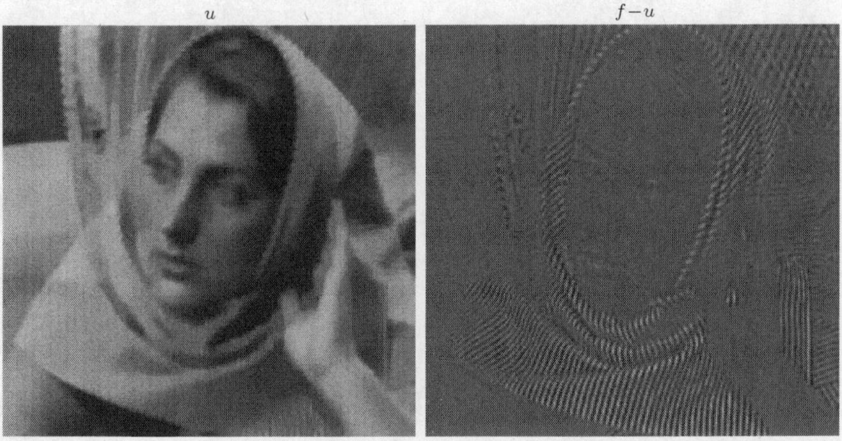

FIGURE 5.5: A decomposition of f from Figure 5.1 using the model (5.11). The oscillatory component $v \in \dot{W}^{-\alpha,1}$, $\alpha = 0.1$, $\lambda = 1.25$.

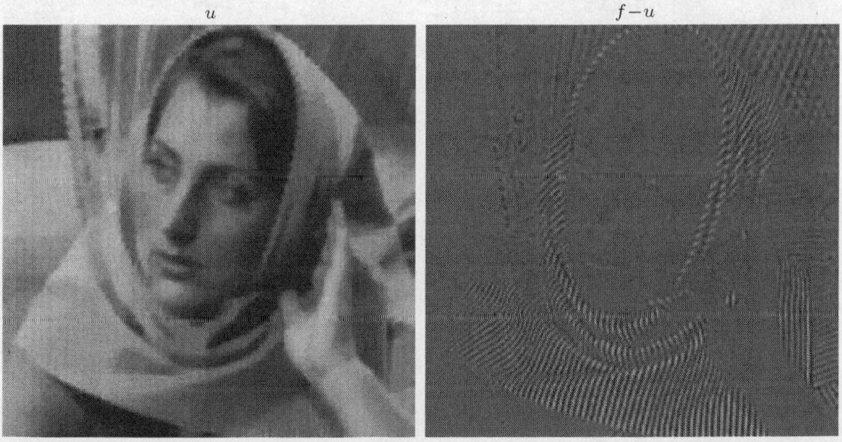

FIGURE 5.6: A decomposition of f from Figure 5.1 using the model (5.11). The oscillatory component $v \in \dot{W}^{-\alpha,1}$, $\alpha = 0.5$, $\lambda = 15$.

Riesz potential K

We use now the Riesz potential I_α to define the kernel K, as in (5.6). Problem (5.10) is equivalent to

$$\int_\Omega |\nabla u| + \lambda \|f - u\|_{\dot{W}^{-\alpha,1}}. \tag{5.11}$$

The following final results are taken from the work of Garnett et al. [151].

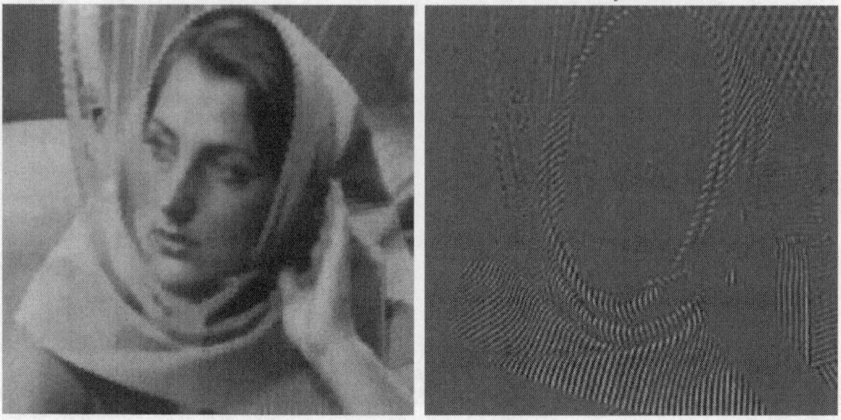

FIGURE 5.7: A decomposition of f from Figure 5.1 using the model (5.11). The oscillatory component $v \in \dot{W}^{-\alpha,1}$, $\alpha = 0.6$, $\lambda = 30$.

Figure 5.5 shows a decomposition of f from Figure 5.1 using the model (5.11). The oscillatory component $v \in \dot{W}^{-\alpha,1}$, $\alpha = 0.1$, $\lambda = 1.25$. Figure 5.6 shows a decomposition of f from Figure 5.1 using the model (5.11). The oscillatory component $v \in \dot{W}^{-\alpha,1}$, $\alpha = 0.5$, $\lambda = 15$. Figure 5.7 shows a decomposition of f from Figure 5.1 using the model (5.11). The oscillatory component $v \in \dot{W}^{-\alpha,1}$, $\alpha = 0.6$, $\lambda = 30$.

5.3 Exercises

Exercise 5.1 *Recall in two dimensions that G denotes the Banach space consisting of all generalized functions v that can be written as:*

$$v(x_1, x_2) = \partial_{x_1} g_1(x_1, x_2) + \partial_{x_2} g_2(x_1, x_2),\ g_1, g_2 \in L^\infty(\mathbb{R}^2),$$

induced by the norm $||\cdot||_G$ defined as the lower bound of all L^∞ norms of the functions $|\vec{g}|$ where $\vec{g} = (g_1, g_2)$, $|\vec{g}(x_1, x_2)| = \sqrt{g_1(x_1, x_2)^2 + g_2(x_1, x_2)^2}$ and where the infimum is computed over all decompositions of v into $v = div\,\vec{g}$.

Recall in two dimensions the cartoon+texture model from (5.3) as an ap-

proximation of Meyer's (BV, G) model:

$$\inf_{u,g_1,g_2} \left\{ G_p(u, g_1, g_2) = \int |Du| + \mu \int |f - u - div\,\vec{g}|^2 \, dx \right.$$
$$\left. + \lambda \left[\int \left(\sqrt{g_1^2 + g_2^2} \right)^p dx \right]^{\frac{1}{p}} \right\}. \tag{5.12}$$

(a) *Compute the Euler–Lagrange equations related to (5.12).*

(b) *In the standard Rudin–Osher–Fatemi model (5.1), when the minimizer is assumed to be in $W^{1,1}(\mathbb{R}^2)$, the residual is given by $f - u = -\frac{1}{2\lambda} K(u)$, with $K(u)$ the curvature operator defined by $K(u) = div\left(\frac{\nabla u}{|\nabla u|}\right)$. Give the residual $f - (u + v) = f - (u + div\vec{g})$ of model (5.12).*

(c) *Prove that for $p = 1$, $|\nabla K(u)| = \lambda$.*

(d) *Prove, for $p > 1$, that:*

$$\lambda \left(\frac{\sqrt{g_1^2 + g_2^2}}{||\sqrt{g_1^2 + g_2^2}||_{L^p}} \right)^{p-1} = |\nabla K(u)|,$$

$$\lambda \, ||\sqrt{g_1^2 + g_2^2}||_{L^p} = \int \sqrt{g_1^2 + g_2^2} |\nabla K(u)|.$$

One can check that if $\vec{g}$ is not of constant magnitude, then $\int \sqrt{g_1^2 + g_2^2} |\nabla K(u)|$ can be viewed as a weighted total variation of the curvature $K(u)$.

(e) *Prove that if $(\hat{u} = 0, \hat{\vec{g}} \in (L^p)^2)$ is a minimizer of (5.12), then*

$$||f - div\,\hat{\vec{g}}||_G \leq \frac{1}{2\mu}.$$

(f) *Prove that if $(\hat{u} \in BV, \hat{\vec{g}} = (0, 0))$ is a minimizer of (5.12), then*

$$||\nabla (\hat{u} - f)||_{L^q} \leq \frac{\lambda}{2\mu},$$

with $\frac{1}{p} + \frac{1}{q} = 1$.

(g) *Prove that $(\hat{u} = 0, \hat{\vec{g}} = (0, 0))$ is a minimizer of (5.12) if and only if $||f||_G \leq \frac{1}{2\mu}$ and $||\nabla f||_{L^q} \leq \frac{\lambda}{2\mu}$.*

Exercise 5.2 *For $f \in L^2(\Omega)$ with $\Omega \subset \mathbb{R}^2$ open, bounded, connected and with Lipschitz boundary $\partial\Omega$, recall the decomposition model (5.4),*

$$\inf_u E(u) = \int_\Omega |Du| + \lambda \int_\Omega |\nabla(\Delta^{-1})(f-u)|^2 \, dx. \tag{5.13}$$

(a) In this question, we check that the last component of (5.13) makes sense. Let $V_0 = \left\{ P \in H^1(\Omega), \int_\Omega P(x)\, dx = 0 \right\}$.

(i) Prove that V_0, as a closed vector subspace of $H^1(\Omega)$, is a Hilbert space for the reduced norm $|P|_{W^{1,2}(\Omega)} = ||\nabla P||_{L^2(\Omega)}$.

(ii) Using Lax–Milgram theorem, deduce that if $v \in L^2(\Omega)$ with $\int_\Omega v(x)\, dx = 0$, then the problem

$$-\Delta P = v, \quad \frac{\partial P}{\partial \vec{n}}_{|\partial\Omega} = 0,$$

with $\vec{n}$ the unit outward normal to $\partial\Omega$, admits a unique solution in V_0.

(b) Assuming that $u \in W^{1,1}(\Omega)$, derive the Euler–Lagrange equation $E_u = 0$ associated with (5.13).

(c) Instead of directly solving the obtained Euler–Lagrange equation with gradient descent (that is, $u_t = -E_u$), apply the Laplacian operator to the Euler–Lagrange equation, to solve $u_t = \Delta E_u$. Derive this evolution problem obtained after applying a gradient descent method. Show in particular that the assumption $\int_\Omega v\, dx = \int_\Omega (f-u)\, dx = 0$ is fulfilled through the descent process.

(d) Prove that the new flow $u_t = \Delta E_u$ decreases the energy.

(e) Let us consider the problem:

$$\inf_{u \in BV(\Omega)} \left\{ F(u) = \int_\Omega |Du| + \lambda \int_\Omega |\nabla(\Delta^{-1})(f-u)|^2 \, dx, \right.$$
$$\left. \int_\Omega (f-u)\, dx = 0 \right\}. \tag{5.14}$$

(i) Justify that the infimum is finite.

(ii) Let (u_n) be a minimizing sequence of (5.14). Using Poincaré–Wirtinger's inequality and the embedding $L^2(\Omega) \subset L^1(\Omega)$, prove

that (u_n) is uniformly bounded in $BV(\Omega)$. Therefore, there is a subsequence of (u_n), still denoted by (u_n) and an element $u \in BV(\Omega)$ such that $u_n \to u$ strongly in $L^1(\Omega)$ and

$$\int_\Omega |Du| \le \liminf_{n\to+\infty} \int_\Omega |Du_n|.$$

(iii) To each u_n, we can associate from the above a unique $P_n \in H^1(\Omega)$ such that $-\Delta P_n = f - u_n$ and $\int_\Omega P_n(x)\,dx = 0$. Prove, still using Poincaré–Wirtinger's inequality, that (P_n) is uniformly bounded in $H^1(\Omega)$. Deduce that there exists a subsequence of (P_n) still denoted by (P_n) and $P \in H^1(\Omega)$ such that $P_n \to P$ in $L^2(\Omega)$. Justify that $\int_\Omega P(x)\,dx = 0$ and that $f - u = -\Delta P$.

Exercise 5.3 *In this exercise, related with examples given in this chapter, and following Chan, Esedoḡlu and Nikolova [88], we illustrate the fact that continuous curve evolution-based gradient descent algorithms can get stuck in a local minimizer according to the chosen initial condition. The assumptions are given next. The observed image f is binary, that is, $f(x) \in \{0,1\}$ for all $x \in \mathbb{R}^N$ and is assumed to be a corrupted version of another binary image $u : \mathbb{R}^N \to \{0,1\}$ to be recovered. f is thus expressed as $f(x) = \chi_\Omega(x)$ with Ω an arbitrary bounded measurable subset of $\mathbb{R}^N$ and we search for u of the form $u(x) = \chi_\Sigma(x)$ with Σ a subset of $\mathbb{R}^N$. We consider the following nonconvex optimization problem:*

$$\min_{\substack{\Sigma\subset\mathbb{R}^N\\ u(x)=\chi_\Sigma(x)}} \int_{\mathbb{R}^N} |Du| + \lambda \int_{\mathbb{R}^N} (u(x) - \chi_\Omega(x))^2\,dx. \tag{5.15}$$

(a) Show that (5.15) can be rephrased as

$$\min_{\substack{\Sigma\subset\mathbb{R}^N\\ u(x)=\chi_\Sigma(x)}} Per(\Sigma) + \lambda|\Sigma \,\triangle\, \Omega|,$$

where $Per(\cdot)$ denotes the perimeter of the set, $|\cdot|$ is the N-dimensional Lebesgue measure and $S_1 \triangle S_2$ denotes the symmetric difference between the two sets S_1 and S_2.

(b) A usual technique of approximation consists in expressing the problem in the level set framework. Let us thus denote by $\Phi : \mathbb{R}^N \to \mathbb{R}$ a Lipschitz continuous function such that $\Sigma = \{x \in \mathbb{R}^N \mid \Phi(x) > 0\}$ and $\partial\Sigma = \{x \in \mathbb{R}^N \mid \Phi(x) = 0\}$.

(i) Show that (5.15) can be reformulated by means of the functional

$$E(\Phi,\lambda) = \int_{\mathbb{R}^N} |DH(\Phi)| + \lambda \int_{\mathbb{R}^N} (H(\Phi(x)) - \chi_\Omega(x))^2\,dx.$$

(ii) *By replacing the one-dimensional Heaviside function H by a smooth approximation H_ϵ, give the Euler–Lagrange equation associated with functional E and derive the gradient flow. Show that the energy decreases in time.*

(iii) *Consider the two-dimensional case $f = \chi_{B_R(0)}$ with $B_R(0)$ the ball of radius R centered at the origin. As initial condition, we take Φ such that $\Sigma = B_R(0)$.*

a *Justify that the evolution equation maintains radial symmetry of Φ. It means that at any given time $t \geq 0$, the candidate for minimization represented through Φ is of the form $\{x \in \mathbb{R}^2 \mid \Phi(x) > 0\} = B_r(0)$ with $r \geq 0$ a radius. As a consequence, $u(x) = \chi_{B_r(0)}(x)$.*

b *Show that the energy involved in minimization problem (5.15) can be rewritten under the form*

$$J(r) = 2\pi r + \lambda \pi \, |R^2 - r^2|.$$

c *Show that if $R < \frac{2}{\lambda}$, the minimum of J is at $r = 0$. In practice, it means that the denoising model removes disks of radius smaller than the value $\frac{2}{R}$.*

d *Show that if $R > \frac{1}{\lambda}$, J has a local maximum at $r = \frac{1}{\lambda}$.*

e *Conclude that if the initial condition is chosen to be $u = f$ with $f = \chi_{B_R(0)}$ with $R \in]\frac{1}{\lambda}, \frac{2}{\lambda}[$, the curve evolution-based gradient descent algorithm remains stuck in the local minimizer, while the global minimizer is $u = 0$.*

This observation motivated Chan, Esedoḡlu and Nikolova to derive algorithms yielding to global minimizers. The key idea is to replace the L^2 norm of the Rudin–Osher–Fatemi model by the L^1 norm. They prove among other things that if the observed image f is the characteristic function of a bounded domain $\Omega \subset \mathbb{R}^N$, then for any $\lambda \geq 0$ there is a minimizer that is also the characteristic function of a possibly different domain.

Part II

Image Segmentation and Boundary Detection

Chapter 6

Mumford and Shah Functional for Image Segmentation

We present in this chapter an overview of the Mumford and Shah model for image segmentation. We discuss its various formulations, some of its properties, and the mathematical framework.

6.1 Description of Mumford and Shah model

An important problem in image analysis and computer vision is the segmentation that aims to partition a given image into its constituent objects or to find boundaries of such objects. This chapter is devoted to the description and analysis of the classical Mumford and Shah functional proposed for image segmentation. In [245, 247, 246], Mumford and Shah formulated an energy minimization problem that allows us to compute optimal piecewise-smooth or piecewise-constant approximations u of a given initial image g. Since then, their model has been analyzed and considered in depth by many authors who studied properties of minimizers, approximations, and applications to image segmentation, image partition, image restoration, and more generally to image analysis and computer vision.

We denote by $\Omega \subset \mathbb{R}^d$ the image domain (an interval if $d = 1$, a rectangle in the plane if $d = 2$, or a rectangular parallelepiped if $d = 3$). More generally, we assume that Ω is open, bounded, and connected. Let $g : \Omega \to \mathbb{R}$ be a given gray-scale image (a signal in one dimension, a planar image in two dimensions, or a volumetric image in three dimensions). It is natural and without losing any generality to assume that g is a bounded function in Ω, $g \in L^\infty(\Omega)$.

As formulated by Mumford and Shah [247], the *segmentation problem* in image analysis and computer vision consists in computing a decomposition

$$\Omega = \Omega_1 \cup \Omega_2 \cup \ldots \cup \Omega_n$$

of the domain Ω of the image g such that

(i) The image g varies smoothly and/or slowly *within* each Ω_i.

(ii) The image g varies discontinuously and/or rapidly across most of the boundary K between different Ω_i.

From the point of view of approximation theory, the segmentation problem may be restated as seeking ways to define and compute *optimal approximations* of a general function $g(x)$ by piecewise-smooth functions $u(x)$, i.e., functions u whose restrictions u_i to the pieces Ω_i of a partition of the domain Ω are continuous or differentiable (or even Sobolev $H^1(\Omega_i)$ functions).

We show in Figure 6.1 an example of partition of a rectangle Ω into eight disjoint regions Ω_1,..., Ω_8 that make up Ω together with their boundaries.

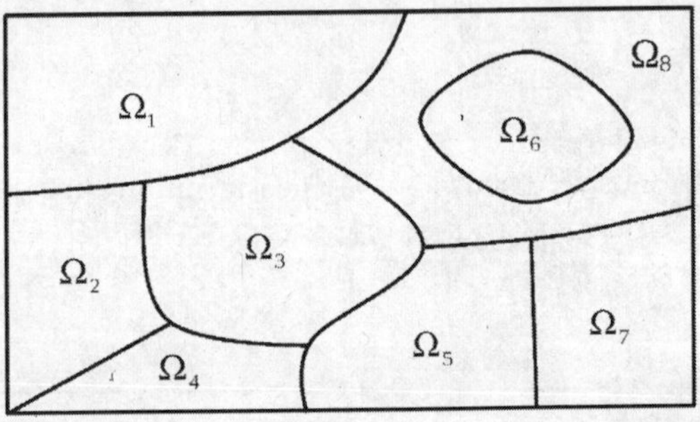

FIGURE 6.1: Example of partition of a rectangle Ω into eight regions Ω_1,..., Ω_8.

In what follows, $\{\Omega_i\}_{i=1}^n$ will be disjoint connected open subsets of a domain Ω, each with a piecewise-smooth boundary, and K will be a closed set, as the union of boundaries of Ω_i inside Ω; thus

$$\Omega = \Omega_1 \cup \Omega_2 \cup ... \cup \Omega_n \cup K, \quad K = \Omega \cap (\partial\Omega_1 \cup ... \cup \partial\Omega_n).$$

The functional E to be minimized for image segmentation is defined by [245, 247, 246],

$$E(u, K) = \mu^2 \int_\Omega (u - g)^2 dx + \int_{\Omega \setminus K} |\nabla u|^2 dx + \nu |K|, \tag{6.1}$$

where $u : \Omega \to \mathbb{R}$ is continuous or even differentiable inside each Ω_i (or $u \in H^1(\Omega_i)$), and may be discontinuous across K. $|K|$ stands for the total surface measure of the hypersurface K (the counting measure if $d = 1$, the length measure if $d = 2$, the area measure if $d = 3$). Later, we will define $|K|$ by $\mathcal{H}^{d-1}(K)$, the $d - 1$ dimensional Hausdorff measure in $\mathbb{R}^d$.

As explained by Mumford and Shah, dropping any of these three terms in (6.1) results in $\inf E = 0$: without the first, take $u = 0$, $K = \emptyset$; without the second, take $u = g$, $K = \emptyset$; without the third, take for example in the discrete case K to be the boundary of all pixels of image g, each Ω_i be a pixel, and u to be the average (value) of g over each pixel. The presence of all three terms leads to non-trivial solutions u, and an optimal pair (u, K) can be seen as a cartoon of the actual image g, providing a simplification of the given image g.

An important result is obtained when we restrict E to piecewise-constant functions u, i.e., $u =$ constant c_i on each open set Ω_i. Multiplying E by μ^{-2}, we have

$$\mu^{-2}E(u,K) = \sum_i \int_{\Omega_i} (g - c_i)^2 dx + \nu_0|K|, \tag{6.2}$$

where $\nu_0 = \nu/\mu^2$. It is easy to verify that this is minimized in the variables c_i by setting

$$c_i = \text{mean}_{\Omega_i}(g) = \frac{\int_{\Omega_i} g(x)dx}{|\Omega_i|},$$

where $|\Omega_i|$ denotes here the Lebesgue measure of Ω_i (e.g., length if $d = 1$, area if $d = 2$, and volume if $d = 3$), so it is sufficient to minimize

$$E_0(K) = \sum_i \int_{\Omega_i} (g - \text{mean}_{\Omega_i}(g))^2 dx + \nu_0|K|.$$

It is possible to interpret E_0 as the limit functional of E as $\mu \to 0$ [247].

We next recall the definition of Sobolev functions $u \in H^1(U) = W^{1,2}(U)$ [3], necessary to properly define a minimizer u when K is fixed (see Chapter 2, Subsection 2.6.1).

Definition 19 *Let $U \subset \mathbb{R}^d$ be an open set. We denote by $W^{1,2}(U)$ (or by $H^1(U)$) the set of functions $u \in L^2(\Omega)$, whose first order distributional partial derivatives belong to $L^2(U)$. This means that there are functions $u_1, ..., u_d \in L^2(U)$ such that*

$$\int_U u(x)\frac{\partial \varphi}{\partial x_i}(x)dx = -\int_U u_i(x)\varphi(x)dx$$

for $1 \le i \le d$ and for all functions $\varphi \in C_c^\infty(U)$.

We may denote by $\frac{\partial u}{\partial x_i}$ the distributional derivative u_i of u, and by $\nabla u = (\frac{\partial u}{\partial x_1}, ..., \frac{\partial u}{\partial x_d})$ its distributional gradient. In what follows, we denote by $|\nabla u|(x)$ the Euclidean norm of the gradient vector at x. $H^1(U) = W^{1,2}(U)$ becomes a Banach space endowed with the norm

$$\|u\|_{W^{1,2}(U)} = \Big[\int_U u^2 dx + \sum_{i=1}^d \int_U \Big(\frac{\partial u}{\partial x_i}\Big)^2 dx\Big]^{1/2}.$$

6.1.1 First variation

In order to better understand and analyze the minimization problem (6.1), it is useful to compute its first variation with respect to each of the unknowns. For simplification of notation, we will use $\|\cdot\|_2$ for the L^2 norm.

6.1.1.1 Minimizing in u with K fixed

Let us assume first that K is fixed, as a closed subset of the open and bounded set $\Omega \subset \mathbb{R}^d$, and denote

$$E(u) = \mu^2 \int_{\Omega\setminus K} (u-g)^2 dx + \int_{\Omega\setminus K} |\nabla u|^2 dx,$$

for $u \in W^{1,2}(\Omega \setminus K)$, where $\Omega \setminus K$ is open and bounded, and $g \in L^2(\Omega \setminus K)$. We have the following classical results obtained as a consequence of the standard method of calculus of variations.

Proposition 4 *There is a unique minimizer of the problem*

$$\inf_{u\in W^{1,2}(\Omega\setminus K)} E(u). \tag{6.3}$$

Proof. [120] First, we note that $0 \leq \inf E < +\infty$, since we can choose $u_0 \equiv 0$ and $E(u_0) = \mu^2 \int_{\Omega\setminus K} g^2(x)dx < +\infty$. Thus we can denote by $m = \inf_u E(u)$ and let $\{u_j\}_{j\geq 1} \in W^{1,2}(\Omega \setminus K)$ be a minimizing sequence, such that $\lim_{j\to\infty} E(u_j) = m$.

Recall that for $u, v \in L^2$,

$$\left\|\frac{u+v}{2}\right\|_2^2 + \left\|\frac{u-v}{2}\right\|_2^2 = \frac{1}{2}\left\|u\right\|_2^2 + \frac{1}{2}\left\|v\right\|_2^2,$$

and so

$$\left\|\frac{u+v}{2}\right\|_2^2 = \frac{1}{2}\left\|u\right\|_2^2 + \frac{1}{2}\left\|v\right\|_2^2 - \left\|\frac{u-v}{2}\right\|_2^2. \tag{6.4}$$

Let $u, v \in W^{1,2}(\Omega \setminus K)$; thus $E(u), E(v) < \infty$. We apply (6.4) to $u - g$ and $v - g$, and then to ∇u and ∇v; we obtain

$$\begin{aligned} E\Big(\frac{u+v}{2}\Big) &= \frac{1}{2}E(u) + \frac{1}{2}E(v) - \frac{\mu^2}{4}\int_{\Omega\setminus K} |u-v|^2 dx - \frac{1}{4}\int_{\Omega\setminus K} |\nabla(u-v)|^2 dx, \\ &= \frac{1}{2}E(u) + \frac{1}{2}E(v) \\ &\quad - \begin{cases} \frac{\mu^2}{4}\|u-v\|^2_{W^{1,2}(\Omega\setminus K)} + (\frac{1}{4} - \frac{\mu^2}{4})\|\nabla(u-v)\|_2^2 & \text{if} \quad \frac{1}{4} \geq \frac{\mu^2}{4} \\ \frac{1}{4}\|u-v\|^2_{W^{1,2}(\Omega\setminus K)} + (\frac{\mu^2}{4} - \frac{1}{4})\|u-v\|_2^2 & \text{if} \quad \frac{1}{4} \leq \frac{\mu^2}{4} \end{cases}. \end{aligned}$$

If we choose $u, v \in W^{1,2}(\Omega \setminus K)$, such that $E(u), E(v) \leq m + \epsilon$, then

$$\begin{aligned} m &\leq E\Big(\frac{u+v}{2}\Big) \\ &\leq m + \epsilon - \begin{cases} \frac{\mu^2}{4}\|u-v\|^2_{W^{1,2}(\Omega\setminus K)} + (\frac{1}{4} - \frac{\mu^2}{4})\|\nabla(u-v)\|_2^2 & \text{if} \quad \frac{1}{4} \geq \frac{\mu^2}{4} \\ \frac{1}{4}\|u-v\|^2_{W^{1,2}(\Omega\setminus K)} + (\frac{\mu^2}{4} - \frac{1}{4})\|u-v\|_2^2 & \text{if} \quad \frac{1}{4} \leq \frac{\mu^2}{4} \end{cases}, \end{aligned}$$

thus

$$\|u-v\|^2_{W^{1,2}(\Omega\setminus K)} \le \begin{cases} \frac{4\epsilon}{\mu^2} \text{ if } \frac{1}{4} \ge \frac{\mu^2}{4} \\ 4\epsilon \text{ if } \frac{1}{4} \le \frac{\mu^2}{4} \end{cases}. \tag{6.5}$$

We let $w_j = u_j - u_1$. From (6.5), $\{w_j\}$ is a Cauchy sequence in $W^{1,2}(\Omega\setminus K)$; let w denote its limit and set $u_0 = u_1 + w$. Then

$$\begin{aligned} E(u_0) &= \mu^2\|u_0-g\|_2^2 + \|\nabla u_0\|_2^2 = \mu^2\|(u_1-g)+w\|_2^2 + \|\nabla u_1 + \nabla w\|_2^2, \\ &= \lim_{j\to+\infty} \Big[\mu^2\|(u_1-g)+w_j\|_2^2 + \|\nabla u_1 + \nabla w_j\|_2^2\Big], \\ &= \lim_{j\to+\infty} E(u_j) = m, \end{aligned}$$

by the continuity of L^2 norms. This shows the existence of minimizers. The uniqueness follows from (6.5) by taking $\epsilon = 0$. □

Proposition 5 *The unique solution u of (6.3) is solution of the elliptic problem*

$$\int_{\Omega\setminus K} \nabla u(x)\cdot\nabla v(x)dx = -\mu^2\int_{\Omega\setminus K}[u(x)-g(x)]v(x)dx, \quad \forall v\in W^{1,2}(\Omega\setminus K), \tag{6.6}$$

or of

$$\triangle u = \mu^2(u-g)$$

in the sense of distributions in $\Omega\setminus K$, with associated boundary condition $\frac{\partial u}{\partial \vec{N}} = 0$ on $\partial(\Omega\setminus K)$, where $\vec{N}$ is the exterior unit normal to the boundary

Proof. Indeed, let $\epsilon \mapsto A(\epsilon) = E(u+\epsilon v)$ for $\epsilon\in\mathbb{R}$ and arbitrary $v \in W^{1,2}(\Omega\setminus K)$. Then A is a quadratic function of ϵ, given by

$$\begin{aligned} A(\epsilon) &= \mu^2\int_{\Omega\setminus K}(u-g)^2dx + \epsilon^2\mu^2\int_{\Omega\setminus K} v^2dx + 2\epsilon\mu^2\int_{\Omega\setminus K}(u-g)vdx \\ &+ \int_{\Omega\setminus K}|\nabla u|^2dx + \epsilon^2\int_{\Omega\setminus K}|\nabla v|^2dx + 2\epsilon\int_{\Omega\setminus K}\nabla u\cdot\nabla vdx, \end{aligned}$$

and we have

$$\begin{aligned} A'(\epsilon) =& 2\epsilon\mu^2\int_{\Omega\setminus K} v^2dx + 2\mu^2\int_{\Omega\setminus K}(u-g)vdx + 2\epsilon\int_{\Omega\setminus K}|\nabla v|^2dx \\ &+ 2\int_{\Omega\setminus K}\nabla u\cdot\nabla vdx, \end{aligned}$$

and

$$A'(0) = 2\mu^2\int_{\Omega\setminus K}(u-g)vdx + 2\int_{\Omega\setminus K}\nabla u\cdot\nabla vdx.$$

Since we must have $E(u) = A(0) \le A(\epsilon) = E(u+\epsilon v)$ for all $\epsilon\in\mathbb{R}$ and all $v\in W^{1,2}(\Omega\setminus K)$, we impose $A'(0) = 0$ for all such v, which yields the weak formulation (6.6).

If in addition u would be a strong classical solution of the problem, or if it would belong to $W^{2,2}(\Omega \setminus K)$, then integrating by parts in the last relation we obtain

$$A'(0) = 2\mu^2 \int_{\Omega\setminus K} (u-g)v dx - 2\int_{\Omega\setminus K} (\triangle u)v dx + \int_{\partial(\Omega\setminus K)} \nabla u \cdot \vec{N} v ds = 0.$$

Taking now $v \in C_0^1(\Omega \setminus K) \subset W^{1,2}(\Omega \setminus K)$, we obtain

$$\triangle u = \mu^2(u-g) \text{ in } \Omega \setminus K.$$

Using this and taking now $v \in C^1(\Omega \setminus K)$, we deduce the associated implicit boundary condition $\nabla u \cdot \vec{N} = \frac{\partial u}{\partial \vec{N}} = 0$ on the boundary of $\Omega \setminus K$ (in other words, on the boundary of Ω and of each Ω_i). □

Assume now that $g \in L^\infty(\Omega \setminus K)$, which is not a restrictive assumption when g represents an image. We can deduce that the unique minimizer u of (6.3) satisfies $\|u\|_\infty \le \|g\|_\infty$ (as expected, due to the smoothing properties of the energy). To prove this, we first state the following classical lemma (see [120], Chapter A3).

Lemma 2 *If $\Omega \setminus K$ is open, and if $u \in W^{1,2}(\Omega \setminus K)$, then $u^+ = \max(u, 0)$ also lies in $W^{1,2}(\Omega \setminus K)$ and $|\nabla u^+(x)| \le |\nabla u(x)|$ almost everywhere.*

Now let $u^*(x) = \max\{-\|g\|_\infty, \min(\|g\|_\infty, u(x))\}$ be the obvious truncation of u. Lemma 2 implies that $u^* \in W^{1,2}(\Omega \setminus K)$ and that $\int_{\Omega\setminus K} |\nabla u^*(x)|^2 dx \le \int_{\Omega\setminus K} |\nabla u(x)|^2 dx$. We also obviously have $\int_{\Omega\setminus K}(u^* - g)^2 dx \le \int_{\Omega\setminus K}(u-g)^2 dx$, and we deduce that $E(u^*) \le E(u)$. But u is the unique minimizer of E, thus $u(x) = u^*(x)$ almost everywhere and we deduce $\|u\|_\infty \le \|g\|_\infty$.

Remark 13 *Several classical regularity results for a weak solution u of (6.3) can be stated:*

- *If $g \in L^\infty(\Omega \setminus K)$, then $u \in C^1_{loc}(\Omega \setminus K)$ ([120], Chapter A3).*
- *If $g \in L^2(\Omega \setminus K)$, then $u \in W^{2,2}_{loc}(\Omega \setminus K) = H^2_{loc}(\Omega \setminus K)$, which implies that u solves the PDE ([139], Chapter 6.3)*

$$\triangle u = \mu^2(u-g) \text{ a.e. in } \Omega \setminus K.$$

6.1.1.2 Minimizing in K

We wish to formally compute the first variation of $E(u, K)$ with respect to K. Let us assume that (u, K) is a minimizer of E from (6.1), and we vary K. Let us assume that locally, $K \cap U$ is the graph of a regular function ϕ, where U is a small neighborhood near a regular, simple point P of K. Without loss of generality, we can assume that $U = D \times I$ where I is an interval in $\mathbb{R}$ and $K \cap U = \{(x_1, x_2, ..., x_d) \in U = D \times I : \ x_d = \phi(x_1, ..., x_{d-1})\}$. Let u^+ denote the restriction of u to

$$U^+ = \{(x_1, x_2, ..., x_d) : \ x_d > \phi(x_1, ..., x_{d-1})\} \cap U,$$

and u^- the restriction of u to

$$U^- = \{(x_1, x_2, ..., x_d): \; x_d < \phi(x_1, ..., x_{d-1})\} \cap U,$$

and choose H^1 extensions of u^+ from U^+ to U, and of u^- from U^- to U. For small ϵ, define a deformation K_ϵ of K inside U as the graph of

$$x_d = \phi(x_1, ..., x_{d-1}) + \epsilon\psi(x_1, ..., x_{d-1}),$$

such that ψ is zero outside D, and $K_\epsilon = K$ outside U. We illustrate in Figure 6.2 the neighborhood U, $U \cap K$ and its deformation.

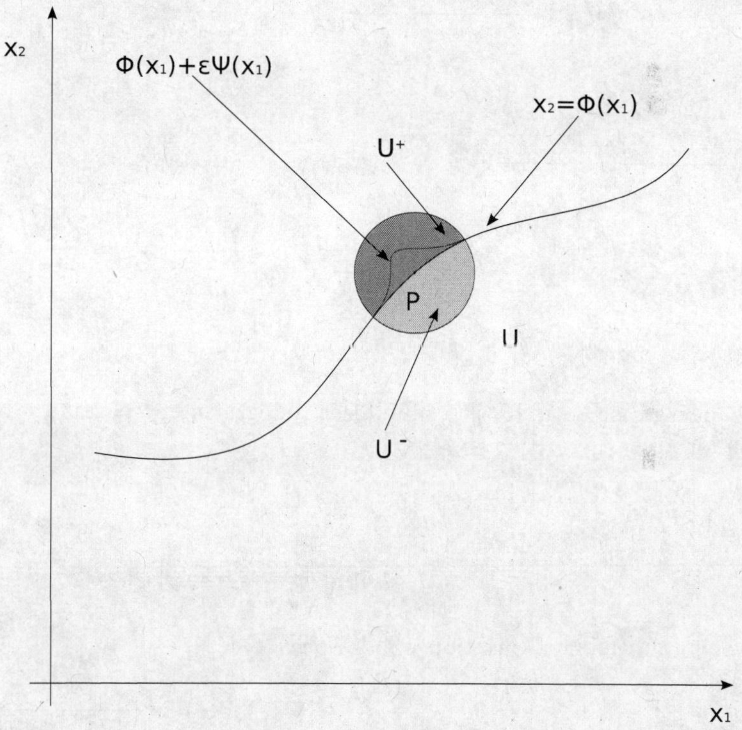

FIGURE 6.2: Neighborhood U near a regular, simple point P of K.

Define

$$u_\epsilon(x) = \begin{cases} u(x) & \text{if} \quad x \notin U, \\ (\text{extension of } u^+)(x) & \text{if} \quad x \in U, \; x \text{ above } K_\epsilon \cap U \\ (\text{extension of } u^-)(x) & \text{if} \quad x \in U, \; x \text{ below } K_\epsilon \cap U. \end{cases}$$

Now, using $z=(x_1,...,x_{d-1})$,

$$\begin{aligned}
E(u_\epsilon,K_\epsilon)-E(u,K) &= \mu^2\int_U\Big[(u_\epsilon-g)^2-(u-g)^2\Big]dx\\
&+\int_{U\setminus K_\epsilon}|\nabla u_\epsilon|^2dx-\int_{U\setminus K}|\nabla u|^2dx\\
&+\nu\Big[|K_\epsilon\cap U|-|K\cap U|\Big],\\
&= \mu^2\int_D\Big(\int_{\phi(z)}^{\phi(z)+\epsilon\psi(z)}\Big[(u^--g)^2-(u^+-g)^2\Big]dx_d\Big)dz\\
&+\int_D\Big(\int_{\phi(z)}^{\phi(z)+\epsilon\psi(z)}\Big[|\nabla u^-|^2-|\nabla u^+|^2\Big]dx_d\Big)dz\\
&+\nu\int_D\Big[\sqrt{1+|\nabla(\phi+\epsilon\psi)|^2}-\sqrt{1+|\nabla\phi|^2}\Big]dz.
\end{aligned}$$

Thus,

$$\begin{aligned}
\lim_{\epsilon\to 0}\frac{E(u_\epsilon,K_\epsilon)-E(u,K)}{\epsilon} &= \mu^2\int_D\Big[(u^--g)^2-(u^+-g)^2\Big]\Big|_{x_d=\phi(z)}\psi(z)dz\\
&+\int_D\Big[|\nabla u^-|^2-|\nabla u^+|^2\Big]\Big|_{x_d=\phi(z)}\psi(z)dz\\
&+\nu\int_D\frac{\nabla\phi\cdot\nabla\psi}{\sqrt{1+|\nabla\phi|^2}}dz,\\
&= 0,
\end{aligned}$$

for all such ψ, since (u,K) is a minimizer. Integrating by parts, we formally obtain for all ψ:

$$\int_D\Big\{\Big[(\mu^2(u^--g)^2+|\nabla u^-|^2)-(\mu^2(u^+-g)^2+|\nabla u^+|^2)\Big]\Big|_{x_d=\phi(z)} -\nu\text{div}\Big(\frac{\nabla\phi}{\sqrt{1+|\nabla\phi|^2}}\Big)\Big\}\psi(z)dz=0,$$

and we obtain the first variation with respect to K,

$$\Big[\mu^2(u^--g)^2+|\nabla u^-|^2\Big]-\Big[\mu^2(u^+-g)^2+|\nabla u^+|^2\Big]-\nu\text{div}\Big(\frac{\nabla\phi}{\sqrt{1+|\nabla\phi|^2}}\Big)=0 \tag{6.7}$$

on $K\cap U$. Noticing that the last term represents the curvature of $K\cap U$, and if we write the energy density as

$$e(u;x)=\mu^2(u(x)-g(x))^2+|\nabla u(x)|^2,$$

we finally obtain

$$e(u^+)-e(u^-)+\nu\text{curv}(K)=0 \text{ on } K$$

(at regular points of K, provided that the traces of u and of $|\nabla u|$ on each side of K are taken in the sense of Sobolev traces).

We conclude this section by stating another important result from [247] regarding the type of singular points of K, when (u, K) is a minimizer of E from (6.1), in two dimensions, $d = 2$. For the rather technical proof of this result, we refer the reader to the instructive and inspiring constructions from [247].

Theorem 29 *Let $d = 2$. If (u, K) is a minimizer of $E(u, K)$ such that K is a union of simple $C^{1,1}$-curves K_i meeting $\partial\Omega$ and meeting each other only at their endpoints, then the only vertices of K are:*

(1) Points P on the boundary $\partial\Omega$ where one K_i meets $\partial\Omega$ perpendicularly
(2) Triple points P where three K_i meet with angles $2\pi/3$
(3) Crack-tips where a K_i ends and meets nothing

6.2 Weak formulation of Mumford and Shah functional: MSH^1

To better study the mathematical properties of the Mumford and Shah functional (6.1), it is necessary to define the measure of K as its $(d-1)$-dimensional Hausdorff measure $\mathcal{H}^{d-1}(K)$ (see Chapter 2), which is the most natural way to extend the notion of length to nonsmooth sets.

We consider a first variant of the functional,

$$F(u, K) = \mu^2 \int_{\Omega \setminus K} (u-g)^2 dx + \int_{\Omega \setminus K} |\nabla u|^2 dx + \nu \mathcal{H}^{d-1}(K). \tag{6.8}$$

The problem in this formulation has been studied [117, 118] in two dimensions. These results were extended to three and more dimensions [227].

However, in order to apply the direct method of calculus of variations for proving existence of minimizers, it is necessary to find a topology for which the functional is lower semi-continuous, while ensuring compactness of minimizing sequences. Unfortunately, the last functional $K \mapsto \mathcal{H}^{d-1}(K)$ is not lower semi-continuous with respect to any compact topology [30], [120], [20].

To overcome this difficulty, the set K is substituted by the jump set S_u of u, thus K is eliminated, and the problem, called the weak formulation, becomes, in its second variant (see [19] and [122]),

$$\inf_u F(u) = \mu^2 \int_{\Omega \setminus S_u} (u-g)^2 dx + \int_{\Omega \setminus S_u} |\nabla u|^2 dx + \nu \mathcal{H}^{d-1}(S_u). \tag{6.9}$$

For illustration, we also give the weak formulation in one dimension for signals. The problem of reconstructing and segmenting a signal u from a degraded input g deriving from a distorted transmission, can be modeled as finding the minimum

$$\inf_u \mu^2 \int_a^b (u-g)^2 dt + \int_{(a,b)\setminus S_u} |u'|^2 dt + \nu\#(S_u),$$

where $\Omega = (a,b)$, S_u denotes the set of discontinuity points of u in the interval (a,b), and $\#(S_u) = \mathcal{H}^0(S_u)$ denotes the counting measure of S_u or its cardinal.

In order to show that (6.9) has a solution, the notion of special functions of bounded variation and the compactness property of the SBV space due to Ambrosio [18, 19] (see also [20], [27]) are necessary (see Chapter 2, Subsection 2.6.2.2).

Theorem 30 *Let $g \in L^\infty(\Omega)$, with $\Omega \subset \mathbb{R}^d$ open, bounded and connected. Then there is a minimizer $u \in SBV(\Omega) \cap L^\infty(\Omega)$ of*

$$F(u) = \mu^2 \int_{\Omega\setminus S_u} (u-g)^2 dx + \int_{\Omega\setminus S_u} |\nabla u|^2 dx + \nu\mathcal{H}^{d-1}(S_u).$$

Proof. We notice that $0 \le \inf_{SBV(\Omega)\cap L^\infty(\Omega)} F < \infty$, because we can take $u_0 = 0 \in SBV(\Omega) \cap L^\infty(\Omega)$ and using the fact that $g \in L^\infty(\Omega) \subset L^2(\Omega)$, $F(u_0) < \infty$. Thus there is a minimizing sequence $u_n \in SBV(\Omega) \cap L^\infty(\Omega)$ satisfying $\lim_{n\to\infty} F(u_n) = \inf F$. We also note that, by the truncation argument from before, we can assume that $\|u_n\|_\infty \le \|g\|_\infty < \infty$. Since $F(u_n) \le C < \infty$ for all $n \ge 0$, and using $g \in L^\infty(\Omega) \subset L^2(\Omega)$, we deduce that $\|u_n\|_2 \le C$ and $\int_{\Omega\setminus S_{u_n}} |\nabla u_n|^2 dx + \mathcal{H}^{d-1}(S_{u_n}) < C$ for some positive real constant C. Using these and Ambrosio's compactness result from Chapter 2, Subsection 2.6.2.2, we deduce that there is a subsequence u_{n_k} of u_n, and $u \in SBV(\Omega)$, such that $u_{n_k} \rightharpoonup u$ in $L^2(\Omega)$, $\nabla u_{n_k} \rightharpoonup \nabla u$ in $L^2(\Omega)^d$. Therefore, $F(u) \le \liminf_{n_k\to\infty} F(u_{n_k}) = \inf F$, and we can also deduce that $\|u\|_\infty \le \|g\|_\infty$. □

For additional existence, regularity results and fine properties of minimizers, and for the connections between problems (6.8) and (6.9), we refer the reader to Dal Maso, Morel, Solimini [117, 118], the important monographs by Morel, Solimini [242], Ambrosio, Fusco, Pallara [20], David [120], and Braides [61]. Existence and regularity of minimizers for the piecewise-constant case can be found in [247], Congedo, Tamanini [212, 234, 312, 313], Larsen [202], and other works.

6.3 Mumford and Shah TV functional: $MSTV$

There is another version of the previous Mumford–Shah regularizer in the weak form $\beta \int_{\Omega\setminus S_u} |\nabla u|^2 dx + \alpha \int_{S_u} d\mathcal{H}^{d-1}$, introduced by Shah [288] and analyzed in [7], where a family of functionals is constructed with their approximations. These are well defined on the space $GBV(\Omega)$ of generalized functions of bounded variation.

Definition 20 *We say that $u \in L^1(\Omega)$ belongs to $GBV(\Omega)$ if its truncations $u_T = (-T) \vee (u \wedge T)$ are in $BV(\Omega)$ for any $T > 0$ (where $a \wedge b$ denotes $min(a, b)$ and $a \vee b$ denotes $max(a, b)$).*

If $u \in GBV(\Omega)$, then $u_T \in BV(\Omega)$ for any $T > 0$. Then we can define $S_u = \cup_{T>0} S_{u_T}$ and the approximate gradient ∇u, the traces u^+ and u^-, and the total variation of the Cantor part $|C_u|$ as the limits of the corresponding quantities defined for u_T.

Then, for $u \in GBV(\Omega) \cap L^2(\Omega)$, the following functional is well defined [7],

$$F^{MSTV}(u) = \int_\Omega (u-g)^2 dx + \Psi^{MSTV}(u), \tag{6.10}$$

where the $MSTV$-like regularizer is

$$\Psi^{MSTV}(u) = \beta \int_{\Omega \setminus S_u} |\nabla u| dx + \alpha \int_{S_u} \frac{|u^+ - u^-|}{1 + |u^+ - u^-|} d\mathcal{H}^{d-1} + |C_u|(\Omega),$$

with $\alpha, \beta > 0$ constants.

Note that there is a non-convex constraint $t \mapsto \frac{|t|}{1+|t|} < |t|$, $t > 0$ on the jump part $u^+ - u^-$. When $u \in BV(\Omega)$, it is interesting to compare $\Psi^{MSTV}(u)$ $(\alpha = \beta = 1)$ with the total variation of u (which is convex),

$$\Psi^{TV}(u) = \int_{\Omega \setminus S_u} |\nabla u| dx + \int_{S_u} |u^+ - u^-| d\mathcal{H}^{d-1} + |C_u|(\Omega).$$

Phase-field and elliptic approximations of the functionals in (6.9) and (6.10) will be presented in the next chapter.

6.4 Exercises

Some of the exercises below were inspired by [120], [242], and [52].

Exercise 6.1 *Let Ω and g be as before and let $\lambda > 0$ be given. Define*

$$E(u, K) = a \int_{\Omega \setminus K} (u-g)^2 dx + b\mathcal{H}^{d-1}(K) + c \int_{\Omega \setminus K} |\nabla u|^2 dx.$$

Set $\tilde{\Omega} = \lambda\Omega$, $\tilde{g}(x) = g(\frac{x}{\lambda})$, and call $\tilde{E}$ the analogue of E on the domain $\tilde{\Omega}$, defined with the function $\tilde{g}$ and the coefficients $\tilde{a} = \lambda^{-d} a$, $\tilde{b} = \lambda^{1-d} b$, and

$\tilde{c} = \lambda^{2-d}c$. *Assume that* (u, K) *is such that* $E(u, K)$ *is finite, and set* $\tilde{K} = \lambda K$ *and* $\tilde{u}(x) = u(\frac{x}{\lambda})$.

(a) Show that $\tilde{E}(\tilde{u}, \tilde{K}) = E(u, K)$.

(b) Let $g^* = \mu\tilde{g}$, $u^* = \mu\tilde{u}$ *for some* $\mu > 0$, *and* E^* *corresponding to* $\tilde{\Omega}$, g^* *with the constants* $a^* = \mu^2\tilde{a} = \mu^2\lambda^{-d}a$, $b^* = \tilde{b} = \lambda^{1-d}b$, *and* $c^* = \mu^2\tilde{c} = \mu^2\lambda^{2-d}c$. *Show that* $E^*(u^*, \tilde{K}) = E(u, K)$.

Exercise 6.2 *Consider the piecewise-constant Mumford–Shah functional*

$$E(c_i, \Omega_i) = \sum_i \int_{\Omega_i} (g - c_i)^2 dx + \nu_0 |K|.$$

Show that, for fixed regions Ω_i, *the optimal constants* c_i *are given by*

$$c_i = \frac{\int_{\Omega_i} g(x)dx}{|\Omega_i|}.$$

Exercise 6.3 *Consider a merging method for image partition consisting of merging iteratively the regions of a segmentation* (Ω_i). *Assume that the criterion for merging* Ω_i *and* Ω_j *is:*

$$a(\Omega_i \cup \Omega_j) - a(\Omega_i) - a(\Omega_j) < 0$$

for some function a depending on Ω_i. *Show that this merging method decreases the energy*

$$E((\Omega_i)) = \sum_i a(\Omega_i).$$

Exercise 6.4 *Consider the graduated non-convexity functions* $G_p : \mathbb{R} \to \mathbb{R}$ *defined by*

$$G_p(t) = \begin{cases} t^2 & \text{if} \quad |t| < \frac{1}{r}, \\ 1 - \frac{(|t|-r)^2}{4p} & \text{if} \quad \frac{1}{r} \le |t| < r, \\ 1 & \text{if} \quad r \le |t|, \end{cases}$$

where $p \ge 0$ *and* $r = \sqrt{4p + 1}$ *are parameters. Show that* G_p *is convex if and only if* $p \ge 1$.

Exercise 6.5 *Consider the graduated non-convexity discrete functionals,*

$$E_p(u) = \sum_{i=1}^{n} \Big[\mu^2(u_i - g_i)^2 + G_p(u_i - u_{i-1})\Big],$$

where the interval $I = (a, b)$ *has been discretized,* $g = (g_i)$ *are the values of the given data at the discrete points of* I, $u = (u_i)$ *is the unknown optimal approximation* $(i = 1, 2, ..., n)$, *and*

$$G_0(t) = \begin{cases} t^2 & \text{if} \quad |t| \le 1, \\ 1 & \text{if} \quad |t| > 1, \end{cases} = \min\{t^2, 1\}$$

with G_p given in Exercise 6.4. A discrete version of the one-dimensional Mumford–Shah functional is E_0 for $p = 0$, setting for simplicity $\nu = 1$, and defining the edge set K as the set of points i for which $|u_i - u_{i-1}| > h = 1$, where $h > 0$ is the space step. In order to find a (local or global) minimizer of E_0, use the graduated non-convexity algorithm by minimizing the sequence of functionals E_p, letting p decrease from 1 to 0. Use a previous minimizer as a starting point for the next minimization with smaller p. Apply the algorithm to a one-dimensional signal.

Exercise 6.6 *Repeat Exercise 6.5 now in two dimensions for images. Consider the corresponding discrete Mumford–Shah functional*

$$E_0(u) = \sum_{i,j} \Big[\mu^2(u_{i,j} - g_{i,j})^2 + G_0(u_{i+1,j} - u_{i,j}) + G_0(u_{i,j+1} - u_{i,j})\Big].$$

Define $E_p(u)$ and apply the graduated non-convexity algorithm.

Chapter 7

Phase-Field Approximations to Mumford and Shah Problem

Since the original Mumford and Shah functional (6.1) (or its weak formulation (6.9)) is non-convex, it has an unknown set K of lower dimension, and it is not the lower-semicontinuous envelope of a more amenable functional, it is difficult to find smooth approximations and to solve the minimization in practice.

Several approximations have been proposed, including the weak membrane model and the graduated non-convexity of Blake and Zisserman [52] (which can be seen as a discrete version of the Mumford and Shah segmentation problem); discrete finite differences approximations starting with the work of Chambolle [82], [84], [83] (also proving the Γ–convergence of Blake-Zisserman approximations to the weak Mumford–Shah functional in one dimension); finite element approximations by Chambolle and Dal Maso [85] and by Chambolle and Bourdin [59], [58]; phase-field elliptic approximations due to Ambrosio and Tortorelli [21], [22] (with generalizations by Braides [61] and extensions by Shah [288], Alicandro, Braides, and Shah [7]); region growing and merging methods proposed by Koepfler, Lopez, and Morel [195], by Morel and Solimini [242], by Dal Maso, Morel, and Solimini [117, 118], level set approximations proposed by Chan and Vese [92, 94, 335, 93, 95], by Samson et al. [280, 279], and by Tsai, Yezzi, and Willsky [328]; and approximations by nonlocal functionals by Braides and Dal Maso [62], among others.

We present in this chapter the phase-field elliptic approximations in more detail, together with their applications to image restoration.

We would like to refer the reader to the monographs and textbooks by Braides [61], by Morel and Solimini [242], and by Ambrosio, Fusco, and Pallara [20] on detailed presentations of approximations to the Mumford and Shah functional. For proving the convergence of some of these approximations to the Mumford and Shah functional, the notion of Γ–convergence is used, which is briefly presented in Chapter 2, Subsection 2.7.7.

Let Ω be an open, bounded and connected subset of $\mathbb{R}^d$, with Lipschitz boundary $\partial\Omega$, and $g : \Omega \to \mathbb{R}$ be a given image belonging to $L^\infty(\Omega)$. The goal is to find optimal piecewise-smooth approximations of g using phase-field and elliptic approximations to the Mumford and Shah functional (6.9), with applications to image segmentation or to image restoration such as denoising and deblurring.

7.1 Ambrosio and Tortorelli phase-field elliptic approximations

A specific strategy, closer to the initial formulation of the Mumford–Shah problem in terms of pairs $(u, K = S_u)$, is based on the approximation by functionals depending on two variables (u, v), the second one related to the set $K = S_u$.

7.1.1 Approximations of the perimeter by elliptic functionals

The Modica–Mortola theorem [240, 241] allows for the variational approximation of the perimeter functional $E \mapsto P(E, \Omega) = \int_\Omega |D\chi_E| < \infty$ of an open subset E of Ω by the quadratic elliptic functionals

$$MM_\epsilon(v) = \int_\Omega \Big(\epsilon|\nabla v|^2 + \frac{W(v)}{\epsilon}\Big)dx, \quad v \in H^1(\Omega),$$

where $W : \mathbb{R} \to \mathbb{R}$ is a "double-well" potential. For instance, choosing $W(t) = t^2(1-t)^2$, assuming that Ω is bounded with Lipschitz boundary and setting $MM_\epsilon(v) = \infty$ if $v \in L^2(\Omega) \setminus H^1(\Omega)$, the functionals $MM_\epsilon(v)$ Γ–converge in $L^2(\Omega)$ to

$$F(v) = \begin{cases} \frac{1}{3}P(E, \Omega) & \text{if } v = \chi_E \text{ for some } E \in \mathcal{B}(\Omega), \\ \infty & \text{otherwise}, \end{cases}$$

where $\mathcal{B}(\Omega)$ denotes the Borel field of Ω.

Minimizing the functional $MM_\epsilon(v)$ with respect to v yields the associated Euler–Lagrange equation with boundary condition,

$$W'(v) = 2\epsilon^2 \Delta v \text{ in } \Omega, \qquad \frac{\partial v}{\partial \vec{N}} = 0 \text{ on } \partial\Omega, \tag{7.1}$$

which can be easily solved in practice by finite differences.

7.1.2 Ambrosio–Tortorelli approximations

In the Mumford and Shah functional, the set $K = S_u$ is not necessarily the boundary of an open and bounded domain, but a construction similar to $MM_\epsilon(v)$ can still be used, with the potential $W(t) = \frac{1}{4}(1-t)^2$ instead. Ambrosio and Tortorelli proposed two elliptic approximations [21], [22] to the weak formulation of the Mumford and Shah problem (6.9). We present the second one [22], being simpler than the first one [21] and commonly used in practice.

Let $X = L^2(\Omega)^2$ and positive coefficients α and β. Let us define for small

$\epsilon > 0$,

$$AT_\epsilon(u,v) = \left| \begin{array}{l} \displaystyle\int_\Omega (u-g)^2 dx + \beta \int_\Omega \Big(v^2 + o(\epsilon)\Big)|\nabla u|^2 dx \\ \displaystyle +\alpha \int_\Omega \Big(\epsilon|\nabla v|^2 + \frac{(v-1)^2}{4\epsilon}\Big) dx, \text{ if } (u,v) \in H^1(\Omega)^2,\ 0 \le v \le 1, \\ +\infty, \text{ otherwise,} \end{array} \right. \tag{7.2}$$

where $o(\epsilon)$ converges faster than ϵ to zero. We also define the limiting Mumford–Shah functional,

$$F(u,v) = \begin{cases} \displaystyle\int_\Omega (u-g)^2 dx + \beta \int_\Omega |\nabla u|^2 + \alpha \mathcal{H}^{d-1}(S_u), & \text{if } u \in SBV(\Omega),\ v \equiv 1, \\ +\infty, & \text{otherwise.} \end{cases}$$

Theorem 31 *AT_ϵ Γ-converges to $F(u,v)$ as $\epsilon \searrow 0$ in X. Moreover, AT_ϵ admits a minimizer (u_ϵ, v_ϵ) such that up to subsequences, u_ϵ converges to some $u \in SBV(\Omega)$ a minimizer of $F(u,1)$ and $\inf AT_\epsilon(u_\epsilon, v_\epsilon) \to F(u,1)$.*

Interesting generalizations of this result are given and proved by Braides in [61].

In practice, the Euler–Lagrange equations associated with the alternating minimization of AT_ϵ with respect to $u = u_\epsilon$ and $v = v_\epsilon$ are used and discretized by finite differences (note that we dropped the subscript ϵ for simplicity of notation, but inside the approximating functionals, u and v still represent u_ϵ and v_ϵ respectively). These are

$$\frac{\partial AT_\epsilon}{\partial u}(u,v) = 2(u-g) - 2\beta \text{div}\Big((v^2 + o(\epsilon))\nabla u\Big) = 0 \text{ in } \Omega \tag{7.3}$$

$$\frac{\partial AT_\epsilon}{\partial v}(u,v) = 2\beta v|\nabla u|^2 - 2\alpha\epsilon\Delta v + \frac{\alpha}{2\epsilon}(v-1) = 0 \text{ in } \Omega, \tag{7.4}$$

combined with corresponding boundary conditions $v^2 \nabla u \cdot \vec{N} = 0$ and $\nabla v \cdot \vec{N} = 0$ on $\partial\Omega$, where $\vec{N}$ denotes the exterior unit normal to the boundary.

We can interpret the Ambrosio–Tortorelli approximations by looking at the above Euler–Lagrange equations. Rewriting them as

$$u = g + \beta \text{div}\Big((v^2 + o(\epsilon))\nabla u\Big), \tag{7.5}$$

$$v = \frac{\frac{\alpha}{2\epsilon} + 2\alpha\epsilon\Delta v}{\frac{\alpha}{2\epsilon} + 2\beta|\nabla u|^2}, \tag{7.6}$$

we can see that, if $v \approx 0$ at some point, there is no diffusion in u at that point that might belong to an edge; if, on the contrary, $v \approx 1$ at some point, there is diffusion in u at that point to obtain a smooth approximation (see equation (7.5)). Moreover, if $|\nabla u| \approx 0$ at some point, then $v \approx 1$ with stronger diffusion on v; otherwise, if $|\nabla u|$ is large, $v \approx 0$ with a very small diffusion coefficient (see equation (7.6)).

One of the finite differences approximations to compute u and v in two dimensions $x = (x_1, x_2)$ is as follows (note that in the numerical calculations, we have dropped the constant $o(\epsilon)$ since it is very small). We use a time-dependent scheme in $u = u(x_1, x_2, t)$ and a stationary semi-implicit fixed-point scheme in $v = v(x_1, x_2)$. Let $\triangle x_1 = \triangle x_2 = h$ be the space step, Δt be the time step, and $g_{i,j}$, $u^n_{i,j}$, $v^n_{i,j}$ be the discrete versions of g, and of u and v at iteration $n \geq 0$, for $1 \leq i \leq M$, $1 \leq j \leq N$. Initialize $u^0 = g$ and $v^0 = 0$.

For $n \geq 1$, compute and repeat to steady state, for $i = 2, ..., M-1$ and $j = 2, ..., N-1$ (combined with Neumann boundary conditions on $\partial\Omega$):

$$
\begin{aligned}
|\nabla u^n|^2_{i,j} &= \Big(\frac{u^n_{i+1,j} - u^n_{i,j}}{h}\Big)^2 + \Big(\frac{u^n_{i,j+1} - u^n_{i,j}}{h}\Big)^2, \\
0 &= 2\beta v^{n+1}_{i,j} |\nabla u^n|^2_{i,j} \\
&\quad - 2\frac{\alpha\epsilon}{h^2}(v^n_{i+1,j} + v^n_{i-1,j} + v^n_{i,j+1} + v^n_{i,j-1} - 4v^{n+1}_{i,j}) \\
&\quad + \frac{\alpha}{2\epsilon}(v^{n+1}_{i,j} - 1), \\
\frac{u^{n+1}_{i,j} - u^n_{i,j}}{\Delta t} &= -(u^n_{i,j} - g_{i,j}) \\
&\quad + \frac{\beta}{h^2}\Big[(v^{n+1}_{i,j})^2(u^n_{i+1,j} - u^n_{i,j}) + (v^{n+1}_{i,j})^2(u^n_{i,j+1} - u^n_{i,j}) \\
&\quad - (v^{n+1}_{i-1,j})^2(u^n_{i,j} - u^n_{i-1,j}) - (v^{n+1}_{i,j-1})^2(u^n_{i,j} - u^n_{i,j-1})\Big],
\end{aligned}
$$

which is equivalent to

$$
\begin{aligned}
|\nabla u^n|^2_{i,j} &= \Big(\frac{u^n_{i+1,j} - u^n_{i,j}}{h}\Big)^2 + \Big(\frac{u^n_{i,j+1} - u^n_{i,j}}{h}\Big)^2, \\
v^{n+1}_{i,j} &= \frac{\frac{\alpha}{2\epsilon} + \frac{2\alpha\epsilon}{h^2}(v^n_{i+1,j} + v^n_{i-1,j} + v^n_{i,j+1} + v^n_{i,j-1})}{\frac{\alpha}{2\epsilon} + 2\beta|\nabla u^n|^2_{i,j} + \frac{8\alpha\epsilon}{h^2}}, \\
u^{n+1}_{i,j} &= u^n_{i,j} + \Delta t\Big\{ -(u^n_{i,j} - g_{i,j}) + \frac{\beta}{h^2}\Big[(v^{n+1}_{i,j})^2(u^n_{i+1,j} - u^n_{i,j}) \\
&\quad + (v^{n+1}_{i,j})^2(u^n_{i,j+1} - u^n_{i,j}) - (v^{n+1}_{i-1,j})^2(u^n_{i,j} - u^n_{i-1,j}) \\
&\quad - (v^{n+1}_{i,j-1})^2(u^n_{i,j} - u^n_{i,j-1})\Big]\Big\}.
\end{aligned}
$$

Experimental results of u_ϵ and v_ϵ obtained using the above Ambrosio–Tortorelli approximations applied to the well-known Barbara image from Figure 7.1 are shown in Figures 7.2, 7.3 and 7.4, varying the regularizing coefficients. Thus $\alpha = \beta = 10$ in Figure 7.2, $\alpha = \beta = 5$ in Figure 7.3, and $\alpha = \beta = 1$ in Figure 7.4. In all cases $\epsilon = 0.0001$ is kept fixed during the iterations, and 150 iterations were performed. We note that, as expected, less regularization (smaller α and β) gives more edges in the image v. We also plot in Figure 7.5 the numerical energy versus iterations for the case $\alpha = \beta = 10$, $\epsilon = 0.0001$.

FIGURE 7.1: Original image g.

7.2 Shah approximation to the MSTV functional

For the Ambrosio–Tortorelli regularizer, we introduce the notation,

$$\Psi_\epsilon^{MSH^1}(u,v) = \beta \int_\Omega v^2 |\nabla u|^2 dx + \alpha \int_\Omega \left(\epsilon |\nabla v|^2 + \frac{(v-1)^2}{4\epsilon} \right) dx. \tag{7.7}$$

Shah [288] suggested a modified version of the above regularizer (7.7) by replacing the norm square of $|\nabla u|$ by the norm in the first term:

$$\Psi_\epsilon^{MSTV}(u,v) = \beta \int_\Omega v^2 |\nabla u| dx + \alpha \int_\Omega \left(\epsilon |\nabla v|^2 + \frac{(v-1)^2}{4\epsilon} \right) dx.$$

This functional Γ–converges to the Ψ^{MSTV} functional [7]

$$\Psi^{MSTV}(u) = \beta \int_{\Omega \setminus K} |\nabla u| dx + \alpha \int_K \frac{|u^+ - u^-|}{1 + |u^+ - u^-|} d\mathcal{H}^{d-1} + |D_c u|(\Omega)$$

given in Chapter 6, Section 6.3, where u^+ and u^- denote the image values on two sides of the jump set $K = S_u$ of u, and $D_c u$ is the Cantor part of the measure-valued derivative Du. The regularizer Ψ_ϵ^{MSTV} and its approximations Ψ_ϵ^{MSTV} are well defined on the space $GBV(\Omega)$ defined in Chapter 6, Section 6.3, called the space of generalized functions of bounded variation.

7.3 Applications to image restoration

In this section, we present applications of the Ambrosio–Tortorelli and Shah approximations to image restoration, following the work of Bar et al.

FIGURE 7.2: Results using the Ambrosio–Tortorelli approximations ($\alpha = \beta = 10$, $\epsilon = 0.0001$): left u_ϵ, right v_ϵ.

FIGURE 7.3: Results using the Ambrosio–Tortorelli approximations ($\alpha = \beta = 5$, $\epsilon = 0.0001$): left u_ϵ, right v_ϵ.

FIGURE 7.4: Results using the Ambrosio–Tortorelli approximations ($\alpha = \beta = 1$, $\epsilon = 0.0001$): left u_ϵ, right v_ϵ.

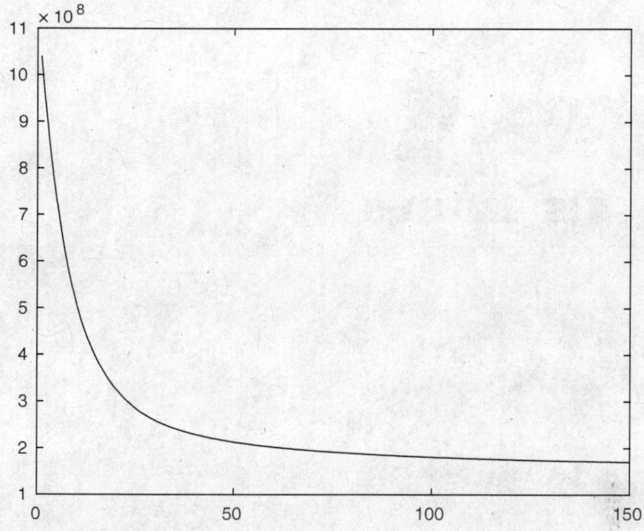

FIGURE 7.5: Numerical energy versus iterations for the Ambrosio–Tortorelli approximations ($\alpha = \beta = 10$, $\epsilon = 0.0001$, 150 iterations).

[40, 39, 42]. We use these approximations of the Mumford and Shah functional to deblur images in the presence of Gaussian or random-valued impulse noise. The numerical results presented in this section have been obtained by Jung and Vese in [183]. We denote here the given degraded image by $f : \Omega \to \mathbb{R}$. Thus, by incorporating the proper fidelity terms depending on the noise model, we design two types of energies as

$$\text{Gaussian noise model:} \quad E^G(u,v) = \eta \int_\Omega (f - k * u)^2 dx + \Psi_\epsilon^{MS}(u,v),$$

$$\text{Impulse noise model:} \quad E^{Im}(u,v) = \eta \int_\Omega |f - k * u| dx + \Psi_\epsilon^{MS}(u,v),$$

where Ψ_ϵ^{MS} is either $\Psi_\epsilon^{MSH^1}$ or Ψ_ϵ^{MSTV}. Minimizing these functionals in u and v, we obtain the Euler–Lagrange equations

$$\frac{\partial E^G}{\partial v} = \frac{\partial E^{Im}}{\partial v} = 2\beta v \phi(|\nabla u|^2) - 2\epsilon\alpha\Delta v + \alpha\left(\frac{v-1}{2\epsilon}\right) = 0,$$

$$\text{Gaussian noise model:} \quad \frac{\partial E^G}{\partial u} = 2\eta\, \tilde{k} * (k * u - f) + L^{MS} u = 0,$$

$$\text{Impulse noise model:} \quad \frac{\partial E^{Im}}{\partial u} = \eta\, \tilde{k} * \text{sign}(k * u - f) + L^{MS} u = 0,$$

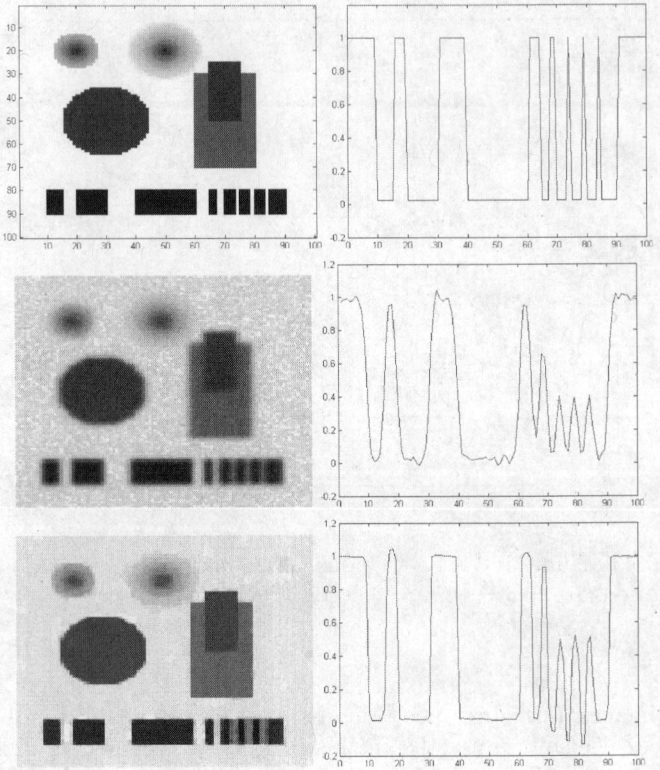

FIGURE 7.6: Image recovery: Gaussian blur kernel with $\sigma_b = 1$ and Gaussian noise with $\sigma_n = 5$. Top: original image and its cross section. Middle: noisy blurry version and its cross section. Bottom: recovered image and its cross section using MSTV (SNR = 33.5629). $\beta = 0.0045$, $\alpha = 0.00000015$, $\epsilon = 0.000001$.

with corresponding boundary conditions, where $\tilde{k}(x, y) = k(-x, -y)$, $\eta > 0$ is a tuning parameter, and

$$L^{MS}u = \begin{cases} -2\beta \operatorname{div}\left(v^2 \nabla u\right) & \text{if} \quad \Psi_\epsilon^{MS} = \Psi_\epsilon^{MSH^1} \\ -\beta \operatorname{div}\left(v^2 \frac{\nabla u}{|\nabla u|}\right) & \text{if} \quad \Psi_\epsilon^{MS} = \Psi_\epsilon^{MSTV} \end{cases},$$

$\phi(s) = s$ in the MSH^1 case and $\phi(s) = \sqrt{s}$ in the $MSTV$ case.

The energy functionals $E^G(u, v)$ and $E^{Im}(u, v)$ are convex in each variable and bounded from below. Therefore, to solve two Euler–Lagrange equations simultaneously, the alternate minimization approach with gradient descent is applied. Note that since both energy functionals are not convex in the joint variable (u, v), when using gradient descent, we may compute only a local

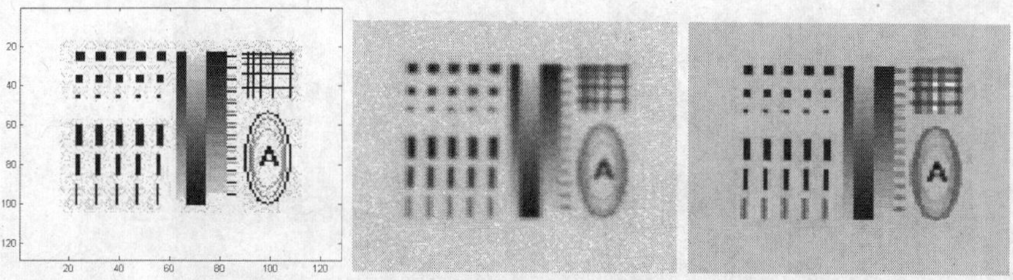

FIGURE 7.7: Original image; noisy blurry version using Gaussian kernel with $\sigma_b = 1$, then contaminated by Gaussian noise with $\sigma_n = 5$; recovered image using $MSTV$ (SNR = 14.4693). $\beta = 0.007$, $\alpha = 0.00000015$, $\epsilon = 0.0000005$.

minimizer. However, this is not a drawback in practice, since the initial guess for u in our experimental results is taken to be the data f.

The MS regularizers presented here were tested on several images with different blur kernels and type of noise. In each case, we show an original image, its noisy and blurry version, and the reconstruction. As a measure of quality of restoration, we use the signal-to-noise ratio (SNR). Higher SNR values indicate better restoration, while the degraded images have lower SNR. First, we test the Gaussian noise model in Figures 7.6 through 7.8. Specifically, in Figure 7.6, we use a simple image and its one-dimensional cross section for the visualization. Next, we recover a blurred image contaminated by impulse noise (salt-and-pepper noise or random-valued impulse noise). Thus we have tested the MSH^1 model on the Einstein image (Figure 7.9) with different blur kernels and impulse noise models (salt-and-pepper noise and random-valued impulse noise), with the same noise density of 0.4.

We mention that the parameter η in the L^1 fidelity term was set to $\eta = 10^{-5}$ for all functionals, and in the MS regularizers, the parameters α, β and ϵ were selected manually to provide higher SNR values. The smoothness parameter β increases with noise level while the other parameters α, ϵ are approximately fixed.

Finally, it would be interesting to compare the numerical results shown in Figures 7.6 through 7.8 and obtained using the Ψ_ϵ^{MSTV} regularizer and the Gaussian noise model with those presented in Chapters 3 and 4. These results are slightly better than those from Chapter 3 obtained using the convex total variation regularization. However, the results obtained in Chapter 4 using the nonlocal total variation appeared to be even better.

FIGURE 7.8: Top: Original image; noisy blurry version using the pill-box kernel of radius 2, then contaminated by Gaussian noise with $\sigma_n = 5$. Middle: recovered image u using $MSTV$ (SNR = 25.1968) and MSH^1 (SNR = 23.1324). Bottom: corresponding residuals $f - k * u$. $\beta = 0.0045$ ($MSTV$), 0.06 (MSH^1), $\alpha = 0.00000001$, $\epsilon = 0.00002$.

7.4 Exercises

Exercise 7.1 *Derive the Euler–Lagrange equation given in (7.1) associated with the minimization of* MM_ϵ. *Implement a finite difference algorithm in two dimensions, based on this partial differential equation, for minimizing the approximated length functional, and apply it to an initial domain* E *for decreasing its perimeter. Plot the numerical energy over iterations.*

Exercise 7.2 *Show existence of minimizers* (u_ϵ, v_ϵ) *for the* AT_ϵ *functional, for fixed* ϵ, *on the space* $V = \{(u,v) \in H^1(\Omega)^2,\ 0 \le v \le 1\}$, *and also that* $\|u_\epsilon\|_{L^\infty(\Omega)} \le \|g\|_{L^\infty(\Omega)}$.

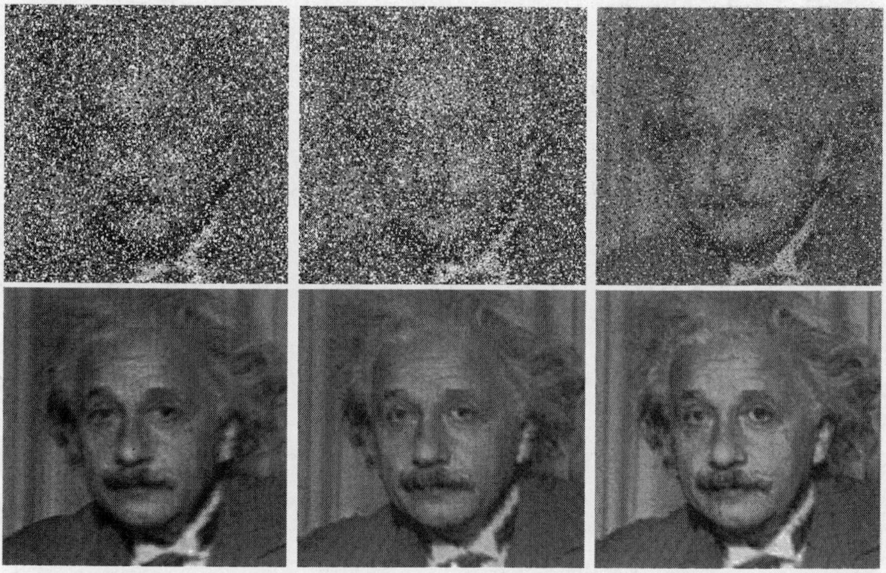

FIGURE 7.9: Restoration using MSH^1 with the image blurred and contaminated by high density ($d = 0.4$) of impulse noise. Top: noisy blurry images (left) using motion blur kernel of length=10, oriented at angle $\theta = 25°$ with respect to the horizon and salt-and-pepper noise with $d = 0.4$, (middle) using Gaussian kernel with $\sigma_b = 1$ and salt-and-pepper noise with $d = 0.4$, (right) using Gaussian kernel with $\sigma_b = 1$ and random-valued impulse noise with $d = 0.4$. Bottom: recovered images using MSH^1, (left) SNR = 17.1106, (middle) SNR = 15.2017, (right) SNR = 16.6960. First and second columns: $\beta = 2$ (MSH^1), $\alpha = 0.001$, $\epsilon = 0.0002$. Third column: $\beta = 2.5$ (MSH^1), $\alpha = 0.000001$, $\epsilon = 0.002$.

Exercise 7.3 *Write the one-dimensional AT_ϵ regularization for an interval (a, b). Find the associated Euler–Lagrange equations and implement a finite-difference numerical approximation for solving them. Apply the obtained algorithm to a one-dimensional signal for segmentation or denoising, and verify numerically the energy decrease over iterations.*

Exercise 7.4 *Derive the Euler–Lagrange equations associated with the minimization of the AT_ϵ energy as given in (7.3) and (7.4).*

Exercise 7.5 *Implement a two-dimensional finite difference algorithm for minimizing a discrete AT_ϵ functional (such as the one presented in this chapter), and apply it to image segmentation or denoising.*

Exercise 7.6 *Using the notations from Section 7.3, derive the Euler–*

Lagrange equation associated with the minimization in (u, v) of

$$E^G(u,v) = \eta \int_\Omega (f - k * u)^2 dx + \Psi_\epsilon^{MSH^1}(u,v).$$

Then derive a finite difference approximation for its minimization in two dimensions and apply it to image restoration in the presence of additive Gaussian noise and blur.

Exercise 7.7 *Using the notations from Section 7.3, derive the Euler–Lagrange equation associated with the minimization in (u, v) of*

$$E^G(u,v) = \eta \int_\Omega |f - k * u|^2 dx + \Psi_\epsilon^{MSTV}(u,v).$$

Then derive a finite difference approximation for its minimization in two dimensions and apply it to image restoration in the presence of additive Laplace noise and blur.

Chapter 8

Region-Based Variational Active Contours

We present in this chapter level set formulations for minimizing the Mumford and Shah functional for segmentation from Chapter 6, as proposed by Chan and Vese in [92], [94], [335], [93], [95] (see also the related work by Tsai, Yezzi, and Willsky [328], Samson et al. [280, 279], Amadieu et al. [17], and Cohen et al. [108, 109]). These works make the link between curve evolution, active contours, and Mumford–Shah segmentation. These models have been proposed by restricting the set of minimizers u to specific classes of functions: piecewise-constant, piecewise-smooth, with the edge set K represented by a union of curves or surfaces that are boundaries of open subsets of Ω. For example, if K is the boundary of an open and bounded subset of Ω, then it can be represented implicitly, as the zero-level line of a level set function ϕ. Thus the set K as an unknown is replaced by an unknown function ϕ that defines it implicitly, and the Euler–Lagrange equations with respect to the unknowns can be easily computed and discretized.

The variational level set terminology due to Zhao et al. [352] presented in Chapter 2, Section 2.9 will be useful to rewrite the Mumford and Shah functional in terms of (u, ϕ), instead of (u, K).

8.1 Piecewise-constant Mumford and Shah segmentation using level sets

We consider level set formulations of the minimal partition problem (6.2).

8.1.1 Two-phase piecewise-constant binary segmentation

The first formulation is for the case when the unknown set of edges K can be represented by $K = \{x \in \Omega : \ \phi(x) = 0\}$ for some (unknown) Lipschitz function $\phi : \Omega \to \mathbb{R}$. In this case we restrict the unknown minimizers u to functions taking two unknown values c_1 and c_2. The corresponding Mumford–

Shah minimization problem can be formulated as [92, 94]

$$\inf_{c_1,c_2,\phi} E(c_1,c_2,\phi) = \int_\Omega (g(x)-c_1)^2 H(\phi)dx + \int_\Omega (g(x)-c_2)^2 H(-\phi)dx \\ + \nu_0 \int_\Omega |DH(\phi)|. \tag{8.1}$$

The unknown minimizer u is expressed as

$$u(x) = c_1 H(\phi(x)) + c_2(1 - H(\phi(x))) = c_1 H(\phi(x)) + c_2 H(-\phi(x)).$$

We replace H by its C^∞ approximation H_ϵ given in Chapter 2, Section 2.9, and we minimize instead

$$E_\epsilon(c_1,c_2,\phi) = \int_\Omega (g(x)-c_1)^2 H_\epsilon(\phi)dx + \int_\Omega (g(x)-c_2)^2 H_\epsilon(-\phi)dx \\ + \nu_0 \int_\Omega |\nabla H_\epsilon(\phi)|dx.$$

The associated Euler–Lagrange equations with respect to c_1, c_2 and ϕ are

$$c_1(\phi) = \frac{\int_\Omega g(x)H_\epsilon(\phi(x))dx}{\int_\Omega H_\epsilon(\phi(x))dx}, \quad c_2(\phi) = \frac{\int_\Omega g(x)H_\epsilon(-\phi(x))dx}{\int_\Omega H_\epsilon(-\phi(x))dx} \tag{8.2}$$

(assuming that the denominators in the expressions for c_1 and c_2 do not vanish), and after simplifications,

$$\delta_\epsilon(\phi)\Big[(g(x)-c_1)^2 - (g(x)-c_2)^2 - \nu_0 \mathrm{div}\Big(\frac{\nabla\phi}{|\nabla\phi|}\Big)\Big] = 0 \text{ in } \Omega, \tag{8.3}$$

with boundary conditions $\nabla\phi \cdot \vec{n} = 0$ on $\partial\Omega$. Since $\delta_\epsilon > 0$ as defined in Chapter 2, Section 2.9, the factor $\delta_\epsilon(\phi)$ can be removed from (8.3), or replaced by $|\nabla\phi|$ to obtain a more geometric motion extended to all level lines of ϕ, as in the standard level set approach.

In the gradient descent approach, we introduce an artificial time $t \geq 0$ and we drive to steady state the following time-dependent partial differential equation with unknown $\phi(t,x)$, $x \in \Omega$, $t > 0$, $\phi(0,x) = \phi_0(x)$ in Ω:

$$\frac{\partial\phi}{\partial t} = \delta_\epsilon(\phi)\Big[(g(x)-c_2)^2 - (g(x)-c_1)^2 + \nu_0 \mathrm{div}\Big(\frac{\nabla\phi}{|\nabla\phi|}\Big)\Big]. \tag{8.4}$$

8.1.2 Multilayer piecewise-constant segmentation

The two-phase binary approach has been generalized by Chung and Vese in [103, 104], where more than one level line of the same level set function ϕ can be used to represent the edge set K. Using m distinct real levels $\{l_1 <$

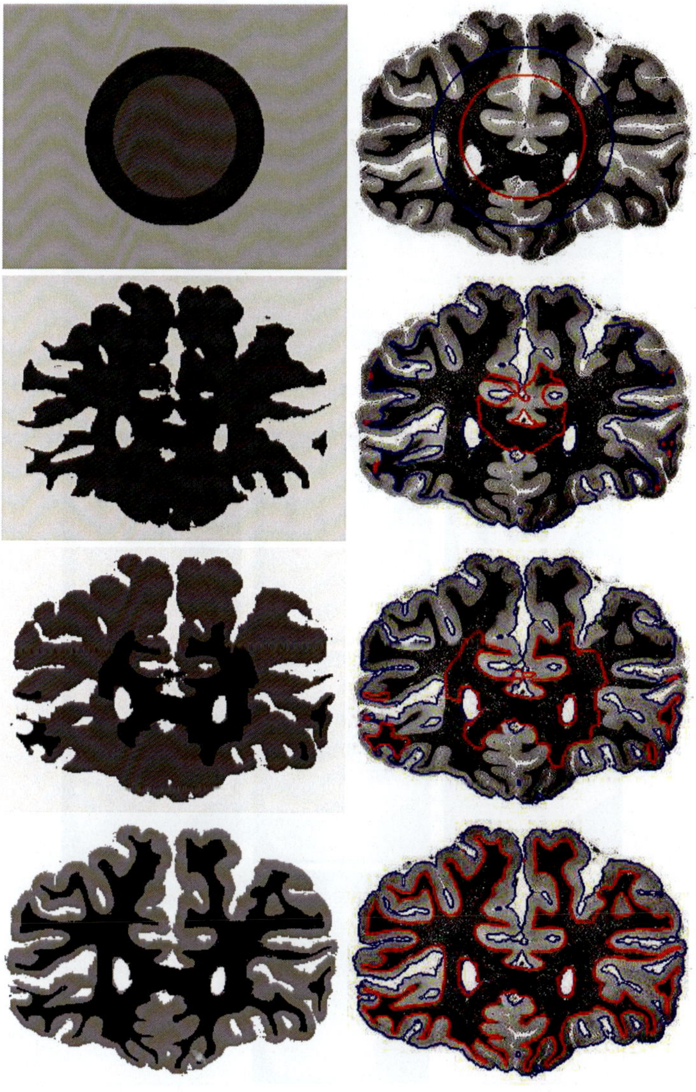

FIGURE 8.5: Segmentation of a brain image using one level set function with two levels.

FIGURE 10.2: Deblurring in the presence of random-valued impulse noise. Top two rows: results related to data f given in Figure 10.1 top row. Bottom two rows: results related to data f given in Figure 10.1 bottom row. First column: (top) data f, (bottom) preprocessed image $\bar{g}$. Second through fourth columns: recovered images using (top) local regularizers (MSH^1, MSTV, TV) and (bottom) nonlocal regularizers (NL/MSH^1, NL/MSTV, NL/TV).

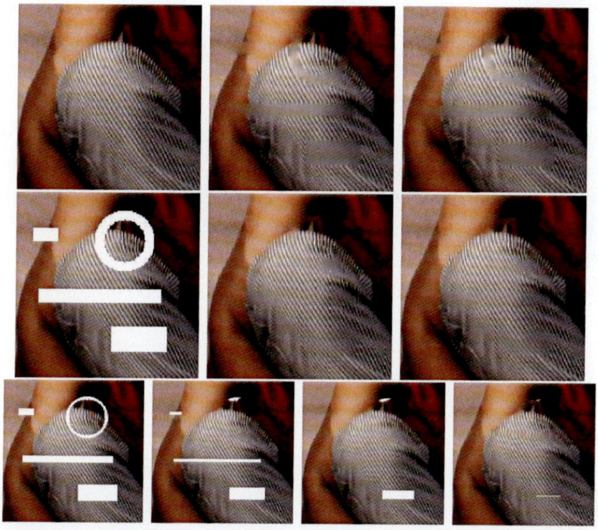

FIGURE 10.8: Inpainting of 150 × 150 size image. First row, (a): original; second row, (a): data f. First row, (b) through (c): recovered images using MSH^1: PSNR = 29.2797 and MSTV: PSNR = 29.4205. Second row, (b) through (c): recovered images using $\mathrm{NL/MSH}^1$: PSNR = 34.4953 and NL/MSTV: PSNR = 34.2406, with 51 × 51 search window and 9 × 9 patch. Bottom row: intermediate steps of the inpainting process with $\mathrm{NL/MSH}^1$ in 50th, 100th, 200th, 350th iterations.

FIGURE 10.9: Super-resolution of still image. Top row, left to right: original image of size 272×272; down-sampled data $f = D_k(h * u)$ of size 68×68 with Gaussian blur kernel h with $\sigma_b = 2$ and $k = 4$; preprocessed (up-sampled) image $\bar{g}$ using bicubic interpolation; deblurred image from applying Lucy–Richardson deconvolution algorithm to $\bar{g}$ at 10th iteration. Bottom row, left to right: recovered images using MSTV with corresponding edge set v and NL/MSTV with corresponding edge set v. $h * u$: PSNR = 21.9423, $\bar{g}$: PSNR = 18.3970, MSTV: PSNR = 23.4551, and NL/MSTV: PSNR = 24.3336.

FIGURE 10.10: Another super-resolution of still image. Top row: original image of size 272×272, blurred image $h * u$ with out-of-focus blur kernel h with radius $r = 3$, down-sampled data $f = D_k(h * u)$ of size 68×68 with $k = 4$, preprocessed (up-sampled) image $\bar{g}$ using bicubic interpolation. Bottom row: recovered images using (left) MSTV, (right) NL/MSTV. $h * u$: PSNR = 21.7459, $\bar{g}$: PSNR = 19.9989, MSTV: PSNR = 21.8869 and NL/MSTV: PSNR = 22.1116.

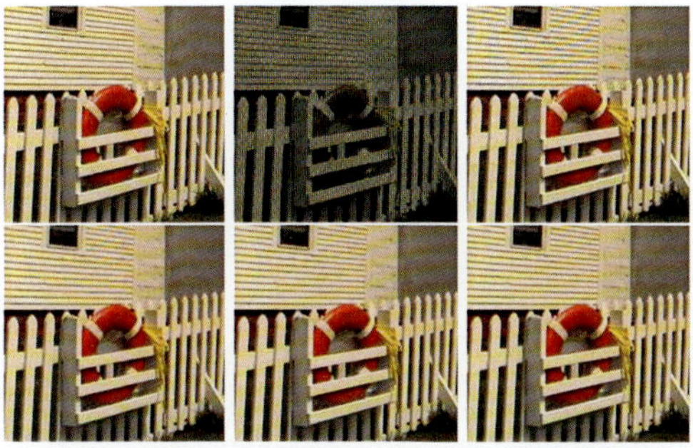

FIGURE 10.11: Demosaicing using NL/MSTV with iterative algorithm. Top row, left to right: original image; data f; interpolated image (PSNR = 27.3076) using Hamilton–Adams for green and bilinear interpolation for $R-G$ and $B-G$. Bottom row, left to right: demosaiced images with decreasing sequence of $h = \{16, 8, 4\}$ and corresponding PSNR values: first = 29.5101, second = 29.7029, third = 29.7128.

FIGURE 10.12: Demosaicing using NL/MSTV with $h = \{16, 8, 2\}$. PSNR values: Hamilton–Adams (top right) = 26.5008, NL/MSTV (bottom row, left to right): 28.1480, 28.2220, and 28.2615.

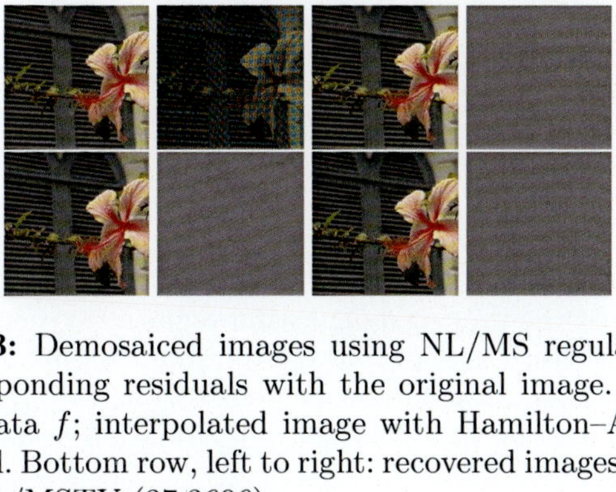

FIGURE 10.13: Demosaiced images using NL/MS regularizers with $h = \{16\}$, and corresponding residuals with the original image. Top row, left to right: original; data f; interpolated image with Hamilton–Adams (PSNR = 36.5672); residual. Bottom row, left to right: recovered images using NL/MSH^1 (37.3598) and NL/MSTV (37.3606).

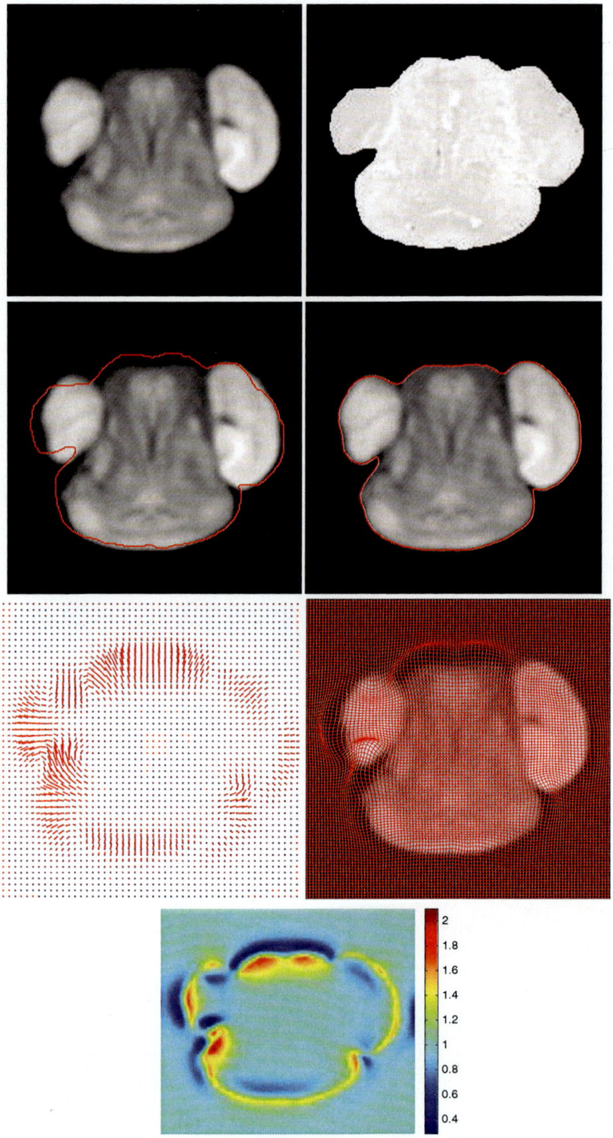

FIGURE 11.2: Application to image registration. First row: left, reference R; right, template T (mouse atlas and gene data). Second row, left to right: contour obtained by the proposed algorithm segmenting template T (starting with Φ_0 defining a disk), superimposed over the reference R; segmented reference, using as Φ_0 the output contour detected at the previous step. Third row, left to right: deformation field; final deformation grid from reference to template. Last row: $\det(\nabla\varphi)$.

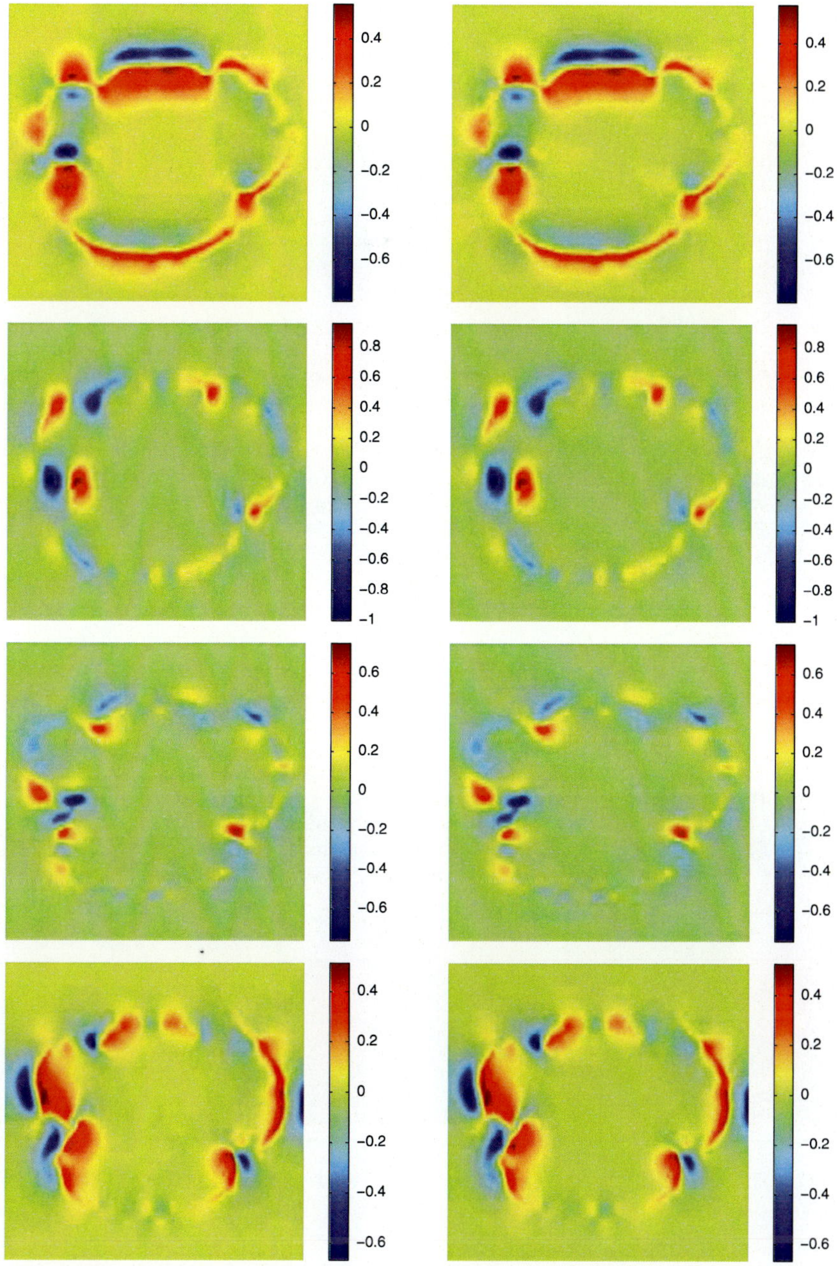

FIGURE 11.3: Components of V and ∇u illustrating good match of two variables.

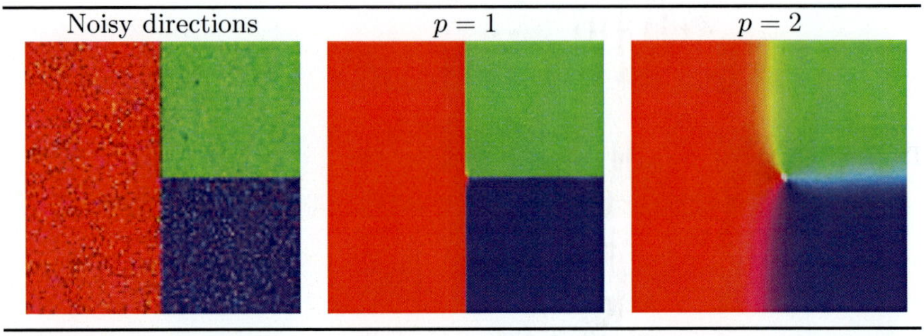

FIGURE 15.10: Directions denoising with $p = 1$ (middle) and $p = 2$ (right). The unit vectors are represented as RGB colors ($\triangle t = 0.01$, $h = 1$).

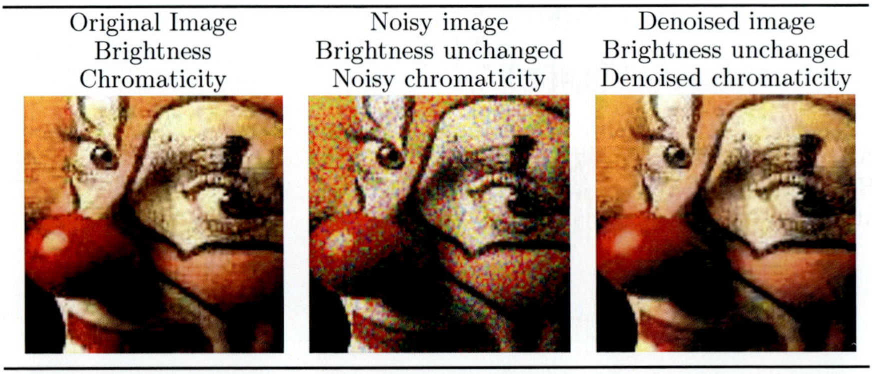

FIGURE 15.11: Chromaticity denoising with $p = 1$. The brightness is kept unchanged from the original image ($\triangle t = 0.01$, $h = 1$, 50 iterations).

$l_2 < ... < l_m\}$, the function ϕ partitions the domain Ω into the following $m+1$ disjoint open regions making up Ω together with their boundaries:

$$\begin{aligned}
\Omega_0 &= \{x \in \Omega : \; -\infty < \phi(x) < l_1\}, \\
\Omega_j &= \{x \in \Omega : \; l_j < \phi(x) < l_{j+1}\}, \; 1 \leq j \leq m-1, \\
\Omega_m &= \{x \in \Omega : \; l_m < \phi(x) < +\infty\}.
\end{aligned}$$

The energy to minimize in this case, depending on $c_0, c_1, ..., c_m, \phi$, will be

$$\begin{aligned}
E(c_0, c_1, ..., c_m, \phi) = &\int_\Omega |g(x) - c_0|^2 H(l_1 - \phi(x))dx \\
&+ \sum_{j=1}^{m-1} \int_\Omega |g(x) - c_j|^2 H(\phi(x) - l_j) H(l_{j+1} - \phi(x))dx \\
&+ \int_\Omega |g(x) - c_m|^2 H(\phi(x) - l_m)dx \\
&+ \nu_0 \sum_{j=1}^{m} \int_\Omega |DH(\phi - l_j)|.
\end{aligned} \tag{8.5}$$

The segmented image will be given by

$$\begin{aligned}
u(x) = &c_0 H(l_1 - \phi(x)) + \sum_{j=1}^{m-1} c_j H(\phi(x) - l_j) H(l_{j+1} - \phi(x)) \\
&+ c_m H(\phi(x) - l_m).
\end{aligned}$$

As before, to minimize the above energy, we approximate and substitute the Heaviside function H by H_ϵ, as $\epsilon \to 0$. The Euler–Lagrange equations associated with the corresponding minimization

$$\inf_{c_0, c_1, ..., c_m, \phi} E_\epsilon(c_0, c_1, ..., c_m, \phi) \tag{8.6}$$

can be expressed as (assuming that the denominators do not vanish),

$$\begin{cases}
c_0(\phi) &= \frac{\int_\Omega g(x) H_\epsilon(l_1 - \phi(x))dx}{\int_\Omega H_\epsilon(l_1 - \phi(x))dx}, \\
c_j(\phi) &= \frac{\int_\Omega g(x) H_\epsilon(\phi(x) - l_j) H_\epsilon(l_{j+1} - \phi(x))dx}{\int_\Omega H_\epsilon(\phi(x) - l_j) H_\epsilon(l_{j+1} - \phi(x))dx}, \\
c_m(\phi) &= \frac{\int_\Omega g(x) H_\epsilon(\phi(x) - l_m)dx}{\int_\Omega H_\epsilon(\phi(x) - l_m)dx},
\end{cases} \tag{8.7}$$

and

$$
\begin{aligned}
0 = & -|g-c_0|^2\delta_\epsilon(l_1-\phi) + \sum_{j=1}^{m-1}|g-c_j|^2\Big[\delta_\epsilon(\phi-l_j)H_\epsilon(l_{j+1}-\phi) \\
& -\delta_\epsilon(l_{j+1}-\phi)H_\epsilon(\phi-l_j)\Big] + |g-c_m|^2\delta_\epsilon(\phi-l_m) \qquad (8.8)\\
& -\nu_0\sum_{j=1}^{m}\Big[\delta_\epsilon(\phi-l_j)\mathrm{div}\Big(\frac{\nabla\phi}{|\nabla\phi|}\Big)\Big],
\end{aligned}
$$

$$\frac{\partial\phi}{\partial\vec{n}}\Big|_{\partial\Omega}=0,$$

where $\vec{n}$ is the exterior unit normal to the boundary $\partial\Omega$.

Again, in the gradient descent approach, we introduce an artificial time $t \geq 0$ and we drive to steady state the time-dependent partial differential equation with unknown $\phi(t,x)$, $x \in \Omega$, $t > 0$, $\phi(0,x) = \phi_0(x)$ in Ω

$$
\begin{aligned}
\frac{\partial\phi}{\partial t} = & |g-c_0|^2\delta_\epsilon(l_1-\phi) + \sum_{j=1}^{m-1}|g-c_j|^2\Big[\delta_\epsilon(l_{j+1}-\phi)H_\epsilon(\phi-l_j) \\
& -\delta_\epsilon(\phi-l_j)H_\epsilon(l_{j+1}-\phi)\Big] - |g-c_m|^2\delta_\epsilon(\phi-l_m) \qquad (8.9)\\
& +\nu_0\sum_{j=1}^{m}\Big[\delta_\epsilon(\phi-l_j)\mathrm{div}\Big(\frac{\nabla\phi}{|\nabla\phi|}\Big)\Big],
\end{aligned}
$$

$$\frac{\partial\phi}{\partial\vec{n}}\Big|_{\partial\Omega}=0.$$

We show in Figure 8.1 examples of partitions of the domain Ω, using m nested level lines of a Lipschitz continuous function ϕ.

We show existence of minimizers of the above functional (8.5). Assume $g \in L^\infty(\Omega)$, that Ω is open, bounded, connected and with Lipschitz boundary $\partial\Omega$. Let us denote by $\chi_j = H(\phi - l_j)$, $1 \leq j \leq m$, where now χ_j must be characteristic functions of sets E_j, with $E_{j+1} \subset E_j$ (this means that if $\chi_{j+1}(x) = 1$ at some point $x \in \Omega$, $\chi_j(x) = 1$ also). This will guarantee that in the new formulation we have

$$1 - \chi_1(x) + \sum_{j=1}^{m-1}\chi_j(x)(1-\chi_{j+1}(x)) + \chi_m(x) \equiv 1$$

for all $x \in \Omega$, i.e., a perfect partition.

Then the problem (8.6) can be reformulated as

$$\inf_{\chi_1,\chi_2,\ldots,\chi_m} E(\chi_1,\chi_2,\ldots,\chi_m), \qquad (8.10)$$

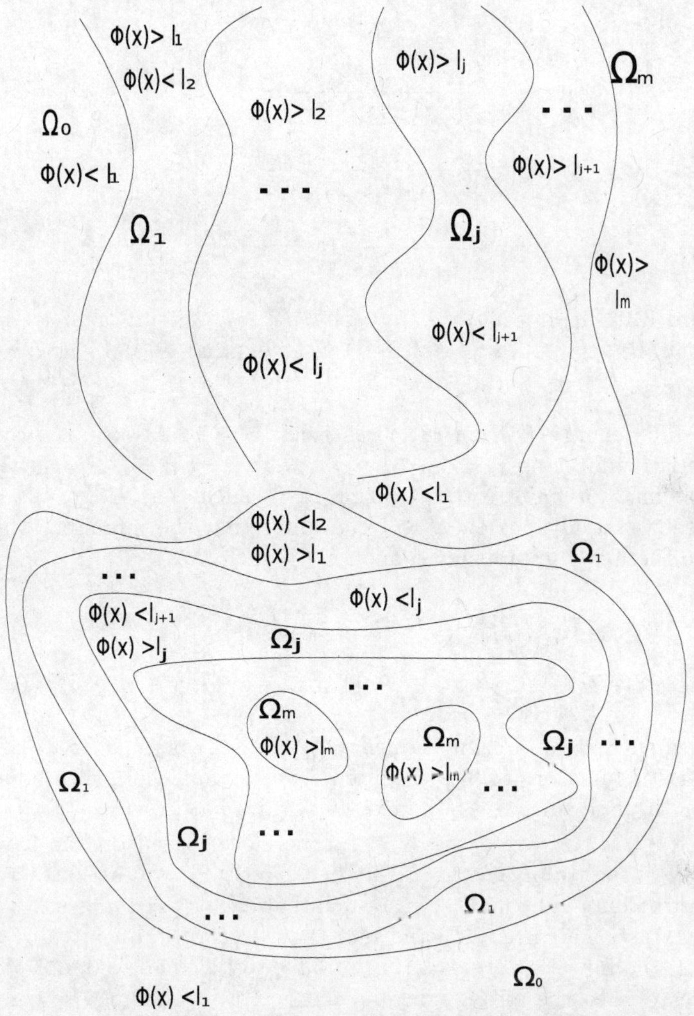

FIGURE 8.1: Two examples of partitions of the domain Ω into $m+1$ disjoint regions using m nested level lines $\{\phi(x) = l_i\}$ of the same function ϕ.

where

$$E(\chi_1, \chi_2, ..., \chi_m) = \int_\Omega (g - c_0(\chi_1))^2 (1 - \chi_1) dx + \sum_{j=1}^{m-1} \int_\Omega (g - c_j(\chi_j, \chi_{j+1}))^2 \chi_j (1 - \chi_{j+1}) dx + \int_\Omega (g - c_m(\chi_m))^2 \chi_m dx + \nu_0 \sum_{j=1}^{m} \int_\Omega |D\chi_j|, \tag{8.11}$$

with

$$\begin{aligned} c_0(\chi_1) =& \frac{\int_\Omega g(x)(1-\chi_1)dx}{\int_\Omega (1-\chi_1)dx}, \\ c_j(\chi_j, \chi_{j+1}) =& \frac{\int_\Omega g(x)\chi_j(1-\chi_{j+1})dx}{\int_\Omega \chi_j(1-\chi_{j+1})dx}, \ 1 \leq j \leq m-1, \\ c_m(\chi_m) =& \frac{\int_\Omega g(x)\chi_m dx}{\int_\Omega \chi_m dx}. \end{aligned}$$

Theorem 32 *The minimization problem (8.10), with E defined in (8.11), has a minimizer $(\chi_1, ..., \chi_m) \in BV(\Omega)^m$, with $\chi_j(x) \in \{0,1\}$ dx-a.e. for $1 \leq j \leq m$, and $\chi_j(x) \geq \chi_{j+1}(x)$ dx-a.e. in Ω.*

Proof. The energy E from (8.11) satisfies $E \geq 0$. Also, it is easy to find characteristic functions $\chi_j = \chi_{E_j}$, $1 \leq j \leq m$, with $E_{j+1} \subset E_j$, E_j with finite perimeter in Ω or finite total variation in Ω, such that $E(\chi_1, \chi_2, ..., \chi_m) < \infty$. These two conditions on E will guarantee that the infimum is finite and therefore there is a minimizing sequence χ_j^k, such that

$$\inf_{\chi_1,...,\chi_m} E(\chi_1, ..., \chi_m) = \lim_{k\to\infty} E(\chi_1^k, \chi_2^k, ..., \chi_m^k),$$

satisfying $\chi_j^k \in BV(\Omega)$, $\chi_j^k(x) \in \{0,1\}$ dx-a.e in Ω, $\chi_j^k(x) \geq \chi_{j+1}^k(x)$ dx-a.e. in Ω.

Taking such a minimizing sequence $(\chi_1^k, ..., \chi_m^k)$ of E, as $k \to \infty$, among characteristic functions of sets of finite perimeter in Ω (i.e., with boundary of finite length), and since Ω is bounded, we obtain that $\|\chi_j^k\|_{BV(\Omega)} = \int_\Omega |D\chi_j^k| + \|\chi_j^k\|_{L^1(\Omega)} \leq M < \infty$ for any $k \geq 1$. Therefore, based on the lower semicontinuity of the total variation ([140] or Chapter 2), we can extract a subsequence, still denoted by $(\chi_1^k, ..., \chi_m^k)$, such that each χ_i^k converges to a function $\chi_i \in BV(\Omega)$ strongly in $L^1(\Omega)$, and such that $\int_\Omega |D\chi_i| \leq \liminf_{k\to\infty} \int_\Omega |D\chi_i^k|$. Moreover, the functions χ_j, $j = 1, ..., m$ have to be equal to 0 or 1 almost everywhere (due to the strong convergence in L^1, thus pointwise convergence dx.-a.e. and that $\chi_j^k \in \{0,1\}$ dx.-a.e.); therefore these must be characteristic functions of sets of finite perimeter in Ω. We also must have in the limit, $\chi_j(x) \geq \chi_{j+1}(x)$, dx.-a.e. in Ω.

On the other hand, it is easy to verify that

$$\begin{aligned} \lim_{k\to\infty} c_0(\chi_1^k) =& c_0(\chi_1), \\ \lim_{k\to\infty} c_j(\chi_j^k, \chi_{j+1}^k) =& c_j(\chi_j, \chi_{j+1}), \quad 1 \leq j \leq m-1, \\ \lim_{k\to\infty} c_m(\chi_m^k) =& c_m(\chi_m). \end{aligned}$$

Then we deduce that

$$E(\chi_1, ..., \chi_m) \leq \liminf_{k\to\infty} E(\chi_1^k, ..., \chi_m^k)$$

and therefore the existence of minimizers among characteristic functions $\chi_1, ..., \chi_m$ of sets of finite perimeter in Ω and with $\chi_j \geq \chi_{j+1}$. □

8.1.3 2^n piecewise-constant segmentation with junctions

The work in [335], [104] shows how the previous Mumford–Shah level set approaches can be extended to piecewise-constant segmentation of images with triple junctions, more than two non-nested regions, or with other complex topologies, by using two or more level set functions that define a perfect partition of the domain Ω. We present here the multiphase approach that allows representation and segmentation of triple junctions.

Thus in this section we show how the two-phase piecewise-constant active contour model without edges from Section 8.1.1 can be generalized to piecewise-constant segmentation of images with more than two segments and junctions, again using (6.2).

We note again that by using only one level set function, we can represent only two phases or segments in the image. Also, other geometrical features, such as triple junctions, cannot be represented using only one level set function. The goal is to look for a multiphase level set model with which we can represent more than two segments or phases, triple junctions and other complex topologies, in an efficient way. We will need only $\log_2 n$ level set functions to represent n phases or segments with complex topologies, such as triple junctions. In addition, the formulation automatically removes the problems of vacuum and overlap, because the partition is a disjoint decomposition and covering of the domain Ω by definition. This is explained next.

Let us consider $m = \log_2 n$ level set functions $\phi_i : \Omega \to \mathbb{R}$, Ω being an open and bounded domain of $\mathbb{R}^2$. The union of the zero-level sets of ϕ_i will represent the edges in the segmented image. We also introduce the "vector level set function" $\Phi = (\phi_1, ..., \phi_m)$, and the "vector Heaviside function" $H(\Phi) = (H(\phi_1), ..., H(\phi_m))$ whose components are only 1 or 0. We can now define the segments or phases in the domain Ω, in the following way: two pixels (x_1, y_1) and (x_2, y_2) in Ω will belong to the same phase or class, if and only if $H(\Phi(x_1, y_1)) = H(\Phi(x_2, y_2))$. In other words, the classes or phases are given by the level sets of the function $H(\Phi)$, i.e. one class is formed by the set

$$\{(x, y) | H(\Phi(x, y)) = \text{constant vector} \in H(\Phi(\Omega))\}$$

(one phase or class contains those pixels (x, y) of Ω having the same value $H(\Phi(x, y))$).

There are up to $n = 2^m$ possibilities for the vector values in the image of $H(\Phi)$. In this way, we can define up to $n = 2^m$ phases or classes in the domain of definition Ω. The classes defined in this way form a disjoint decomposition and covering of Ω. Therefore, each pixel $(x, y) \in \Omega$ will belong to one and only one class by definition, and there is no vacuum or overlap among the phases. This is an important advantage, compared with the classical multiphase representation introduced in Zhao et al. [352] and used in Samson et al. [280, 279].

The set of curves C is represented by the union of the zero level sets of the functions ϕ_i.

We label the classes by I, with $1 \leq I \leq 2^m = n$. Now, let us introduce a constant vector of averages $c = (c_1, ..., c_n)$, where $c_I = mean(g)$ in the class I (note that g denotes the given image to be segmented) and the characteristic function χ_I for each class I. Then the reduced Mumford–Shah energy (6.2) can be written as:

$$F_n^{MS}(c, \Phi) = \sum_{1 \leq I \leq n = 2^m} \int_\Omega (g(x, y) - c_I)^2 \chi_I dxdy + \frac{\nu_0}{2} \sum_{1 \leq I \leq n = 2^m} \int_\Omega |D\chi_I|. \tag{8.12}$$

In order to simplify the model, we will replace the total length term $\frac{1}{2}\sum_{1 \leq I \leq n = 2^m} \int_\Omega |D\chi_I|$ by $\sum_{i=1}^m \int_\Omega |DH(\phi_i)|$. Thus, in some cases, some parts of the curves will be counted more than once in the total length term, or in other words, some edges will have a different weight in the total length term. We will see that with this slight modification and simplification, we still obtain very satisfactory results. Such simplification may have only a very small effect in most cases, because the fitting term is the dominant one.

Therefore, the energy that we will minimize is given by:

$$F_n(c, \Phi) = \sum_{1 \leq I \leq n = 2^m} \int_\Omega (g - c_I)^2 \chi_I dxdy + \sum_{1 \leq i \leq m} \nu_0 \int_\Omega |DH(\phi_i)|. \tag{8.13}$$

Clearly, for $n = 2$ (and therefore $m = 1$), we obtain the two-phase energy (8.1) considered in the active contour model without edges. For the purpose of illustration, let us write the above energy for $n = 4$ phases or classes (and therefore using $m = 2$ level set functions; see Figure 8.2 top):

$$\begin{aligned} F_4(c, \Phi) = & \int_\Omega (g - c_{11})^2 H(\phi_1) H(\phi_2) dxdy \\ & + \int_\Omega (g - c_{10})^2 H(\phi_1)(1 - H(\phi_2)) dxdy \\ & + \int_\Omega (g - c_{01})^2 (1 - H(\phi_1)) H(\phi_2) dxdy \\ & + \int_\Omega (g - c_{00})^2 (1 - H(\phi_1))(1 - H(\phi_2)) dxdy \\ & + \nu_0 \int_\Omega |DH(\phi_1)| + \nu_0 \int_\Omega |DH(\phi_2)|, \end{aligned} \tag{8.14}$$

where $c = (c_{11}, c_{10}, c_{01}, c_{00})$ is a constant vector, and $\Phi = (\phi_1, \phi_2)$.

With these notations, we can express the image-function u as:

$$\begin{aligned} u = & c_{11} H(\phi_1) H(\phi_2) + c_{10} H(\phi_1)(1 - H(\phi_2)) + c_{01}(1 - H(\phi_1)) H(\phi_2) \\ & + c_{00}(1 - H(\phi_1))(1 - H(\phi_2)). \end{aligned}$$

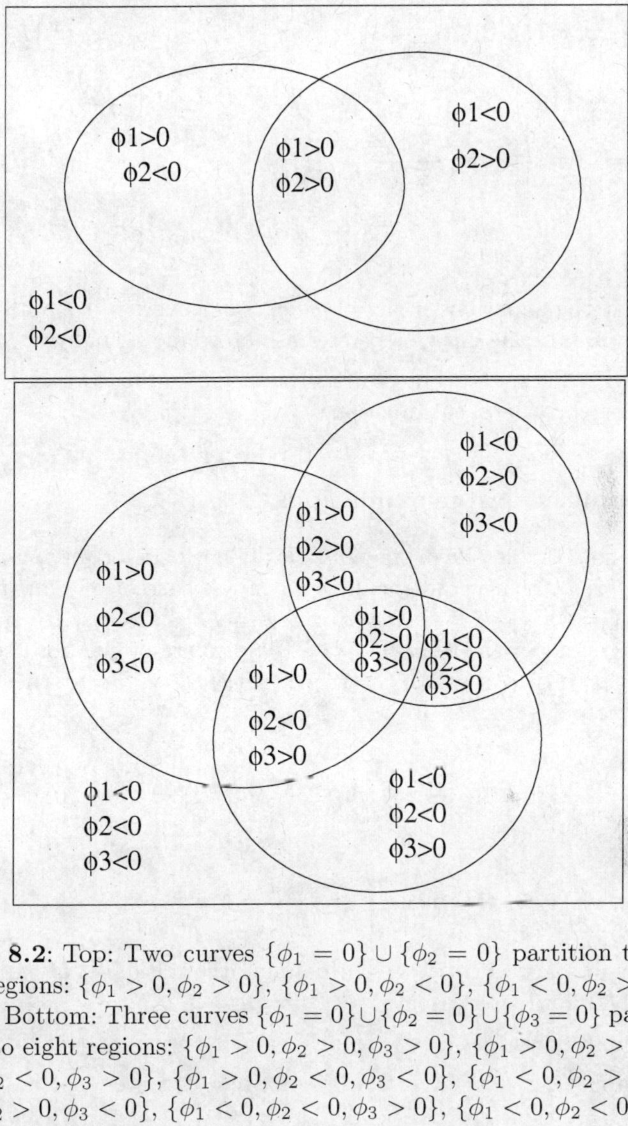

FIGURE 8.2: Top: Two curves $\{\phi_1 = 0\} \cup \{\phi_2 = 0\}$ partition the domain into four regions: $\{\phi_1 > 0, \phi_2 > 0\}$, $\{\phi_1 > 0, \phi_2 < 0\}$, $\{\phi_1 < 0, \phi_2 > 0\}$, $\{\phi_1 < 0, \phi_2 < 0\}$. Bottom: Three curves $\{\phi_1 = 0\} \cup \{\phi_2 = 0\} \cup \{\phi_3 = 0\}$ partition the domain into eight regions: $\{\phi_1 > 0, \phi_2 > 0, \phi_3 > 0\}$, $\{\phi_1 > 0, \phi_2 > 0, \phi_3 < 0\}$, $\{\phi_1 > 0, \phi_2 < 0, \phi_3 > 0\}$, $\{\phi_1 > 0, \phi_2 < 0, \phi_3 < 0\}$, $\{\phi_1 < 0, \phi_2 > 0, \phi_3 > 0\}$, $\{\phi_1 < 0, \phi_2 > 0, \phi_3 < 0\}$, $\{\phi_1 < 0, \phi_2 < 0, \phi_3 > 0\}$, $\{\phi_1 < 0, \phi_2 < 0, \phi_3 < 0\}$.

The Euler–Lagrange equations obtained by minimizing (8.14) with respect to c and Φ, embedded in a dynamical scheme, are: given $\phi_1(0, x, y) = \phi_{1,0}(x, y)$, $\phi_2(0, x, y) = \phi_{2,0}(x, y)$,

$$\begin{aligned}
c_{11}(\Phi) &= mean(g) \text{ in } \{(x, y) : \phi_1(t, x, y) > 0, \phi_2(t, x, y) > 0\}, \\
c_{10}(\Phi) &= mean(g) \text{ in } \{(x, y) : \phi_1(t, x, y) > 0, \phi_2(t, x, y) < 0\}, \\
c_{01}(\Phi) &= mean(g) \text{ in } \{(x, y) : \phi_1(t, x, y) < 0, \phi_2(t, x, y) > 0\}, \\
c_{00}(\Phi) &= mean(g) \text{ in } \{(x, y) : \phi_1(t, x, y) < 0, \phi_2(t, x, y) < 0\},
\end{aligned} \tag{8.15}$$

$$\frac{\partial \phi_1}{\partial t} = \delta_\varepsilon(\phi_1)\Big\{\nu_0 \mathrm{div}\Big(\frac{\nabla\phi_1}{|\nabla\phi_1|}\Big) - \Big[\Big((g-c_{11})^2 - (g-c_{01})^2\Big)H_\varepsilon(\phi_2)$$
$$+ \Big((g-c_{10})^2 - (g-c_{00})^2\Big)(1 - H_\varepsilon(\phi_2))\Big]\Big\}, \tag{8.16}$$
$$\frac{\partial \phi_2}{\partial t} = \delta_\varepsilon(\phi_2)\Big\{\nu_0 \mathrm{div}\Big(\frac{\nabla\phi_2}{|\nabla\phi_2|}\Big) - \Big[\Big((g-c_{11})^2 - (g-c_{10})^2\Big)H_\varepsilon(\phi_1)$$
$$+ \Big((g-c_{01})^2 - (g-c_{00})^2\Big)(1 - H_\varepsilon(\phi_1))\Big]\Big\}. \tag{8.17}$$

We note that the equations in $\Phi = (\phi_1, \phi_2)$ are governed by both mean curvature and jump of the data energy terms across the boundary.

We show in Figure 8.2 bottom, the partition of the domain Ω into eight regions, using three level set functions.

8.1.4 Numerical approximations

We give here the details of the numerical algorithm for solving (8.8) in two dimensions (x, y), using gradient descent, in the case of one function ϕ with two levels $l_1 = 0$, $l_2 = l > 0$. Let $h = \triangle x = \triangle y$ be the space steps, $\triangle t$ be the time step, and $\epsilon = h$. Let (x_i, y_j) be the discrete points, for $1 \le i, j \le M$, and $g_{i,j} = g(x_i, y_j)$, $\phi^n_{i,j} \approx \phi(n\triangle t, x_i, y_j)$, with $n \ge 0$. Recall the usual finite differences formulas

$$\triangle^x_+\phi_{i,j} = \phi_{i+1,j} - \phi_{i,j}, \quad \triangle^x_-\phi_{i,j} = \phi_{i,j} - \phi_{i-1,j},$$

$$\triangle^y_+\phi_{i,j} = \phi_{i,j+1} - \phi_{i,j}, \quad \triangle^y_-\phi_{i,j} = \phi_{i,j} - \phi_{i,j-1}.$$

Set $n = 0$, and start with $\phi^0_{i,j}$ given (defining the initial set of curves). Then, for each $n > 0$ until steady state:

(1) Compute averages $c_0(\phi^n)$, $c_1(\phi^n)$, and $c_2(\phi^n)$.

(2) Compute $\phi^{n+1}_{i,j}$, derived from the finite differences scheme

$$\frac{\phi^{n+1}_{i,j} - \phi^n_{i,j}}{\triangle t} = \delta_\epsilon(\phi^n_{i,j})\Big[\frac{\nu_0}{h^2}\Big(\triangle^x_-\Big(\frac{\phi^n_{i+1,j} - \phi^{n+1}_{i,j}}{|\nabla\phi^n_{i,j}|}\Big) + \triangle^y_-\Big(\frac{\phi^n_{i,j+1} - \phi^{n+1}_{i,j}}{|\nabla\phi^n_{i,j}|}\Big)\Big)$$
$$+ |g_{i,j} - c_0|^2 - |g_{i,j} - c_1|^2 H_\epsilon(l - \phi^n_{i,j})\Big]$$
$$+ \delta_\epsilon(\phi^n_{i,j} - l)\Big[\frac{\nu_0}{h^2}\Big(\triangle^x_-\Big(\frac{\phi^n_{i+1,j} - \phi^{n+1}_{i,j}}{|\nabla\phi^n_{i,j}|}\Big) + \triangle^y_-\Big(\frac{\phi^n_{i,j+1} - \phi^{n+1}_{i,j}}{|\nabla\phi^n_{i,j}|}\Big)\Big)$$
$$- |g_{i,j} - c_2|^2 + |g_{i,j} - c_1|^2 H_\epsilon(\phi^n_{i,j})\Big],$$

where $|\nabla\phi^n_{i,j}| = \sqrt{\left(\frac{\phi^n_{i+1,j}-\phi^n_{i,j}}{h}\right)^2 + \left(\frac{\phi^n_{i,j+1}-\phi^n_{i,j}}{h}\right)^2}$. Let

$$
\begin{aligned}
C_1 &= \frac{1}{\sqrt{\left(\frac{\phi^n_{i+1,j}-\phi^n_{i,j}}{h}\right)^2 + \left(\frac{\phi^n_{i,j+1}-\phi^n_{i,j}}{h}\right)^2}},\\
C_2 &= \frac{1}{\sqrt{\left(\frac{\phi^n_{i,j}-\phi^n_{i-1,j}}{h}\right)^2 + \left(\frac{\phi^n_{i-1,j+1}-\phi^n_{i-1,j}}{h}\right)^2}},\\
C_3 &= \frac{1}{\sqrt{\left(\frac{\phi^n_{i+1,j}-\phi^n_{i,j}}{h}\right)^2 + \left(\frac{\phi^n_{i,j+1}-\phi^n_{i,j}}{h}\right)^2}},\\
C_4 &= \frac{1}{\sqrt{\left(\frac{\phi^n_{i+1,j-1}-\phi^n_{i,j-1}}{h}\right)^2 + \left(\frac{\phi^n_{i,j}-\phi^n_{i,j-1}}{h}\right)^2}}.
\end{aligned}
$$

Let $m_1 = \frac{\triangle t}{h^2}(\delta_\epsilon(\phi^n_{i,j}) + \delta_\epsilon(\phi^n_{i,j} - l))\nu_0$, $C = 1 + m_1(C_1 + C_2 + C_3 + C_4)$. The main update equation for ϕ becomes

$$
\begin{aligned}
\phi^{n+1}_{i,j} =& \frac{1}{C}\Big[\phi^n_{i,j} + m_1(C_1\phi^n_{i+1,j} + C_2\phi^n_{i-1,j} + C_3\phi^n_{i,j+1} + C_4\phi^n_{i,j-1})\\
&+ \triangle t\delta_\epsilon(\phi^n_{i,j})(-(g_{i,j} - c_1)^2 H_\epsilon(l - \phi^n_{i,j})\\
&+ (g_{i,j} - c_0)^2) + \triangle t\delta_\epsilon(\phi^n_{i,j} - l)(-(g_{i,j} - c_2)^2 + (g_{i,j} - c_1)^2 H_\epsilon(\psi^n_{i,j}))\Big],
\end{aligned}
$$

and we repeat these steps until steady state is reached.

8.1.5 Numerical results

We now present several experimental results obtained using the piecewise-constant segmentation models with the level set method discussed above. These models act as denoising, segmentation and active contours. In Figures 8.3 and 8.4, we show experimental results taken from [94] obtained using the binary piecewise-constant model (8.1); we notice in particular how interior contours can be automatically detected. In Figure 8.5, we show an experimental result using the multilayer model (8.5), with $m = 2$ and two levels l_1, l_2, applied to the segmentation of a brain image [104].

We show numerical results using the four-phase model from Subsection 8.1.3. The only varying parameter is ν_0, the coefficient of the length term. We show in particular that triple junctions can be represented and detected using only two level set functions, that interior contours are automatically detected and also that the model is robust in the presence of noise and complex topologies. We begin with a noisy synthetic image with four regions in Figure 8.6. The image contains three objects of distinct intensities, all correctly detected and segmented. Because the energy which is minimized is not convex

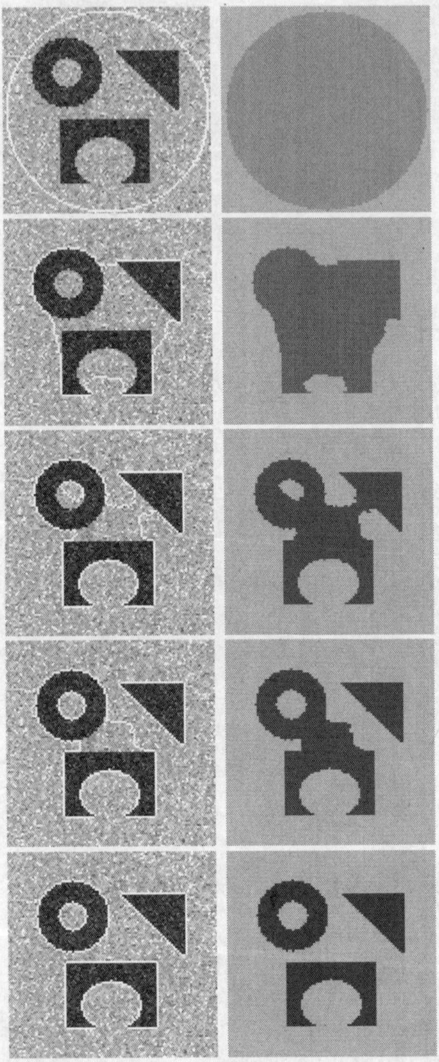

FIGURE 8.3: Detection of different objects in a noisy image, with various convexities and with an interior contour that is automatically detected using only one initial curve. After a short time, an interior contour appears inside the torus and then expands. Top: g and the evolving contours. Bottom: the piecewise-constant approximations u of g over time, given by $u = c_1 H(\phi) + c_2(1 - H(\phi))$.

and there is no uniqueness of minimizers, the algorithm may not converge to a global minimizer for a given initial condition.

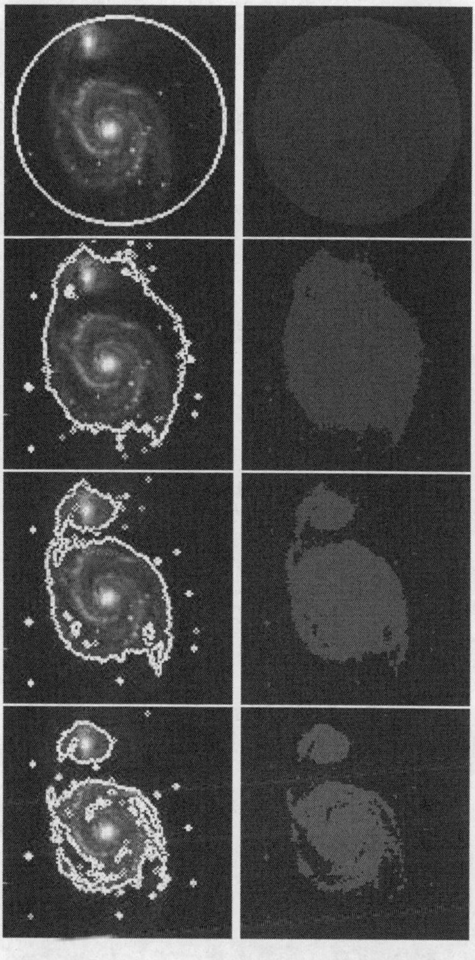

FIGURE 8.4: Numerical result using the binary active contour model without edges (compare with Figure 8.11).

In Figure 8.7 we show a noisy synthetic image with a triple junction. Using only one level set function, the triple junction cannot be represented. Most of the models need three level set functions, as in Zhao et al. [352] and Samson et al. [280, 279]. Here, only two level set functions are needed to represent the triple junction. We show their zero level sets which have to overlap on a segment of the triple junction.

We show next numerical results on a real picture (an MRI brain image), in Figures 8.8 and 8.9. We use here two level set functions, detecting four phases. We also show the final four segments detected by the algorithm. We see how

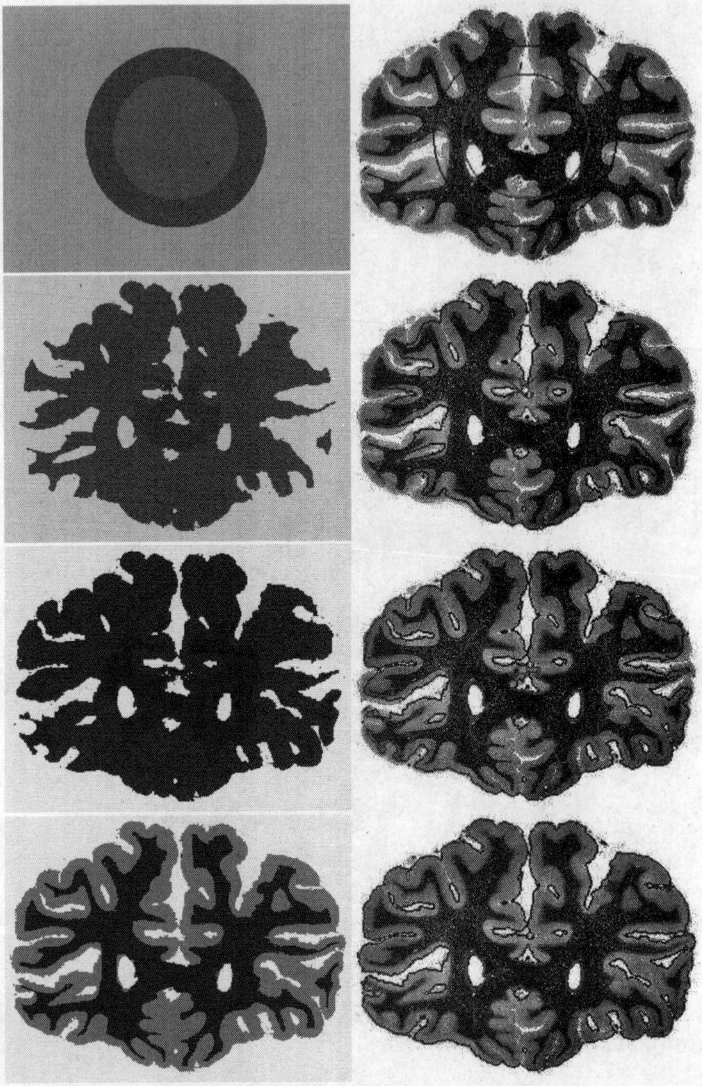

FIGURE 8.5: Segmentation of a brain image using one level set function with two levels.

the model can handle complex topologies, and also that the four phases in Figure 8.9 identify well the gray matter, white matter, and the cerebrospinal fluid.

FIGURE 8.6: Segmentation of a noisy synthetic image, using the four-phase piecewise-constant model. Left: the evolving contours overlay on the original image. Right: computed averages of the four segments $c_{11}, c_{10}, c_{01}, c_{00}$.

8.1.6 Level set method links active contours, Mumford–Shah segmentation, and total variation restoration

If we minimize the Mumford–Shah (6.1) and the Rudin–Osher–Fatemi total variation (3.2) functionals (denoted respectively by F^{MS} and G^{TV}) with data g, restricted to the subset $\{u = H(\phi) \,|\, \phi : \Omega \to \mathbb{R} \text{ is Lipschitz}\}$, i.e., of

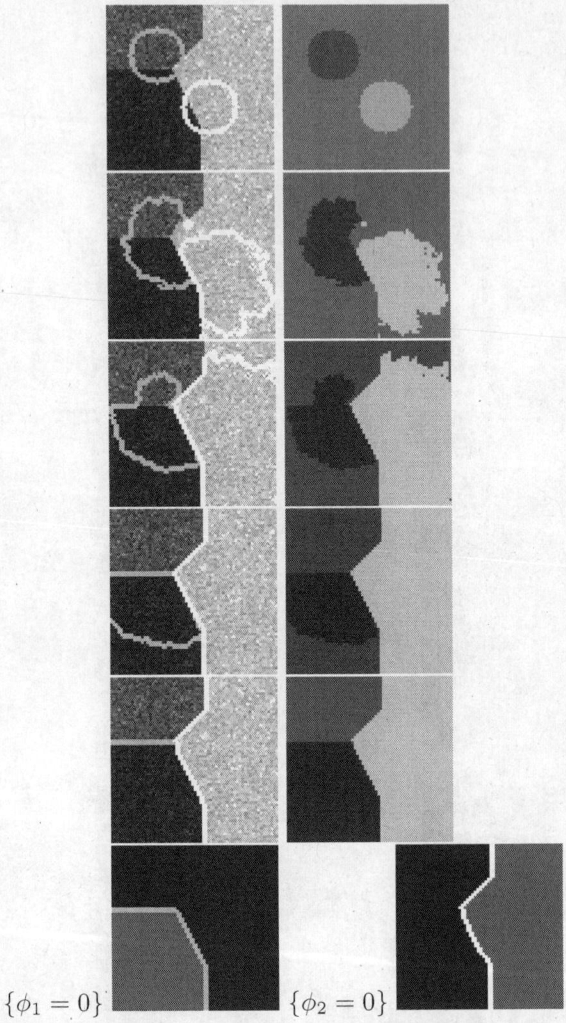

FIGURE 8.7: Results on a synthetic image with a triple junction using the four-phase piecewise-constant model with two level set functions. We also show the zero level sets of ϕ_1 and ϕ_2.

characteristic functions, then the two functionals are the same

$$\begin{aligned}
&F^{MS}(u = H(\phi), K = \{\phi = 0\}) = G^{TV}(u = H(\phi)) \\
&= \int_\Omega (g-1)^2 H(\phi)dx + \int_\Omega (g-0)^2 H(-\phi)dx + \nu_0 \int_\Omega |DH(\phi)|.
\end{aligned}$$

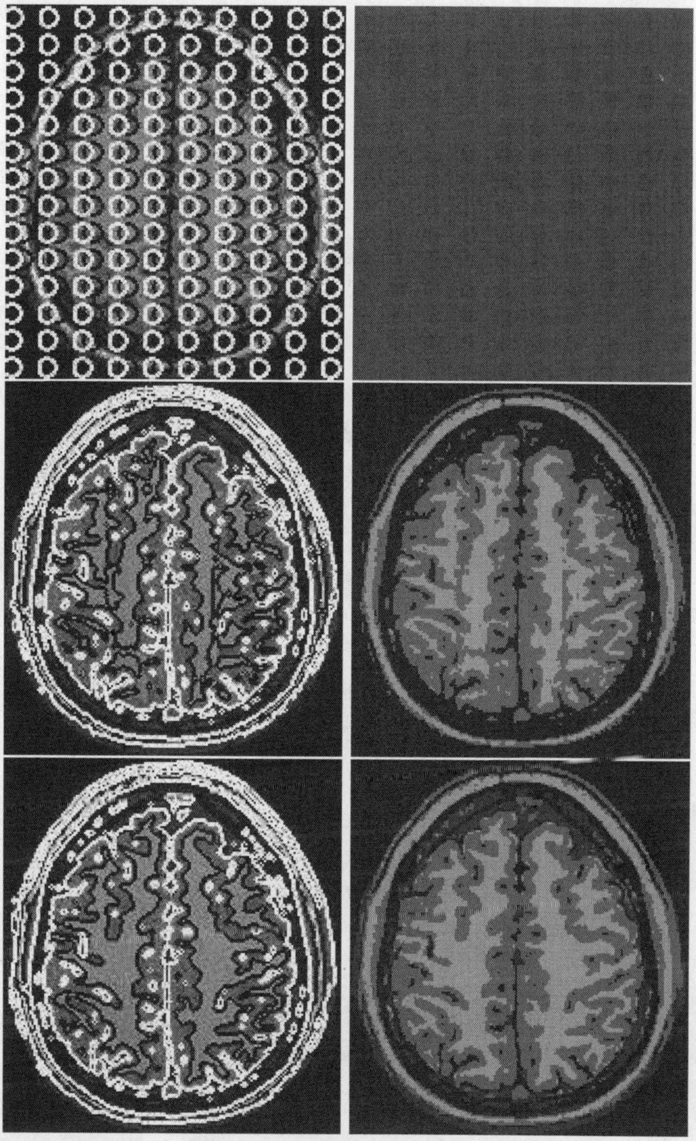

FIGURE 8.8: Segmentation of an MRI brain image, using two level set functions and four constant phases.

We want to present another binary segmentation model based on the convex total variation minimization, and implemented using the level set method. This is motivated by the fact that the Mumford and Shah minimization prob-

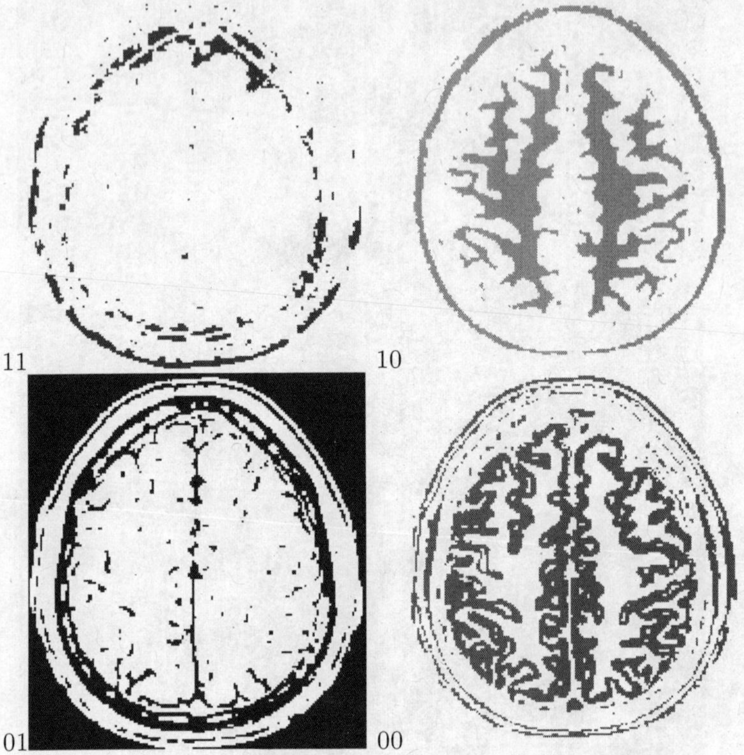

FIGURE 8.9: The algorithm depicts well the final four segments from the previous result (white matter, gray matter, etc.). The final averages are $c_{11} = 45$, $c_{10} = 159$, $c_{01} = 9$, $c_{00} = 103$.

lem is non-convex; due to the isotropic length term present in the Mumford and Shah energy which approximates the edges in the image, the contours are limited to some features, but do not depend on the data g: for instance, if an edge meets the boundary of the image, it has to be at a right angle. By using the total variation minimization in a level set framework, we try to remove these limitations and work with a less non-convex minimization problem. In addition, a new feature appears, by having the scaling coefficient depending on the jump magnitude of the image. Finally, this TV-based level set model has all the advantages of the model presented in Subsection 8.1.1, such as automatic detection of interior contours and detection of contours with or without gradient. A short presentation of this work by Osher and Vese appeared in [253].

If we restrict the Rudin–Osher–Fatemi energy from (3.2) to

$$\Big\{u(x) = c^+ H(\phi(x)) + c^-(1 - H(\phi(x))),\ c^+, c^- \in \mathbb{R},\ \phi : \Omega \to \mathbb{R} \text{ Lipschitz}\Big\},$$

we can write the above energy function of c^+, c^-, ϕ as

$$G(c^+, c^-, \phi) = \int_\Omega (g - c^+)^2 H(\phi)dx + \int_\Omega (g - c^-)^2(1 - H(\phi))dx + \nu_0|c^+ - c^-| \int_\Omega |DH(\phi)|. \tag{8.18}$$

We consider now $C^1(\mathbb{R})$ approximations and regularizations H_ε and δ_ε of the Heaviside function and the Dirac Delta function δ_0, as $\varepsilon \to 0$, such that $H'_\varepsilon = \delta_\varepsilon$.

Minimizing the above energy, and embedding the gradient descent into a dynamic scheme, we obtain

$$c^+(t) = \frac{\int_\Omega g(x)H_\varepsilon(\phi(t,x))dx}{\int_\Omega H_\varepsilon(\phi(t,x))dx} - \Big(\frac{c^+(t) - c^-(t)}{|c^+(t) - c^-(t)|}\Big)\Big(\frac{\nu_0 \int_\Omega \delta_\varepsilon(\phi(t,x))|\nabla\phi(t,x)|dx}{2\int_\Omega H_\varepsilon(\phi(t,x))dx}\Big), \tag{8.19}$$

$$c^-(t) = \frac{\int_\Omega g(x)\Big(1 - H_\varepsilon(\phi(t,x))\Big)dx}{\int_\Omega \Big(1 - H_\varepsilon(\phi(t,x))\Big)dx} - \Big(\frac{c^-(t) - c^+(t)}{|c^-(t) - c^+(t)|}\Big)\Big(\frac{\nu_0 \int_\Omega \delta_\varepsilon(\phi(t,x))|\nabla\phi(t,x)|dx}{2\int_\Omega(1 - H_\varepsilon(\phi(t,x)))dx}\Big), \tag{8.20}$$

$$\frac{\partial\phi}{\partial t} = \delta_\varepsilon(\phi)\Big[\nu_0|c^+ - c^-|\mathrm{div}\Big(\frac{\nabla\phi}{|\nabla\phi|}\Big) - (g - c^+)^2 + (g - c^-)^2\Big]. \tag{8.21}$$

Now the values c^+ and c^- are no longer the averages of the image g on the corresponding regions. In addition, there is an extra factor $|c^+ - c^-|$ in the regularization term, multiplied by ν_0. As we will show in the numerical results, this extra factor, the jump, has the following role: if we apply both models to the same image and with the same regularizing parameter ν_0, by the TV-based model there is a stronger constraint on the total length; therefore less noise will be kept compared with the active contour model without edges (8.1) based on Mumford–Shah.

We give now the main steps of the numerical algorithm. We denote by $t^n = n\triangle t$, $n \geq 0$ the time discretization. We will use the notations

$$A^+(t^n) = \frac{\int_\Omega g(x)H_\varepsilon(\phi(t^n,x))dx}{\int_\Omega H_\varepsilon(\phi(t^n,x))dx}, \; A^-(t^n) = \frac{\int_\Omega g(x)\Big(1 - H_\varepsilon(\phi(t^n,x))\Big)dx}{\int_\Omega \Big(1 - H_\varepsilon(\phi(t^n,x))\Big)dx}$$

for the averages of g on each corresponding region, function of the evolving curve, and

$$B^+(t^n) = \frac{\int_\Omega \delta_\varepsilon(\phi(t^n,x))|\nabla\phi(t^n,x)|dx}{\int_\Omega H_\varepsilon(\phi(t^n,x))dx},$$

$$B^-(t^n) = \frac{\int_\Omega \delta_\varepsilon(\phi(t^n,x))|\nabla\phi(t^n,x)|dx}{\int_\Omega(1-H_\varepsilon(\phi(t^n,x)))dx}.$$

Then for the TV-based model, we use the following time-discretizations

$$\begin{aligned}
c^+(t^{n+1}) &= A^+(t^n) - \left(\tfrac{c^+(t^{n+1})-c^-(t^{n+1})}{|c^+(t^n)-c^-(t^n)|}\right)\tfrac{\nu_0}{2}B^+(t^n),\\
c^-(t^{n+1}) &= A^-(t^n) - \left(\tfrac{c^-(t^{n+1})-c^+(t^{n+1})}{|c^+(t^n)-c^-(t^n)|}\right)\tfrac{\nu_0}{2}B^-(t^n),\\
\frac{\phi^{n+1}-\phi^n}{\triangle t} &= \delta_\varepsilon(\phi^n)\Big[\nu_0|c^+(t^{n+1})-c^-(t^{n+1})|\mathrm{div}\Big(\tfrac{\nabla\phi^{n+1}}{|\nabla\phi^n|}\Big)\\
&\quad -(g-c^+(t^{n+1}))^2+(g-c^-(t^{n+1}))^2\Big].
\end{aligned}$$

The linear system in $c^+(t^{n+1})$, $c^-(t^{n+1})$ can be re-written as:

$$\begin{aligned}
c^+(t^{n+1})\Big(1+\tfrac{\nu_0 B^+(t^n)}{2|c^+(t^n)-c^-(t^n)|}\Big) - c^-(t^{n+1})\Big(\tfrac{\nu_0 B^+(t^n)}{2|c^+(t^n)-c^-(t^n)|}\Big) &= A^+(t^n),\\
c^-(t^{n+1})\Big(1+\tfrac{\nu_0 B^-(t^n)}{2|c^+(t^n)-c^-(t^n)|}\Big) - c^+(t^{n+1})\Big(\tfrac{\nu_0 B^-(t^n)}{2|c^+(t^n)-c^-(t^n)|}\Big) &= A^-(t^n).
\end{aligned}$$

Some numerical applications are presented in Figures 8.10 through 8.13.

8.1.7 Two-phase piecewise-constant binary segmentation with multiplicative noise

The binary segmentation model presented in Subsection 8.1.1 is appropriate for images corrupted by additive Gaussian noise, due to the L^2 data fidelity terms. In that case, the unknown constants c_1 and c_2 became the averages of the data on each side of the contour. If the data g is corrupted by a different type of noise, it is better to adapt the data fidelity terms to the noise statistics. In this direction, an extensive amount of work has been done in [211]. We discuss below one simple case, while for another one we refer the reader to Exercise 8.7 inspired from [191].

In the case of multiplicative noise following a Gaussian distribution of mean 1, let $g = u\cdot\eta$ be a given noisy image containing multiplicative noise η. Assume $g \in L^\infty(\Omega)$, and the unknown u is piecewise-constant. We further assume that $|\Omega| = 1$, $\int_\Omega \eta = 1$, and the convention $\frac{0}{0} = 0$. Inspired by the image restoration model in the case of multiplicative noise [276] for the data fidelity term, we present here a piecewise-constant segmentation model for images corrupted by multiplicative noise, introduced in [204]. The energy minimization in the two-phase case is

$$\inf_{c_1,c_2,\phi} \nu_0\int_\Omega |DH(\phi)| + \int_\Omega(\frac{g}{c_1}-1)^2H(\phi)dx + \int_\Omega(\frac{g}{c_2}-1)^2H(-\phi)dx. \quad (8.22)$$

Keeping ϕ fixed and minimizing with respect to c_1 and c_2, we obtain explicit expressions

$$c_1(\phi) = \frac{\int_\Omega g^2H(\phi)dx}{\int_\Omega gH(\phi)dx},\ c_2(\phi) = \frac{\int_\Omega g^2H(-\phi)dx}{\int_\Omega gH(-\phi)dx}.$$

FIGURE 8.10: Numerical result using the TV-based active contour model.

Replacing H with H_ϵ in (8.22), we obtain the Euler–Lagrange equation for ϕ, parameterizing the gradient descent direction and with free boundary conditions,

$$\frac{\partial \phi}{\partial t} = \delta_\epsilon(\phi)\big[\nu_0 \mathrm{div}\big(\frac{\nabla \phi}{|\nabla \phi|}\big) - (\frac{g}{c_1} - 1)^2 + (\frac{g}{c_2} - 1)^2\big].$$

In Figure 8.14, we successfully use this multiplicative model to denoise

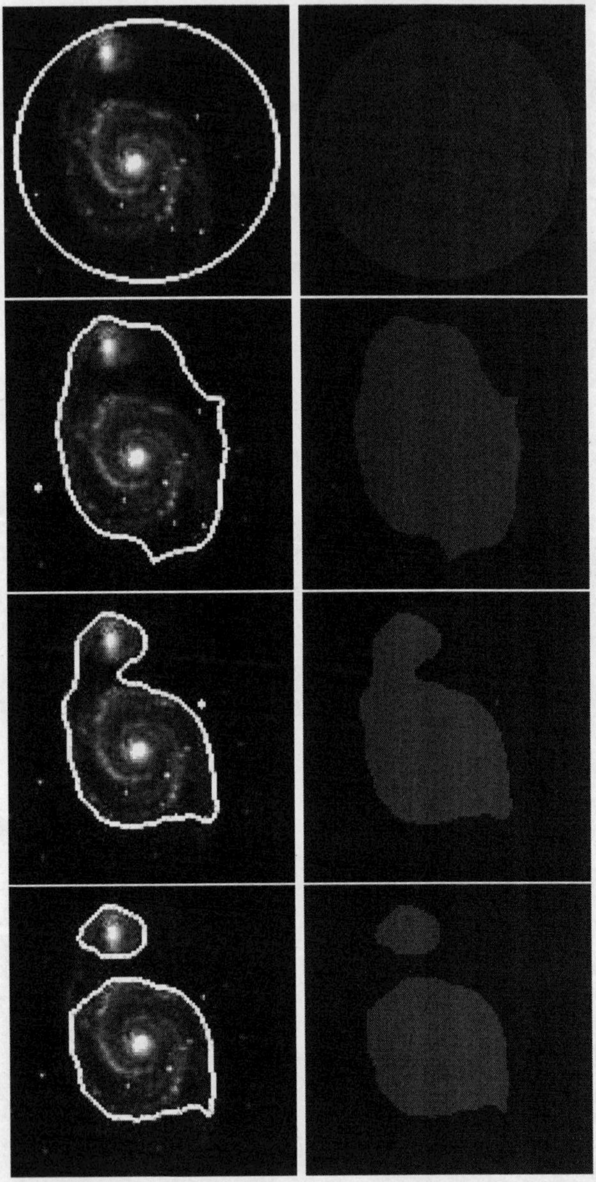

FIGURE 8.11: Another numerical result using the TV-based active contour model.

and segment two images corrupted by multiplicative noise of mean 1 (we have constructed the multiplicative noise using the formula $\eta = 1 + noise$, where

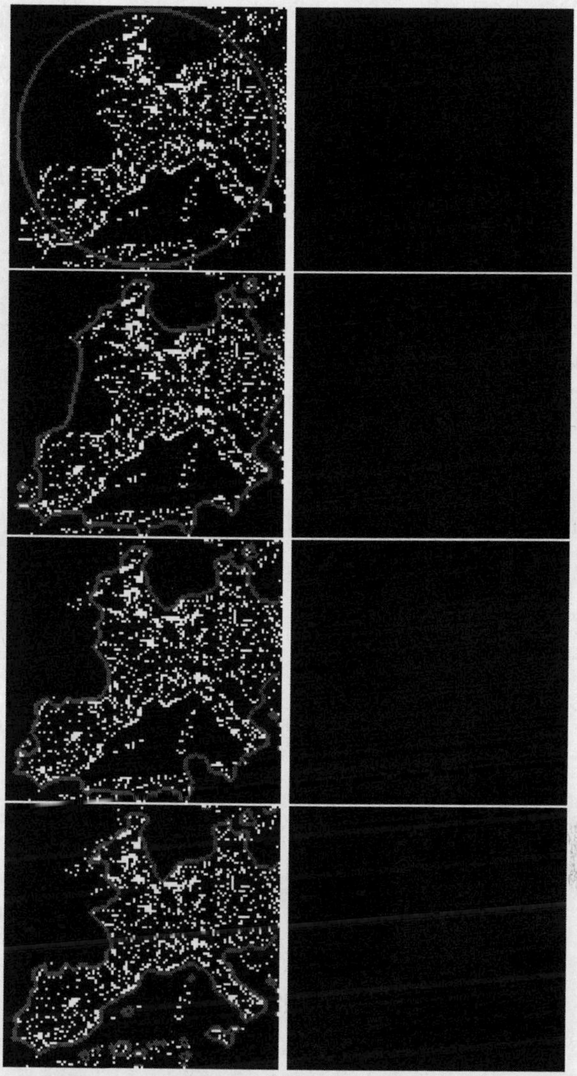

FIGURE 8.12: Numerical result using the TV-based active contour model on a real picture (Europe night lights). As the active contour model without edges [92, 94], it can also detect contours without gradient, also called cognitive contours following G. Kanizsa [185].

noise is white Gaussian noise of zero mean and variance 20). It is natural that the multiplicative noise does not appear that strong in the darker areas.

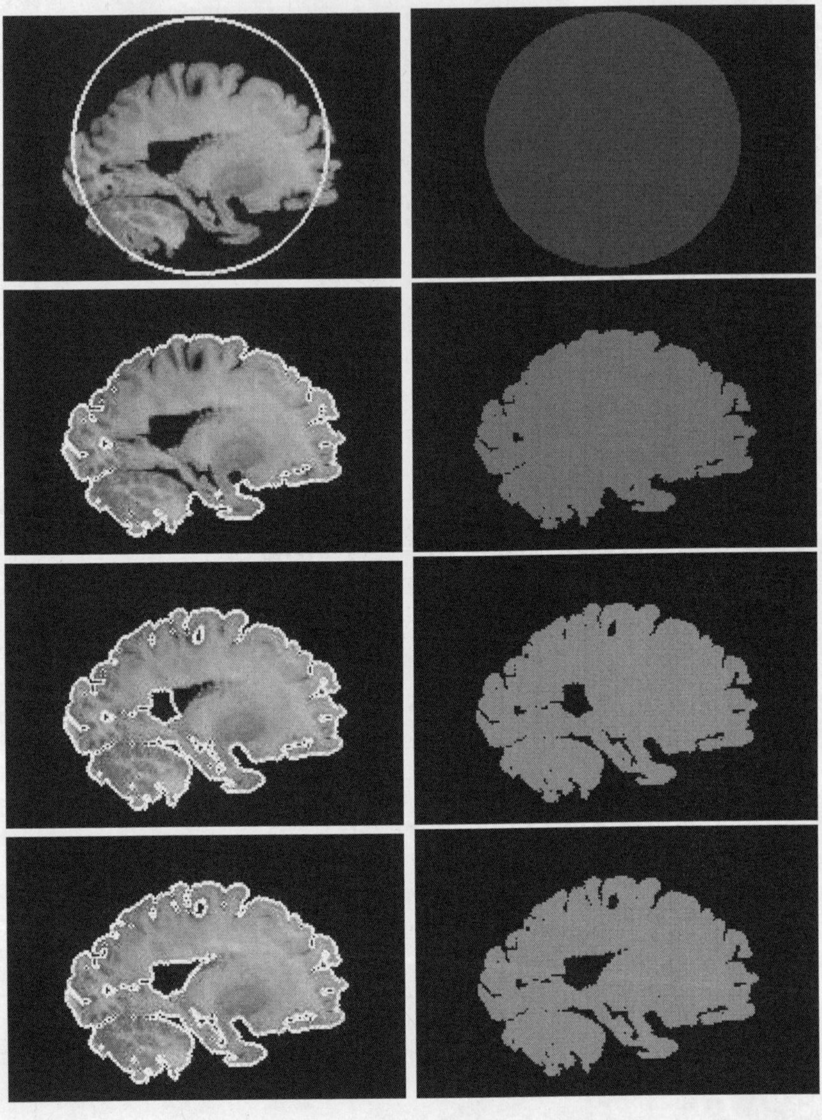

FIGURE 8.13: Numerical result using the TV-based active contour model in three dimensions. Here, only two-dimensional cross-sections of the three-dimensional calculation are shown.

8.2 Piecewise-smooth Mumford and Shah segmentation using level sets

We consider in this section level set formulations of the piecewise-smooth Mumford and Shah segmentation model (6.1).

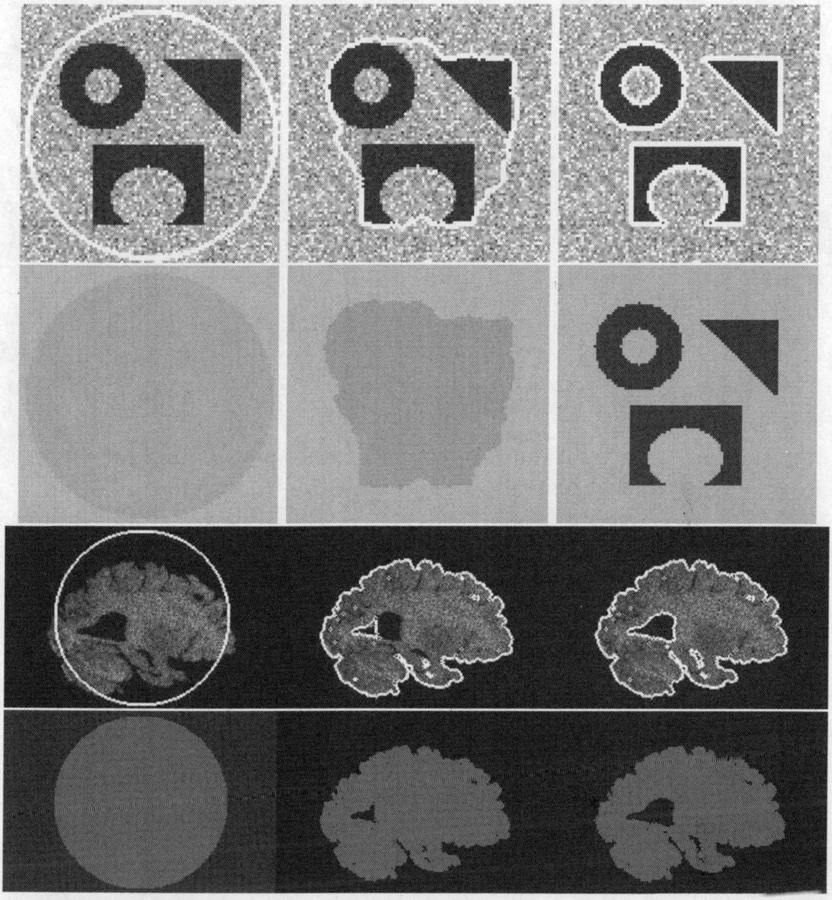

FIGURE 8.14: Two-phase piecewise constant segmentation with multiplicative noise model. Denoising and segmentation of two images corrupted by multiplicative noise of mean 1. Evolution of contours displayed over the initial noisy image and the segmented image $u = c_1 H(\phi) + c_2 H(-\phi)$ over time.

8.2.1 One-dimensional case

In many applications, we deal with a source signal u on $\Omega = (a, b)$. The problem of reconstructing u from a disturbed input g deriving from a distorted transmission, can be modeled as finding the minimum

$$\min_{u,C} \Big\{ \mu \int_{(a,b)\setminus C} |u'|^2 dx + \int_a^b |u - g|^2 dx + \nu \#(C) \Big\}, \tag{8.23}$$

where C denotes the set of discontinuity points of u, and $\#(C)$ denotes the cardinal of C (the counting measure).

We let $C = \Big\{x \in (a,b) |\ \phi(x) = 0\Big\}$, with ϕ being a one dimensional level set function, and we introduce two functions u^+ and u^-, such that $u(x) = u^+(x)H\Big(\phi(x)\Big) + u^-(x)\Big(1 - H\Big(\phi(x)\Big)\Big)$. These two Sobolev functions replace the two unknown constants used in (8.1), and are such that $u^+ \in H^1(\{\phi \geq 0\})$, and $u^- \in H^1(\{\phi \leq 0\})$. Then the energy (8.23) can be written in the level set formulation as

$$\min_{u^+,u^-,\phi} \Big\{\mu \int_a^b |(u^+)'|^2 H(\phi)dx + \mu \int_a^b |(u^-)'|^2 (1 - H(\phi))dx$$
$$+ \int_a^b |u^+ - g|^2 H(\phi)dx + \int_a^b |u^- - g|^2(1 - H(\phi))dx + \nu \int_a^b |H(\phi)'|\Big\}.$$

Minimizing this energy with respect to u^+, u^-, and ϕ, we obtain the associated Euler–Lagrange equations, embedded in a dynamical scheme

$$\begin{aligned}
u^+ &= g + \mu(u^+)'' \text{ in } \{x : \phi(t,x) > 0\},\ (u^+)' = 0 \text{ on } \{x : \phi(t,x) = 0\},\\
u^- &= g + \mu(u^-)'' \text{ in } \{x : \phi(t,x) < 0\},\ (u^-)' = 0 \text{ on } \{x : \phi(t,x) = 0\},\\
\frac{\partial \phi}{\partial t} &= \delta_\varepsilon(\phi)\Big[\nu\Big(\frac{\phi'}{|\phi'|}\Big)' - |u^+ - g|^2 + |u^- - g|^2 - \mu|(u^+)'|^2 + \mu|(u^-)'|^2\Big].
\end{aligned}$$

Note that in one dimension for signal segmentation, only one level set function ϕ suffices to represent the jump set of the piecewise smooth function u.

We show in Figure 8.15 an original signal and its noisy version, together with two points, where $\phi = 0$ at the initial time. In Figure 8.15, we show the segmented signal, and the detected set of jumps given by $\phi = 0$ at the steady state, using the presented level set algorithm in dimension 1. Note that piecewise smooth regions are very well reconstructed by the model, and that the jumps are well located and without smearing.

8.2.2 Two-dimensional case

We now consider the corresponding two-dimensional case under the assumption that the edges denoted by K in the image can be represented by one level set function ϕ, i.e. $K = \{x \in \Omega \,|\, \phi(x) = 0\}$, and we follow the approaches developed in parallel by Chan and Vese [95, 335] and by Tsai, Yezzi, Willsky [328] to minimize the general Mumford and Shah model. As in [335], the link between the unknowns u and ϕ can be expressed by introducing two functions u^+ and u^- (see Figure 8.16), such that

$$u(x) = \begin{cases} u^+(x) \text{ if } \phi(x) \geq 0,\\ u^-(x) \text{ if } \phi(x) \leq 0.\end{cases}$$

We assume that u^+ and u^- are Sobolev H^1 functions on $\phi \geq 0$ and on

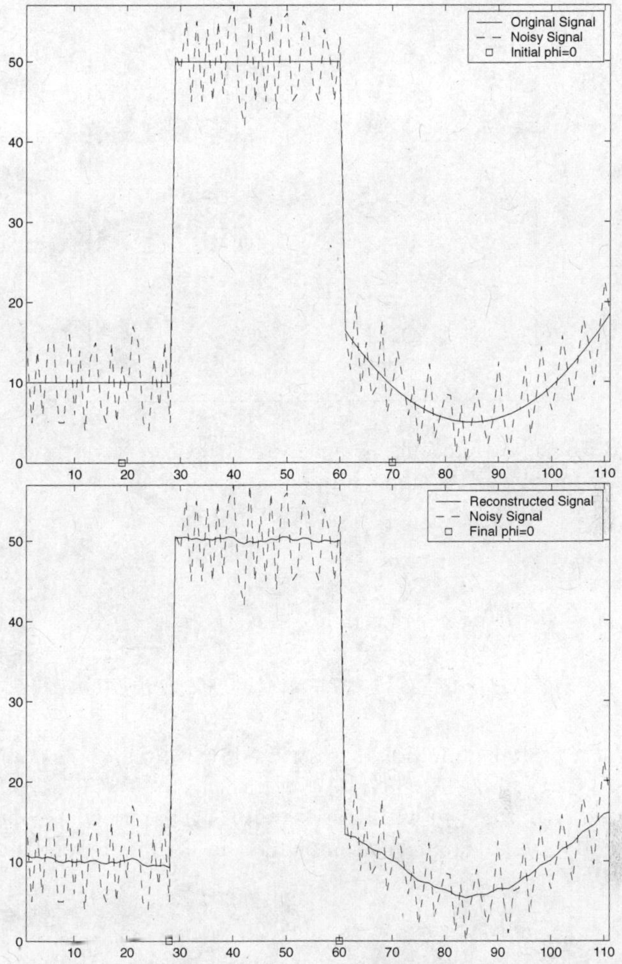

FIGURE 8.15: Top: original and noisy signal, together with the set of points where $\phi = 0$ at the initial time, represented by squares on the x-axis. Bottom: reconstructed signal, noisy signal, and the set of points where $\phi = 0$ at steady state (the jumps).

$\phi \leq 0$, respectively, (with Sobolev traces up to all boundary points, i.e., up to the boundary $\{\phi = 0\}$). We can write the following minimization problem

$$\inf_{u^+, u^-, \phi} E(u^+, u^-, \phi),$$

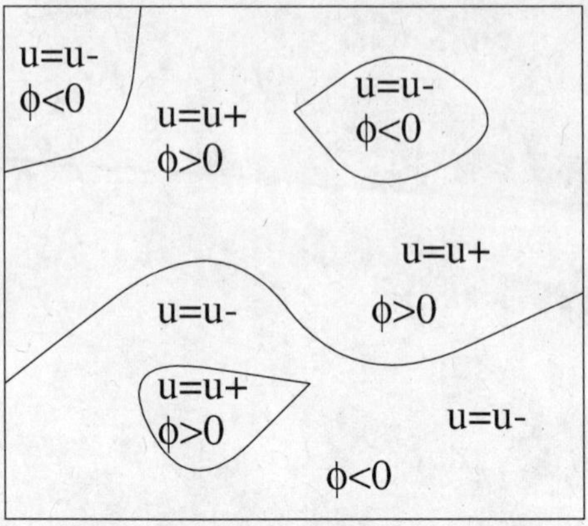

FIGURE 8.16: The functions u^+, u^-, and the zero-level lines of the level set function ϕ for piecewise smooth image partition.

where

$$E(u^+, u^-, \phi) = \mu^2 \int_\Omega |u^+ - g|^2 H(\phi) dx + \mu^2 \int_\Omega |u^- - g|^2 (1 - H(\phi)) dx$$
$$+ \int_\Omega |\nabla u^+|^2 H(\phi) dx + \int_\Omega |\nabla u^-|^2 (1 - H(\phi)) dx + \nu \int_\Omega |DH(\phi)|$$

is the Mumford–Shah functional restricted to $u(x) = u^+(x)H(\phi(x)) + u^-(x)(1 - H(\phi(x)))$.

Minimizing $E(u^+, u^-, \phi)$ with respect to u^+, u^-, and ϕ, we obtain the following Euler–Lagrange equations (embedded in a time-dependent dynamical scheme for ϕ)

$$\mu^2(u^+ - g) = \triangle u^+ \text{ in } \{x : \phi(t,x) > 0\},$$
$$\frac{\partial u^+}{\partial \vec{n}} = 0 \text{ on } \{x : \phi(t,x) = 0\} \cup \partial\Omega, \tag{8.24}$$
$$\mu^2(u^- - g) = \triangle u^- \text{ in } \{x : \phi(t,x) < 0\},$$
$$\frac{\partial u^-}{\partial \vec{n}} = 0 \text{ on } \{x : \phi(t,x) = 0\} \cup \partial\Omega, \tag{8.25}$$

$$\frac{\partial \phi}{\partial t} = \delta_\varepsilon(\phi)\Big[\nu \text{div}\Big(\frac{\nabla \phi}{|\nabla \phi|}\Big) - \mu^2 |u^+ - g|^2$$
$$- |\nabla u^+|^2 + \mu^2 |u^- - g|^2 + |\nabla u^-|^2\Big], \tag{8.26}$$

where $\partial/\partial\vec{n}$ denotes the partial derivative in the normal direction $\vec{n}$ at the corresponding boundary. We also associate the boundary condition $\frac{\partial \phi}{\partial \vec{n}} = 0$ on $\partial\Omega$ to equation (8.26).

We show in Figures 8.17 and 8.18 experimental results taken from [335] obtained with the piecewise smooth two-phase model.

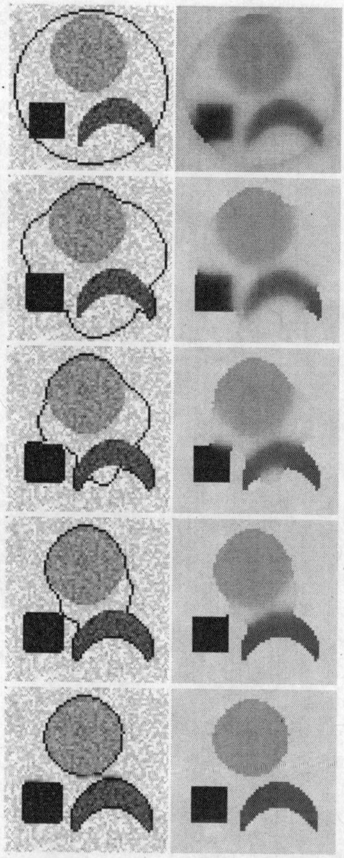

FIGURE 8.17: Results on a noisy image using the level set algorithm for the piecewise smooth Mumford–Shah model with one level set function. The algorithm performs as active contours, denoising and edge detection.

There are cases when the boundaries K of regions forming a partition of the image cannot be represented by the boundary of an open domain. To overcome this, several solutions have been proposed in this framework and we mention two of them: (i) in the work by Tsai, Yezzi, Willsky [328], the minimization of $E(u^+, u^-, \phi)$ is repeated inside each of the two regions previously computed; (ii) in the work of Chan and Vese [335], two or more level set functions are used. For example, in two dimensions, the problem can be solved using only two level set functions, and we do not have to know a priori how many gray levels the image has (or how many segments). The idea is based on the four-color theorem. Based on this observation, we can "color" all the regions in a partition using only four colors, such that any two adjacent regions have

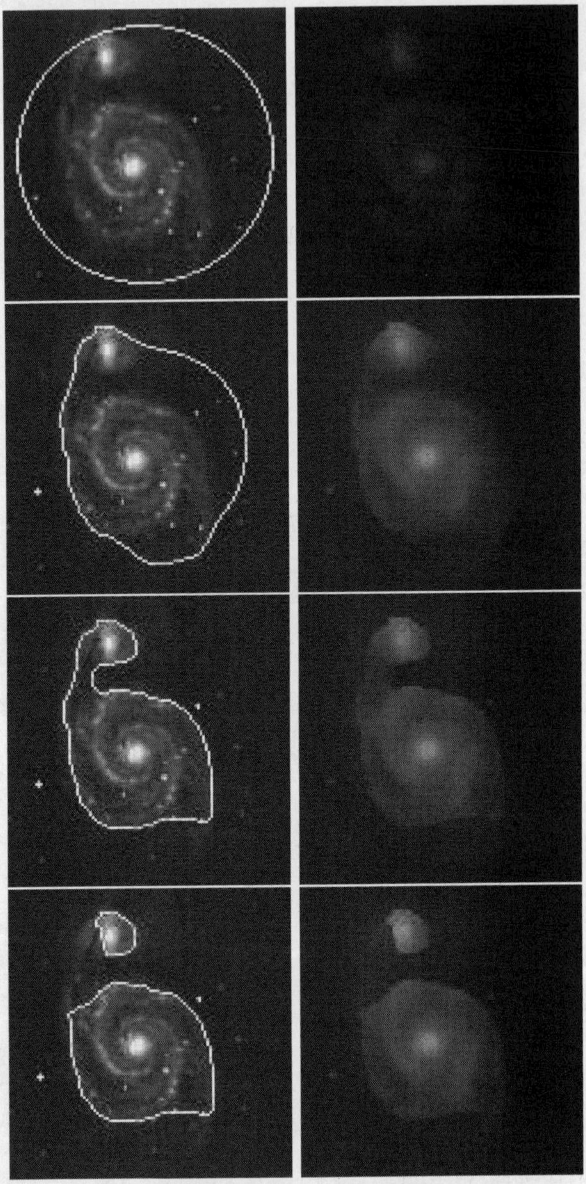

FIGURE 8.18: Numerical result using the piecewise smooth Mumford–Shah level set algorithm with one level set function on a piecewise smooth real galaxy image.

different colors. Therefore, using two level set functions, we can identify the four colors by the following (disjoint) sets: $\{\phi_1 > 0, \phi_2 > 0\}$, $\{\phi_1 < 0, \phi_2 < 0\}$, $\{\phi_1 < 0, \phi_2 > 0\}$, $\{\phi_1 > 0, \phi_2 < 0\}$. The boundaries of the regions forming the partition will be given by $\{\phi_1 = 0\} \cup \{\phi_2 = 0\}$, and this will be the set of curves K. Note that, in this particular multiphase formulation of the problem, we do not have the problems of overlapping or vacuum (i.e., the phases are disjoint, and their union is the entire domain Ω, by definition).

As before, the link between the function u and the four regions can be made by introducing four functions $u^{++}, u^{+-}, u^{-+}, u^{--}$, which are in fact the restrictions of u to each of the four phases, as follows (see Figure 8.19):

$$u(x) = \begin{cases} u^{++}(x), & \text{if } \phi_1(x) > 0 \text{ and } \phi_2(x) > 0, \\ u^{+-}(x), & \text{if } \phi_1(x) > 0 \text{ and } \phi_2(x) < 0, \\ u^{-+}(x), & \text{if } \phi_1(x) < 0 \text{ and } \phi_2(x) > 0, \\ u^{--}(x), & \text{if } \phi_1(x) < 0 \text{ and } \phi_2(x) < 0. \end{cases}$$

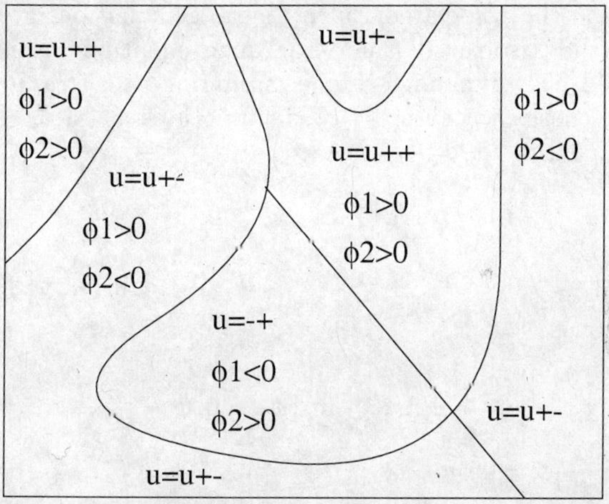

FIGURE 8.19: The functions u^{++}, u^{+-}, u^{-+}, u^{--} and the zero-level lines of the level set functions ϕ_1, ϕ_2 for piecewise smooth image partition.

Again, using the Heaviside function, the relation between u, the four functions u^{++}, u^{+-}, u^{-+}, u^{--} and the level set functions ϕ_1 and ϕ_2 can be expressed by

$$\begin{aligned} u = & u^{++}H(\phi_1)H(\phi_2) + u^{+-}H(\phi_1)(1 - H(\phi_2)) \\ & + u^{-+}(1 - H(\phi_1))H(\phi_2) + u^{--}(1 - H(\phi_1))(1 - H(\phi_2)). \end{aligned}$$

We then introduce an energy in a level set formulation based on the

Mumford–Shah functional:

$$
\begin{aligned}
&E(u, \phi_1, \phi_2) = \\
&\int_\Omega |u^{++} - g|^2 H(\phi_1)H(\phi_2)dx + \mu \int_\Omega |\nabla u^{++}|^2 H(\phi_1)H(\phi_2)dx \\
&+ \int_\Omega |u^{+-} - g|^2 H(\phi_1)(1 - H(\phi_2))dx + \mu \int_\Omega |\nabla u^{+-}|^2 H(\phi_1)(1 - H(\phi_2))dx \\
&+ \int_\Omega |u^{-+} - g|^2 (1 - H(\phi_1))H(\phi_2)dx + \mu \int_\Omega |\nabla u^{-+}|^2 (1 - H(\phi_1))H(\phi_2)dx \\
&+ \int_\Omega |u^{--} - g|^2 (1 - H(\phi_1))(1 - H(\phi_2))dx \\
&+ \mu \int_\Omega |\nabla u^{--}|^2 (1 - H(\phi_1))(1 - H(\phi_2))dx + \nu \int_\Omega |DH(\phi_1)| + \nu \int_\Omega |DH(\phi_2)|.
\end{aligned}
$$

Note that the expression $\int_\Omega |DH(\phi_1)| + \int_\Omega |DH(\phi_2)|$ is not exactly the total length of $K = \{x \in \Omega : \ \phi_1(x) = 0 \text{ or } \phi_2(x) = 0\}$; it is just an approximation and simplification. In practice, satisfactory results using the above formula are obtained, and the associated Euler–Lagrange equations are simplified.

We obtain the associated Euler–Lagrange equations as in the previous cases, embedded in a dynamic scheme, assuming $\phi_i : (t, x, y) \mapsto \phi_i(t, x, y)$. Minimizing the energy with respect to the functions u^{++}, u^{+-}, u^{-+}, u^{--}, we have, for each fixed t

$$
\begin{aligned}
&(u^{++} - g) = \mu \triangle u^{++} \text{ in } \{\phi_1 > 0, \phi_2 > 0\}, \\
&\frac{\partial u^{++}}{\partial \vec{n}} = 0 \text{ on } \{\phi_1 = 0, \phi_2 \geq 0\}, \{\phi_1 \geq 0, \phi_2 = 0\};
\end{aligned}
$$

$$
\begin{aligned}
&(u^{+-} - g) = \mu \triangle u^{+-} \text{ in } \{\phi_1 > 0, \phi_2 < 0\}, \\
&\frac{\partial u^{+-}}{\partial \vec{n}} = 0 \text{ on } \{\phi_1 = 0, \phi_2 \leq 0\}, \{\phi_1 \geq 0, \phi_2 = 0\};
\end{aligned}
$$

$$
\begin{aligned}
&(u^{-+} - g) = \mu \triangle u^{-+} \text{ in } \{\phi_1 < 0, \phi_2 > 0\}, \\
&\frac{\partial u^{-+}}{\partial \vec{n}} = 0 \text{ on } \{\phi_1 = 0, \phi_2 \geq 0\}, \{\phi_1 \leq 0, \phi_2 = 0\};
\end{aligned}
$$

$$
\begin{aligned}
&(u^{--} - g) = \mu \triangle u^{--} \text{ in } \{\phi_1 < 0, \phi_2 < 0\}, \\
&\frac{\partial u^{--}}{\partial \vec{n}} = 0 \text{ on } \{\phi_1 = 0, \phi_2 \leq 0\}, \{\phi_1 \leq 0, \phi_2 = 0\}.
\end{aligned}
$$

The Euler–Lagrange equations evolving ϕ_1 and ϕ_2, embedded in a dynamic scheme, are formally

$$
\begin{aligned}
\frac{\partial \phi_1}{\partial t} =&\delta_\varepsilon(\phi_1)\Big[\nu \mathrm{div}\Big(\frac{\nabla \phi_1}{|\nabla \phi_1|}\Big) - |u^{++} - g|^2 H_\varepsilon(\phi_2) - \mu|\nabla u^{++}|^2 H_\varepsilon(\phi_2) \\
&- |u^{+-} - g|^2(1 - H_\varepsilon(\phi_2)) - \mu|\nabla u^{+-}|^2(1 - H_\varepsilon(\phi_2)) \\
&+ |u^{-+} - g|^2 H_\varepsilon(\phi_2) + \mu|\nabla u^{-+}|^2 H_\varepsilon(\phi_2) + |u^{--} - g|^2(1 - H_\varepsilon(\phi_2)) \\
&+ \mu|\nabla u^{--}|^2(1 - H_\varepsilon(\phi_2))\Big], \\
\frac{\partial \phi_2}{\partial t} =&\delta_\varepsilon(\phi_2)\Big[\nu \mathrm{div}\Big(\frac{\nabla \phi_2}{|\nabla \phi_2|}\Big) - |u^{++} - g|^2 H_\varepsilon(\phi_1) - \mu|\nabla u^{++}|^2 H_\varepsilon(\phi_1) \\
&+ |u^{+-} - g|^2 H_\varepsilon(\phi_1) + \mu|\nabla u^{+-}|^2 H_\varepsilon(\phi_1) \\
&- |u^{-+} - g|^2(1 - H_\varepsilon(\phi_1)) - \mu|\nabla u^{-+}|^2(1 - H_\varepsilon(\phi_1)) \\
&+ |u^{--} - g|^2(1 - H_\varepsilon(\phi_1)) + \mu|\nabla u^{--}|^2(1 - H_\varepsilon(\phi_1))\Big].
\end{aligned}
$$

We can show, by standard techniques of the calculus of variations on the space $SBV(\Omega)$ (special functions of bounded variations, Chapter 2), and a compactness result due to Ambrosio [18], that the proposed minimization problems from this subsection, in the level set formulation, have a minimizer. Finally, because there is no uniqueness of minimizers and because the problems are non-convex, the numerical results may depend on the initial choice of the curves, and we may compute a local minimum only. We think that by using the seed initialization (see [335]) the algorithms have the tendency to compute a global minimum most of the time. Additional experimental results are shown in [335].

8.3 Applications to variational image restoration with segmentation-based regularization and level sets

This section focuses on the challenging task of edge-preserving variational image restoration. In this context, restoration means image deblurring and denoising, where we deal with Gaussian and impulsive noise models. Terms from the Mumford–Shah segmentation functional are used as regularizers, reflecting the model of piecewise-constant or piecewise-smooth images.

We present here other joint formulations for denoising, deblurring and piecewise-constant segmentation introduced in [181] that can be seen as applications and modifications of the piecewise-constant Mumford–Shah model in level set formulation presented in Section 8.1. For related work we refer the reader to [189, 39, 42, 41, 215]. We use a minimization approach and we consider the gradient descent method. Let $g = h * u + n$ be a given blurred noisy image, where h is a known blurring kernel (such as a Gaussian kernel) and n represents Gaussian additive noise of zero mean. We assume that the contours or jumps in the image u can be represented by the m distinct levels $\{-\infty = l_0 < l_1 < l_2 < \cdots < l_m < l_{m+1} = \infty\}$ of the same implicit (Lipschitz continuous) function $\phi : \Omega \to \mathbb{R}$ partitioning Ω into $m+1$ disjoint open regions $R_j = \{x \in \Omega : l_{j-1} < \phi(x) < l_j\}$, $1 \leq j \leq m+1$. Thus we can recover the denoised and deblurred image $u = c_1 H(\phi - l_m) + \sum_{j=2}^{m} c_j H(\phi - l_{m-j+1}) H(l_{m-j+2} - \phi) + c_{m+1} H(l_1 - \phi)$ by minimizing the following energy functional ($\mu \geq 0$):

$$E(c_1, c_2, \ldots, c_m, c_{m+1}, \phi) = \int_\Omega \bigg| g - h * \Big(c_1 H(\phi - l_m)$$
$$+ \sum_{j=2}^{m} c_j H(\phi - l_{m-j+1}) H(l_{m-j+2} - \phi) + c_{m+1} H(l_1 - \phi) \Big) \bigg|^2 dx$$
$$+ \mu \sum_{j=1}^{m} \int_\Omega |DH(\phi - l_j)|.$$

In the binary case (one level $m = 1$, $l_1 = 0$) we assume the degradation model $g = h * \Big(c_1 H(\phi) + c_2(1 - H(\phi)) \Big) + n$, and we wish to recover $u = c_1 H(\phi) + c_2(1 - H(\phi))$ in Ω together with a segmentation of g. The modified binary segmentation model incorporating the blur becomes

$$E(c_1, c_2, \phi) =$$
$$\inf_{c_1, c_2, \phi} \int_\Omega \Big| g - h * \Big(c_1 H(\phi) + c_2(1 - H(\phi)) \Big) \Big|^2 dx + \mu \int_\Omega |DH(\phi)|.$$

We compute the Euler–Lagrange equations minimizing this energy with respect to c_1, c_2, and ϕ. Using alternating minimization, keeping first ϕ fixed and minimizing the energy with respect to the unknown constants c_1 and c_2, we obtain the following linear system of equations:

$$c_1 \int_\Omega h_1^2 dx + c_2 \int_\Omega h_1 h_2 dx = \int g h_1 dx,$$
$$c_1 \int_\Omega h_1 h_2 dx + c_2 \int_\Omega h_2^2 dx = \int g h_2 dx,$$

with the notations $h_1 = h * H(\phi)$ and $h_2 = h * (1 - H(\phi))$. Note that the linear system has a unique solution because the determinant of the coefficient matrix is not zero due to the Cauchy–Schwarz inequality $\Big(\int_\Omega h_1 h_2 dx \Big)^2 \leq \int_\Omega h_1^2 dx \int_\Omega h_2^2 dx$, where the equality holds if and only if $h_1 = h_2$ for a.e. $x \in \Omega$. But clearly $h_1 = h * H(\phi)$ and $h_2 = h * (1 - H(\phi))$ are distinct; thus we have strict inequality.

Keeping the constants c_1 and c_2 fixed and minimizing the energy with respect to ϕ, we obtain the evolution equation by introducing an artificial time for the gradient descent in $\phi(t,x)$, $t>0$, $x\in\Omega$

$$\frac{\partial\phi}{\partial t} = \delta(\phi)\Big[2\Big(\tilde{h}*g - c_1\tilde{h}*(h*H(\phi)) - c_2\tilde{h}*(h*(1-H(\phi)))\Big)(c_1-c_2) + \mu\mathrm{div}\Big(\frac{\nabla\phi}{|\nabla\phi|}\Big)\Big]$$

where $\tilde{h}(x) = h(-x)$.

We show in Figure 8.20 a numerical result for joint denoising, deblurring and segmentation of a synthetic image, in a binary level set approach.

In the case of two distinct levels $l_1 < l_2$ of the level set function ϕ ($m=2$), we wish to recover a piecewise-constant image of the form $u = c_1H(\phi-l_2) + c_2H(l_2-\phi)H(\phi-l_1) + c_3H(l_1-\phi)$ and a segmentation of g, assuming the degradation model $g = h*\Big(c_1H(\phi-l_2)+c_2H(l_2-\phi)H(\phi-l_1)+c_3H(l_1-\phi)\Big)+n$, by minimizing

$$\inf_{c_1,c_2,c_3,\phi} E(c_1,c_2,c_3,\phi) = \int_\Omega \Big|g - h*\Big(c_1H(\phi-l_2) + c_2H(l_2-\phi)H(\phi-l_1) + c_3H(l_1-\phi)\Big)\Big|^2 dx + \mu\sum_{j=1}^{2}\int_\Omega |DH(\phi-l_j)|. \tag{8.27}$$

Similar to the previous binary model with blur, for fixed ϕ, the unknown constants are computed by solving the linear system of three equations

$$\begin{aligned} c_1\int h_1^2dx + c_2\int h_1h_2dx + c_3\int h_1h_3dx &= \int gh_1dx,\\ c_1\int h_1h_2dx + c_2\int h_2^2dx + c_3\int h_2h_3dx &= \int gh_2dx,\\ c_1\int h_1h_3dx + c_2\int h_2h_3dx + c_3\int h_3^2dx &= \int gh_3dx, \end{aligned}$$

where $h_1 = h*H(\phi-l_2)$, $h_2 = h*\Big(H(l_2-\phi)H(\phi-l_1)\Big)$, and $h_3 = h*H(l_1-\phi)$.

For fixed c_1, c_2 and c_3, by minimizing the functional E with respect to ϕ, we obtain the gradient descent for $\phi(t,x)$, $t>0$, $x\in\Omega$:

$$\begin{aligned} \frac{\partial\phi}{\partial t} =& 2\Big(\tilde{h}*(g - h*(c_1H(\phi-l_2) + c_2H(l_2-\phi)H(\phi-l_1)\\ & c_3H(l_1-\phi)))\Big)\,(c_1\delta(\phi-l_2) + c_2H(l_2-\phi)\delta(\phi-l_1)\\ & -c_2H(\phi-l_1)\delta(l_2-\phi) - c_3\delta(l_1-\phi))\\ & + \mu\mathrm{div}\Big(\frac{\nabla\phi}{|\nabla\phi|}\Big)(\delta(\phi-l_1)+\delta(\phi-l_2)). \end{aligned} \tag{8.28}$$

We show in Figures 8.21 and 8.22 a numerical result for joint denoising, deblurring and segmentation of the brain image, in a multilayer level set approach.

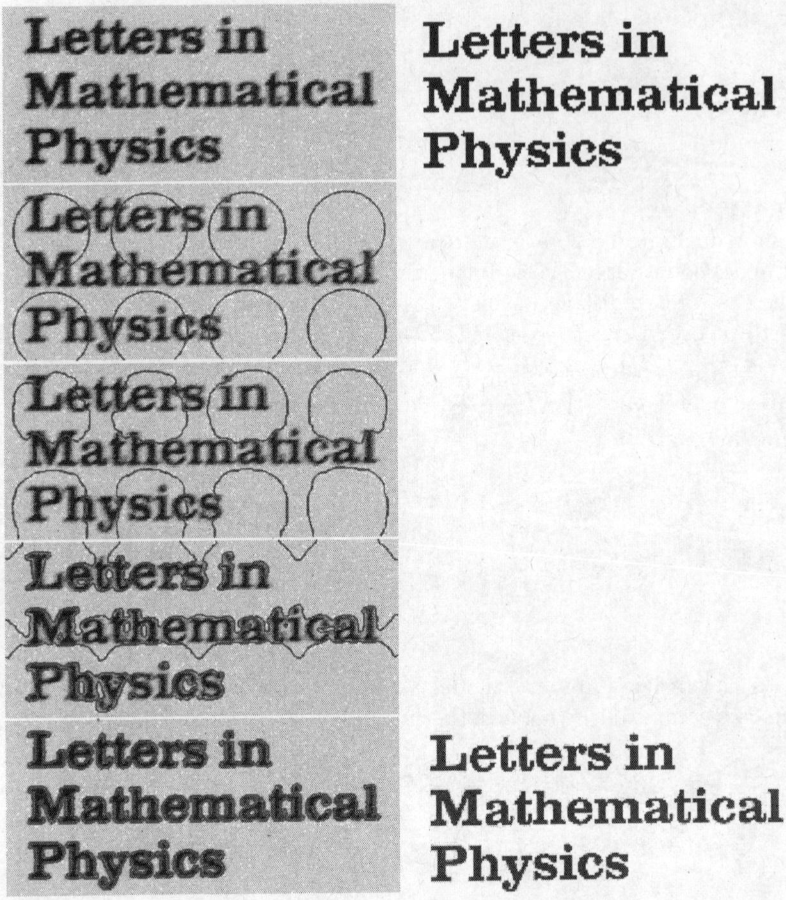

FIGURE 8.20: Joint segmentation, denoising, and deblurring using the binary level set model. Top row: degraded image g (blurred with motion blur kernel of length 10, oriented at an angle $\theta = 25°$ w.r.t. the horizon and contaminated by Gaussian noise with $\sigma_n = 10$), original image. Rows 2 through 5: initial curves, curve evolution at iterations 50, 100, 300 with $\mu = 5 \cdot 255^2$, and the restored image u (SNR = 28.1827). (c_1, c_2): original image $\approx (62.7525, 259.8939)$, restored u, $(61.9194, 262.7795)$.

8.4 Exercises

Exercise 8.1 *Prove the validity of formulas (8.2) and (8.3).*

Exercise 8.2 *Assume that for $t \geq 0$, ϕ solves the partial differential equation*

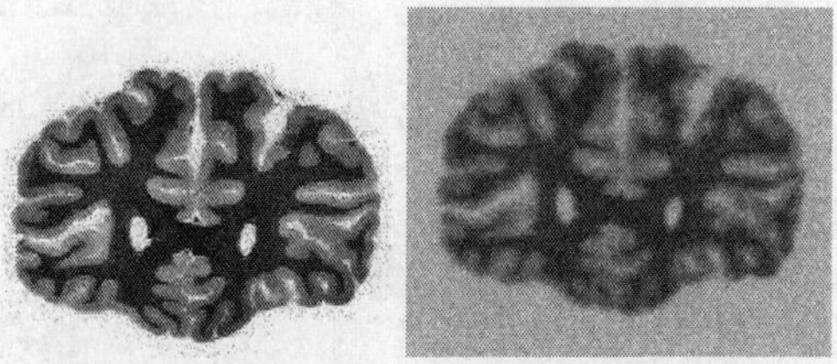

FIGURE 8.21: Original image (left) and its noisy, blurry version (right) blurred with Gaussian kernel with $\sigma_b = 1$ and contaminated by Gaussian noise $\sigma_n = 20$.

(8.4) with $c_1 = c_1(t)$ and $c_2 = c_2(t)$ given by (8.2) applied to $\phi(t, \cdot)$ with the given boundary conditions. Show that the energy

$$\begin{aligned} E(t) &= \int_\Omega (g(x) - c_1(t))^2 H_\epsilon(\phi(t,x))dx + \int_\Omega (g(x) - c_2(t))^2 H_\epsilon(-\phi(t,x))dx \\ &+ \nu_0 \int_\Omega |\nabla H_\epsilon(\phi(t,x))|dx \end{aligned}$$

decreases as t increases.

Exercise 8.3 *Prove the validity of formulas (8.7) and (8.8).*

Exercise 8.4 *Implement the two-phase piecewise-constant active contour model using finite differences and (8.2), (8.4) in two dimensions and apply the algorithm to a given image. Compare the numerical results and the rate of convergence with those obtained using (8.2) and one of the alternative partial differential equations in ϕ:*

$$(g(x) - c_1)^2 - (g(x) - c_2)^2 = \nu_0 \, div\Big(\frac{\nabla\phi}{|\nabla\phi|}\Big) \text{ in } \Omega,$$

that was obtained by dropping $\delta_\epsilon(\phi) > 0$ from (8.3), and

$$\frac{\partial\phi}{\partial t} = |\nabla\phi|\Big[(g(x) - c_2)^2 - (g(x) - c_1)^2 + \nu_0 \, div\Big(\frac{\nabla\phi}{|\nabla\phi|}\Big)\Big],$$

that was obtained replacing $\delta_\epsilon(\phi)$ in (8.4) by $|\nabla\phi|$.

Exercise 8.5 *Assume that for $t \geq 0$, ϕ solves the partial differential equation (8.9) with $c_j = c_j(t)$, $0 \leq j \leq m$, given by (8.7) applied to $\phi(t, \cdot)$ with the*

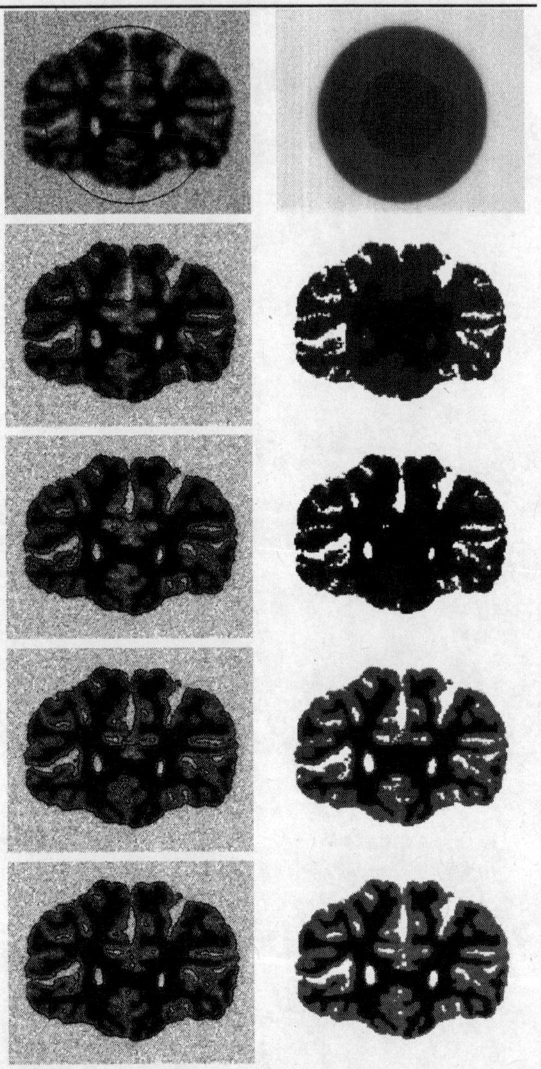

FIGURE 8.22: Curve evolution and restored u using (8.28), $\mu = 0.02 \cdot 255^2$, (c_1, c_2, c_3): original image $\approx$ $(12.7501, 125.3610, 255.6453)$, restored u using $(22.4797, 136.9884, 255.0074)$.

given boundary conditions. Show that the energy

$$E(t) = \int_\Omega |g(x) - c_0(t)|^2 H_\epsilon(l_1 - \phi(t,x))dx$$
$$+ \sum_{j=1}^{m-1} \int_\Omega |g(x) - c_j(t)|^2 H_\epsilon(\phi(t,x) - l_j) H_\epsilon(l_{j+1} - \phi(t,x))dx$$
$$+ \int_\Omega |g(x) - c_m(t)|^2 H_\epsilon(\phi(t,x) - l_m)dx + \nu_0 \sum_{j=1}^{m} \int_\Omega |\nabla H_\epsilon(\phi(t,x) - l_j)|dx$$

decreases as t increases.

Exercise 8.6 *Prove the validity of formulas (8.15), (8.16) and (8.17) as Euler–Lagrange equations associated with the minimization of the four-phase energy (8.14).*

Exercise 8.7 *If the data g is corrupted by additive Laplace noise or by impulse noise, then we would substitute the L^2 data fidelity terms by the corresponding L^1 data fidelity terms, known to penalize outliers. It is well known that these types of noise are effectively removed by the median filter; indeed, the medians c_1 and c_2 of the data g appear instead of the averages. We are led to minimize the regularized energy*

$$\inf_{c_1,c_2,\phi} \int_\Omega |g(x) - c_1| H_\epsilon(\phi) + \int_\Omega |g(x) - c_2| H_\epsilon(-\phi) + \nu_0 \int_\Omega |\nabla H_\epsilon(\phi)|dx,$$

leading to the Euler–Lagrange equations

$$c_1 = median\ g(x)|_{\{\phi(x)\geq 0\}}, \quad c_2 = median\ g(x)|_{\{\phi(x)\leq 0\}},$$

and

$$\delta_\epsilon(\phi)\Big[|g(x) - c_1| - |g(x) - c_2| - \nu_0\, div\Big(\frac{\nabla\phi}{|\nabla\phi|}\Big)\Big] = 0 \ in\ \Omega. \tag{8.29}$$

Discretize these Euler–Lagrange equations by finite differences and gradient descent, and apply the algorithm to binary segmentation of a piecewise-constant image corrupted by Laplace noise or impulse noise. Compare the results to those obtained using (8.2) and (8.4) applied to the same noisy image.

Exercise 8.8 *Prove the validity of formulas (8.19), (8.20), and (8.21).*

Chapter 9

Edge-Based Variational Snakes and Active Contours

The basic idea of snakes or active contour models is to evolve a curve, subject to constraints, from a given image I, to detect objects in that image. For instance, starting with a curve around the objects to be detected, the curve moves toward its interior normal and stops on boundaries of objects, as shown in Figure 9.1.

Applications can be found in medical imaging (for segmentation of anatomical regions or for tumor detection and delineation), in target detection, tracking, high-level vision, and in many other areas. We limit ourselves here to *variational* active contour models based on local information (such as gradient or Laplacian).

Although snakes and active contours can be extended to active surfaces in three dimensions, we limit the presentation to two-dimensional images and evolving curves. Let Ω be a bounded open subset of $\mathbb{R}^2$, with $\partial\Omega$ its boundary. Let $I : \overline{\Omega} \to \mathbb{R}$ be a given image, and $C = (C_1, C_2) : [0, 1] \to \mathbb{R}^2$, $q \mapsto C(q)$, be a parametrized smooth curve as a closed contour (thus $C(0) = C(1)$ and partial derivatives of C_i at 0 and 1 also agree).

Snakes or active contours are divided into two categories when we take into account the representation of the curve:

(1) parametric active contours (as in [187, 346]) when the evolving snake or curve is represented explicitly using splines; and

(2) geometric active contours (as in [80], [188], [94], among other work) when the curve is represented implicitly as the level line of a function.

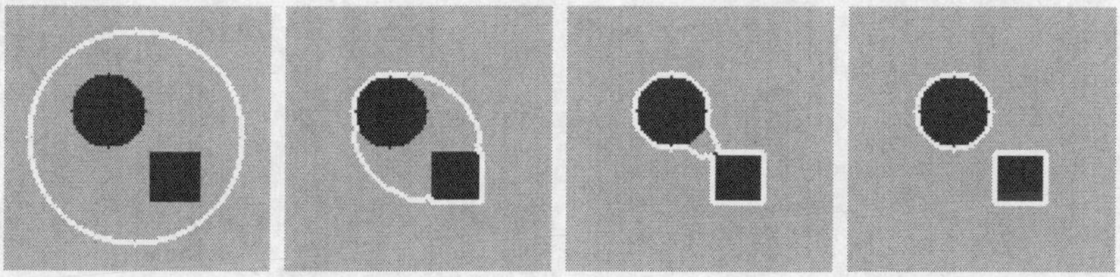

FIGURE 9.1: Active contour evolution for object and boundary detection with change of topology.

We start the presentation with the parametric snake model, the basis of many active contour models.

9.1 Snake model

The influential and versatile snake model of Kass, Witkin and Terzopoulos [187] is based on the minimization of an energy $E_{snake}(C)$ composed of three terms: an internal energy to the curve (curve regularization), an external or potential energy to the curve (image forces depending on I), and a third term composed of other external constraint forces. We consider here only the first two terms. The main energy is described by the following equations:

$$E_{snake}(C) = E_{int}(C) + E_{ext}(C),$$

with the internal energy of the spline-curve given by

$$E_{int}(C) = \frac{\alpha}{2}\int_0^1 |C'(q)|^2 dq + \frac{\beta}{2}\int_0^1 |C''(q)|^2 dq,$$

with weights $\alpha > 0$, $\beta \geq 0$, and the external potential energy (image forces)

$$E_{ext}(C) = w_{line}E_{line} + w_{edge}E_{edge} + w_{term}E_{term},$$

composed of three functionals that help to attract the snake to salient features of image I. The authors in [187] use for the line functional the image itself, $E_{line} = \int_0^1 I(C(q))dq$ (depending on the sign of w_{line}, the snake will be attracted either to light or dark lines). As an edge functional they use $E_{edge} = -\int_0^1 |\nabla I(C(q))|^2 dq$ (based on the gradient edge detector, the snake is attracted to contours with large image gradients) or $E_{edge} = -\int_0^1 (G_\sigma * \Delta I(C(q)))^2 dq = -\int_0^1 (\Delta G_\sigma * I(C(q)))^2 dq$ (minima of this functional lie on zero-crossings of the Laplacian of Gaussian of I which define edges in the Marr-Hildreth scale-space theory [231]). Finally, an additional term functional can be $E_{term} = \int_0^1 curv(G_\sigma * I)(C(q))dq$ (based on the curvature of level lines of a smoothed version of the image I to detect terminations with high curvature).

In this section we consider the simplified snake model given by

$$E_{snake}(C) = \frac{\alpha}{2}\int_0^1 |C'(q)|^2 dq + \frac{\beta}{2}\int_0^1 |C''(q)|^2 dq - \lambda\int_0^1 |\nabla I(C(q))|^2 dq, \quad (9.1)$$

where α, β and λ are positive parameters. The first two terms control the smoothness of the contour (internal energy), while the third term attracts the contour toward an object in the image (external energy). Observe that by minimizing the energy (9.1), we are trying to locate the curve at the points of maxima $|\nabla I|$ acting as an edge detector, while keeping smoothness in the curve (object boundary). Note that in [187], the weights α and β are functions $q \mapsto \alpha(q)$ and $q \mapsto \beta(q)$ depending on the parametrization, to better control the snake.

In order to find local minima of E_{snake} in (9.1), a time-dependent gradient descent method is used, working with a family of smooth closed curves $(t, q) \mapsto C(t, q) = (C_1(t, q), C_2(t, q))$, $q \in [0, 1]$, $t \geq 0$, such that $C(t, 0) = C(t, 1)$ for all $t \geq 0$ (similarly for the first and second order derivatives of C in q). Generalizing the external potential energy to $E_{ext}(C) = \lambda \int_0^1 P(C(q))dq$, it yields to

$$\frac{\partial C_i}{\partial t}(t, q) = \alpha C_{i_{qq}}(t, q) - \beta C_{i_{qqqq}}(t, q) + \lambda F_i(C(t, q)), \; i = 1, 2, \tag{9.2}$$

with, for simplicity of notations, $F(C) = \begin{pmatrix} F_1(C) \\ F_2(C) \end{pmatrix} = -\nabla P(C)$ and $C_{iq} = \dfrac{\partial C_i}{\partial q}$.

Let us set $q_j = jh$, $j \in \{0, \cdots, m\}$ with h the space step and $h = \frac{1}{m}$. Let $T > 0$ be fixed. Let us introduce a subdivision of the interval $[0, T]$ into sub-intervals of length τ. One has $t_n = n\tau$, $\forall n \in \{0, \cdots, \frac{T}{\tau}\}$. Let us then denote by $C_{i,j}^n$ an approximation of $C_i(t_n, q_j = jh)$.

Let us denote by D^+ the finite difference operator such that $(D^+ v)_j = \frac{v_{j+1} - v_j}{h}$ and by D^- the finite difference operator such that $(D^- v)_j = \frac{v_j - v_{j-1}}{h}$. Note that $(D^+ D^- v)_j = (D^- D^+ v)_j$. An implicit finite difference scheme is employed to discretize (9.2), yielding to

$$\begin{aligned}
\frac{C_{i,j}^{n+1} - C_{i,j}^n}{\tau} =& \alpha \left(D^+ D^- C_i^{n+1}\right)_j - \beta \left(D^+ D^- D^+ D^- C_i^{n+1}\right)_j + \lambda F_i(C_j^n), \\
=& \alpha \frac{C_{i,j+1}^{n+1} - 2C_{i,j}^{n+1} + C_{i,j-1}^{n+1}}{h^2} - \beta \left[\frac{C_{i,j+2}^{n+1} - 2C_{i,j+1}^{n+1} + C_{i,j}^{n+1}}{h^4} \right. \\
& \left. -2 \frac{C_{i,j+1}^{n+1} - 2C_{i,j}^{n+1} + C_{i,j-1}^{n+1}}{h^4} + \frac{C_{i,j}^{n+1} - 2C_{i,j-1}^{n+1} + C_{i,j-2}^{n+1}}{h^4} \right] \\
& + \lambda F_i(C_j^n),
\end{aligned}$$

or equivalently,

$$(Id + \tau A)\, C_i^{n+1} = C_i^n + \tau \lambda F_i(C_j^n),$$

with A a pentadiagonal matrix whose coefficients depend on α, β and h defined by

$$A = \begin{pmatrix}
\frac{2\alpha}{h^2}+\frac{6\beta}{h^4} & -\frac{\alpha}{h^2}-\frac{4\beta}{h^4} & \frac{\beta}{h^4} & 0 & \cdots & \cdots & 0 \\
-\frac{\alpha}{h^2}-\frac{4\beta}{h^4} & \frac{2\alpha}{h^2}+\frac{6\beta}{h^4} & -\frac{\alpha}{h^2}-\frac{4\beta}{h^4} & \frac{\beta}{h^4} & \ddots & & \vdots \\
\frac{\beta}{h^4} & -\frac{\alpha}{h^2}-\frac{4\beta}{h^4} & \ddots & \ddots & \ddots & \ddots & \vdots \\
0 & \frac{\beta}{h^4} & \ddots & \ddots & \ddots & \frac{\beta}{h^4} & 0 \\
\vdots & \ddots & \ddots & \ddots & \ddots & -\frac{\alpha}{h^2}-\frac{4\beta}{h^4} & \frac{\beta}{h^4} \\
\vdots & & \ddots & \frac{\beta}{h^4} & -\frac{\alpha}{h^2}-\frac{4\beta}{h^4} & \frac{2\alpha}{h^2}+\frac{6\beta}{h^4} & -\frac{\alpha}{h^2}-\frac{4\beta}{h^4} \\
0 & \cdots & \cdots & 0 & \frac{\beta}{h^4} & -\frac{\alpha}{h^2}-\frac{4\beta}{h^4} & \frac{2\alpha}{h^2}+\frac{6\beta}{h^4}
\end{pmatrix}.$$

At each discrete time t_n, the curve is submitted to external forces computed at time t_{n-1}. Also, with constant parameters α and β, the LU decomposition of $(Id + \tau A)$ is made once and in this case, the orders of magnitude for α and β are respectively h^2 and h^4 so that the coefficients in the linear system matrix are of the same order.

A more general edge detector can be defined using a positive and decreasing function g, depending on the gradient of the image I, such that $\lim_{z\to\infty} g(z) = 0$. A popular choice is $g(t) = \frac{1}{1+|t|^p}$, $p \geq 1$, to obtain

$$g(|\nabla I(x)|) = \frac{1}{1 + |\nabla(G_\sigma * I)(x)|^p},\ p \geq 1,$$

where $G_\sigma * I$, a smoother version of I, is the convolution of the image I with the Gaussian $G_\sigma(x) = \frac{1}{2\pi\sigma^2} e^{-\|x\|^2/2\sigma^2}$. The function $g(|\nabla I|)$ is strictly positive in homogeneous regions, and near zero on edges (see Figure 9.2). This generalizes the negative gradient norm $-|\nabla I|$ as an edge detector, and the general snake model without the elastic term becomes

$$E(C) = \int_0^1 |C'(q)|^2 dq + \lambda \int_0^1 g(|\nabla I(C(q))|)^2 dq. \tag{9.3}$$

The above snake model has some limitations. The proposed snake energy is not intrinsic; in other words, different parameterizations of the curve can lead to different values of the energy. Moreover, due to the explicit representation of the spline curve (or spline surface in three dimensions) reparameterization of the curve (or surface) may be needed during the numerical evolution. Finally, the evolving curve cannot automatically change topology (merging or breaking). The models discussed in the next sections allow for level set or implicit representation of the curve with discretization on a fixed rectangular grid, and automatic changes of topology can happen without need of reparameterization. An improved version of the classical snake model is the gradient vector flow [346] of Xu and Prince.

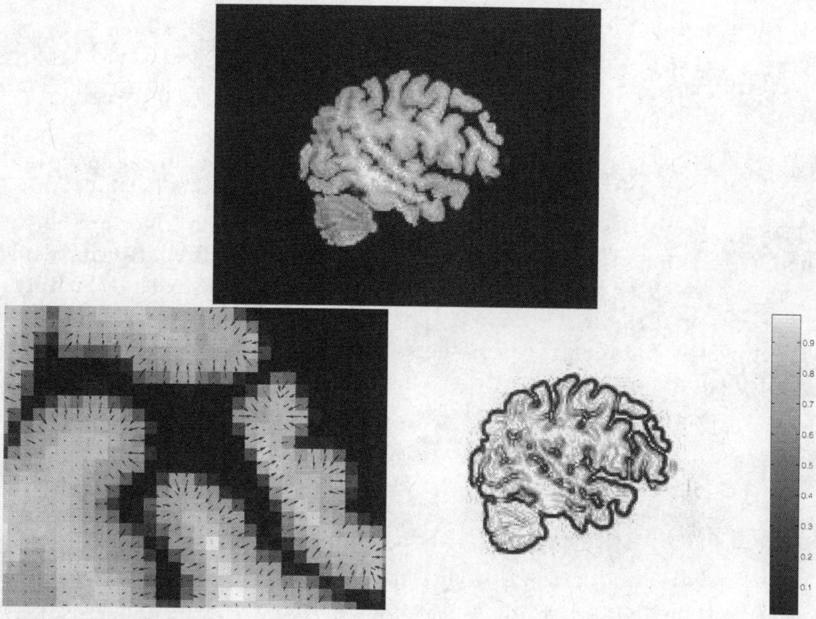

FIGURE 9.2: Top: a given brain slice image I. Bottom left to right: zoom on the related gradient map ∇I, and associated edge-map $g(|\nabla I|)$.

9.1.1 Gradient vector flow snake

Xu and Prince [346] pointed out two limitations of the classical snake model of Kass, Witkin and Terzopoulos and of other related active contour models using a local edge detector: problems with initialization (the initial contour must be close to the true boundary and, for instance, it should surround it), and poor convergence to boundary concavities. The gradient vector flow [346] is insensitive to initialization (the initialization can be made inside, outside, or across the object boundary) and it can move into boundary concavities. The main idea is to modify the time-dependent Euler–Lagrange equation

$$\frac{\partial C_i}{\partial t} = \alpha C_{i_{qq}} - \beta C_{i_{qqqq}} - \lambda \frac{\partial E_{ext}}{\partial C_i}(C), \ i = 1, 2, \tag{9.4}$$

associated with the snake energy into

$$\frac{\partial C_i}{\partial t} = \alpha C_{i_{qq}} - \beta C_{i_{qqqq}} + \lambda v_i(C), \ i = 1, 2, \tag{9.5}$$

where $v(x) = (v_1(x), v_2(x))$ is an external force field called gradient vector flow (GVF). Vector v is chosen so that the snake can move into boundary concavities and the detected boundary is independent of the initial curve.

This is defined as the minimizer of the "interpolating" energy

$$E(v) = \int_\Omega \left[\mu(|\nabla v_1|^2 + |\nabla v_2|^2) + |\nabla f|^2 |v - \nabla f|^2\right] dx, \tag{9.6}$$

where $f(x)$ is the negative of one of the external forces used in the snake model: $f(x)$ is $|\nabla I(x)|^2$, $|\nabla(G_\sigma * I)(x)|^2$, $-I(x)$, or $-(G_\sigma * I)(x)$. In particular, the vectors ∇f point toward the edges and are normal to the edges at the edges. When $|\nabla f|$ is large, v will be close to ∇f due to the second term, while in regions where $|\nabla f| \approx 0$, v will be slowly varying due to the H^1 diffusion regularization of the first term.

Vector v is obtained by time-dependent gradient descent of the Euler–Lagrange equations associated with (9.6), with unknown $v(t, x)$,

$$\frac{\partial v_i}{\partial t} = \mu \triangle v_i - (v_i - f_{x_i})|\nabla f|^2, \tag{9.7}$$

where f_{x_i} denotes $\dfrac{\partial f}{\partial x_i}$.

To conclude this part, we would like to point out the work by Jifeng et al. ([177]) which provides an improved external force field for active contour models. They propose improving the diffusion properties of the gradient vector flow model by replacing the Laplacian operator appearing in (9.7) by its diffusion term in the normal direction, that is, the normalized infinity Laplacian operator defined (when x is such that $\frac{\nabla v_i(x)}{|\nabla v_i(x)|} \neq 0$) by $v_{i\vec{N}\vec{N}} = \vec{N}^T D^2 v_i \vec{N} = \dfrac{1}{|\nabla v_i|^2}\left(v_{ix_1}^2 v_{ix_1x_1} + v_{ix_2}^2 v_{ix_2x_2} + 2 v_{ix_1x_2} v_{ix_1} v_{ix_2}\right)$, $\vec{N} = \frac{\nabla v_i}{|\nabla v_i|}$. Unlike the gradient vector flow model, this new field called NGVF is anisotropic. The model is stable for larger time steps, slightly improves segmentation results, and allows faster detection of long and thin concavities.

9.2 Geodesic active contours

The geodesic active contour model proposed by Caselles, Kimmel and Sapiro in [80] and Kichenassamy et al. in [188] is:

$$\inf_C J(C) = \int_0^1 |C'(q)| \cdot g(|\nabla I(C(q))|) dq. \tag{9.8}$$

This is a problem of geodesic computation in a Riemannian space, according to a metric induced by the image I. Solving the minimization problem (9.8) consists of finding the path of minimal new length in that metric.

Using the steepest descent method to deform an initial curve $C(0, \cdot)$ toward a (local) minima of J, we should follow a curve evolution equation given by the time-dependent Euler–Lagrange equation associated with (9.8):

$$\frac{\partial C}{\partial t} = g\kappa\vec{N} - (\nabla g \cdot \vec{N})\vec{N}, \tag{9.9}$$

with $\vec{N}$ the unit inward normal vector to the curve, $(\cdot, \cdot)$ denoting the Euclidean inner product in $\mathbb{R}^2$ and where we have set $g := g(|\nabla I|)$ for the sake of conciseness.

The equation $\frac{\partial C}{\partial t} = \kappa\vec{N}$ corresponds to the Euclidean heat flow equation and exhibits regularizing and shortening properties. As stressed in [80], this flow diminishes the total curvature as well as the number of zero crossings and the value of maxima/minima curvature. Gage and Hamilton [150] proved that any closed curve evolving according to this flow remains convex and becomes a point. Grayson [161] proved that any closed curve moving according to this flow, whatever shape it may have, evolves to become a point. In the geodesic active contour model, the curvature component κ is weighted by the edge detector function so that the moving curve stops on the boundaries.

The model is complemented by the component $-(\nabla g \cdot \vec{N})\vec{N}$ (effective when ∇g and $\vec{N}$ are collinear) that guarantees a better attraction to the boundary, in particular when the image gradient varies abruptly along the edges. As depicted in Figure 9.3, the vectors $-\nabla g$ are directed toward the middle of the boundaries and the component $-(\nabla g \cdot \vec{N})\vec{N}$ allows us to detect non-convex shapes, the weighted Euclidean heat flow alone being unable to handle such topological changes. An additional component of the form $\nu\, g\, \vec{N}$ can be included in the model, moving the curve inward or outward (according to the sign of ν), to increase the speed of convergence. It is also of great interest to allow convex initial curves to detect non-convex shapes and to initialize the contour inside the object to be detected. The general flow for C is thus defined by

$$\frac{\partial C}{\partial t} = g\kappa\vec{N} - (\nabla g \cdot \vec{N})\vec{N} + \nu\, g\, \vec{N}. \tag{9.10}$$

The curve is then embedded in a Lipschitz continuous function $\Phi : \Omega \times [0, +\infty[\to \mathbb{R}$ with $(x, t) \mapsto \Phi(x, t)$, such that $C(t, \cdot) = \{x = (x_1, x_2) \in \Omega \,|\, \Phi(x, t) = 0\}$ and

$$\begin{cases} \Phi < 0 \text{ on } w \text{ the interior of } C \\ \Phi > 0 \text{ on } \Omega \setminus \bar{w}. \end{cases}$$

In practice, the function Φ is preferred to be a signed distance function for the stability of numerical computations.

Assuming that C is parameterized such that for increasing $q \in [0, 1]$, C is to the left, the unit inward normal to the curve C at point $C(q)$ is given by

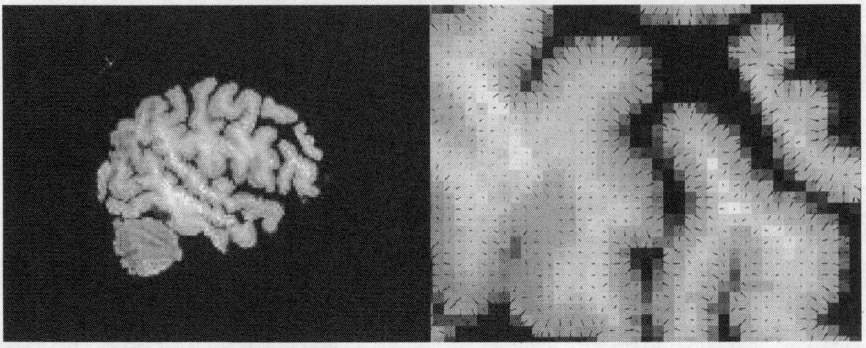

FIGURE 9.3: From left to right: a given brain slice image I and zoom on the vector field $-\nabla g(|\nabla I|)$.

$\vec{N}(q) = \begin{pmatrix} -\frac{C_2'(q)}{\sqrt{(C_1'(q))^2+(C_2'(q))^2}} \\ \frac{C_1'(q)}{\sqrt{(C_1'(q))^2+(C_2'(q))^2}} \end{pmatrix}$. If

$$C(q) = \{(C_1(q), C_2(q)) \in \Omega,\ \Phi(C_1(q), C_2(q)) = 0\},$$

then differentiating the relation $\Phi(C_1(q), C_2(q)) = 0$ with respect to q yields

$$\frac{\partial \Phi}{\partial x_1}(C_1(q), C_2(q))C_1'(q) + \frac{\partial \Phi}{\partial x_2}(C_1(q), C_2(q))C_2'(q) = 0. \tag{9.11}$$

The vectors $\vec{T}(q) = \begin{pmatrix} C_1'(q) \\ C_2'(q) \end{pmatrix}$ (tangent vector to the curve C at point $C(q) = (C_1(q), C_2(q)))$ and $\nabla\Phi(C_1(q), C_2(q))$ are thus orthogonal (which means that $\vec{N}(q)$ and $\nabla\Phi(C_1(q), C_2(q))$ are collinear), and with the above conventions on Φ, there exists a positive λ such that

$$\begin{cases} C_1'(q) = -\lambda \dfrac{\partial \Phi}{\partial x_2}(C_1(q), C_2(q)), \\ C_2'(q) = \lambda \dfrac{\partial \Phi}{\partial x_1}(C_1(q), C_2(q)). \end{cases} \tag{9.12}$$

Differentiating relation (9.11) again with respect to q leads to

$$\begin{aligned} &\frac{\partial \Phi}{\partial x_1}(C_1(q), C_2(q))C_1''(q) + \frac{\partial \Phi}{\partial x_2}(C_1(q), C_2(q))C_2''(q) \\ &+ \frac{\partial^2 \Phi}{\partial x_1^2}(C_1(q), C_2(q))(C_1'(q))^2 + 2\frac{\partial^2 \Phi}{\partial x_1 \partial x_2}(C_1(q), C_2(q))C_1'(q)C_2'(q) \\ &+ \frac{\partial^2 \Phi}{\partial x_2^2}(C_1(q), C_2(q))(C_2'(q))^2 = 0. \end{aligned}$$

Now including the expressions (9.12) and denoting by $\frac{\partial \Phi}{\partial x_i} := \frac{\partial \Phi}{\partial x_i}(C_1(q), C_2(q))$, it follows, from the above relation, that

$$\lambda^2 \left[\frac{\partial^2 \Phi}{\partial x_1^2}(\frac{\partial \Phi}{\partial x_2})^2 - 2\frac{\partial^2 \Phi}{\partial x_1 \partial x_2}\frac{\partial \Phi}{\partial x_1}\frac{\partial \Phi}{\partial x_2} + \frac{\partial^2 \Phi}{\partial x_2^2}(\frac{\partial \Phi}{\partial x_1})^2\right]$$
$$+ \frac{1}{\lambda}\left[C_2'(q)C_1''(q) - C_1'(q)C_2''(q)\right] = 0.$$

Still from relation (9.12), $\lambda = \frac{|C'(q)|}{|\nabla \Phi|}$ and

$$\kappa(q) = \frac{C_1'(q)C_2''(q) - C_2'(q)C_1''(q)}{|C'(q)|^3},$$
$$= \frac{\frac{\partial^2 \Phi}{\partial x_1^2}(\frac{\partial \Phi}{\partial x_2})^2 - 2\frac{\partial^2 \Phi}{\partial x_1 \partial x_2}\frac{\partial \Phi}{\partial x_1}\frac{\partial \Phi}{\partial x_2} + \frac{\partial^2 \Phi}{\partial x_2^2}(\frac{\partial \Phi}{\partial x_1})^2}{|\nabla \Phi|^3} = \operatorname{div}\left(\frac{\nabla \Phi}{|\nabla \Phi|}\right).$$

The idea is now to determine the partial differential equation satisfied by Φ in order that the curve C evolves according to (9.10). By definition,

$$\forall q \in [0,1],\ \forall t \in [0, +\infty[,\ \Phi(C(t), t) = 0.$$

Differentiating this relation with respect to t yields

$$\left(\nabla \Phi(C(t), t) \cdot \frac{\partial C}{\partial t}(t)\right) + \frac{\partial \Phi}{\partial t}(C(t), t) = 0. \tag{9.13}$$

Including the expression (9.10) of $\frac{\partial C}{\partial t}$ in (9.13) as well as the level set counterparts of the inward unit normal to the curve ($\vec{N} = -\frac{\nabla \Phi}{|\nabla \Phi|}$) and the curvature ($\kappa = \operatorname{div}\left(\frac{\nabla \Phi}{|\nabla \Phi|}\right)$) leads to

$$\frac{\partial \Phi}{\partial t} = g|\nabla \Phi|\operatorname{div}\left(\frac{\nabla \Phi}{|\nabla \Phi|}\right) + (\nabla g \cdot \nabla \Phi) + \nu\, g\, |\nabla \Phi|,$$
$$= |\nabla \Phi|\operatorname{div}\left(g\,\frac{\nabla \Phi}{|\nabla \Phi|}\right) + \nu\, g\, |\nabla \Phi|.$$

This equation is a priori valid only on the zero level set of Φ. One of the advantages of the level set method is that the equation can be extended to the whole domain $\Omega \times [0, +\infty[$. The evolution equation is then equipped with an initial condition $\Phi(x,t) = \Phi_0(x)$ and with Neumann boundary conditions. One is thus led to solve

$$\begin{cases} \frac{\partial \Phi}{\partial t} = g|\nabla \Phi|\operatorname{div}\left(\frac{\nabla \Phi}{|\nabla \Phi|}\right) + (\nabla g \cdot \nabla \Phi) + \nu\, g\, |\nabla \Phi| \text{ on } \Omega \times [0, +\infty[, \\ \frac{\partial \Phi}{\partial \vec{n}} = 0 \text{ on } \partial\Omega,\ \vec{n} \text{ denoting the unit outward normal to } \partial\Omega, \\ \Phi(x, 0) = \Phi_0(x). \end{cases}$$

Alternatively, we could have straightforwardly phrased the segmentation problem (9.8) in the level set framework by minimizing

$$\inf_{\phi} J(\phi) = \int_\Omega g(|\nabla I(x)|)|DH(\phi)| + \nu \int_\Omega g(|\nabla I|)\, H(\phi)dx, \tag{9.14}$$

where the second balloon force term is added to increase the evolution speed and to attract the curve toward the boundary, and constitutes in fact an extra area-based speed.

First we substitute the Heaviside function H by a smooth approximation H_ϵ (respectively δ, the Dirac measure, by δ_ϵ) as discussed in Chapter 2, Section 2.9, and we minimize instead

$$\begin{aligned}\inf_{\phi} J(\phi) &= \int_\Omega g(|\nabla I(x)|)|\nabla H_\epsilon(\phi(x))|dx + \nu \int_\Omega g(|\nabla I|)\, H_\epsilon(\phi)dx,\\ &= \int_\Omega g(|\nabla I(x)|)\delta_\epsilon(\Phi(x))|\nabla\phi(x)|dx + \nu \int_\Omega g(|\nabla I|)\, H_\epsilon(\phi)dx. \end{aligned} \tag{9.15}$$

Let us assume that Φ is a minimizer of J and let us define by $G(\varepsilon) = J(\Phi + \varepsilon\Psi)$ for $\varepsilon \in \mathbb{R}$ and Ψ, a test function. As Φ minimizes J, $G'(0) = 0$. By differentiating G with respect to ε, we obtain (still setting $g := g(|\nabla I|)$ for the sake of conciseness)

$$\begin{aligned} G'(\varepsilon) = &\int_\Omega g\, \Big[\delta'_\epsilon(\Phi + \varepsilon\Psi)|\nabla\Phi + \varepsilon\nabla\Psi|\Psi \\ &+ \delta_\epsilon(\Phi + \varepsilon\Psi)\left(\frac{\nabla\Phi + \varepsilon\nabla\Psi}{|\nabla\Phi + \varepsilon\nabla\Psi|}\cdot\nabla\Psi\right)\Big]\, dx \\ &+ \nu \int_\Omega g\, \delta_\epsilon(\Phi + \varepsilon\Psi)\, \Psi\, dx. \end{aligned}$$

Then

$$G'(0) = \int_\Omega g\, \left[\delta'_\epsilon(\Phi)|\nabla\Phi|\Psi + \delta_\epsilon(\Phi)\left(\frac{\nabla\Phi}{|\nabla\Phi|}\cdot\nabla\Psi\right)\right]\, dx + \nu \int_\Omega g\, \delta_\epsilon(\Phi)\, \Psi\, dx.$$

Making an integration by parts in the second term of the first integral yields to

$$\begin{aligned}
G'(0) =& \int_\Omega g\,\delta_\epsilon'(\Phi)|\nabla\Phi|\Psi\,dx - \int_\Omega \mathrm{div}\left(g\delta_\epsilon(\Phi)\frac{\nabla\Phi}{|\nabla\Phi|}\right)\Psi\,dx \\
&+ \int_{\partial\Omega} g\frac{\delta_\epsilon(\Phi)}{|\nabla\Phi|}\frac{\partial\Phi}{\partial\vec{n}}\Psi\,d\sigma + \nu\int_\Omega g\,\delta_\epsilon(\Phi)\,\Psi\,dx, \\
=& \int_\Omega g\,\delta_\epsilon'(\Phi)|\nabla\Phi|\Psi\,dx - \int_\Omega \delta_\epsilon(\Phi)\,\mathrm{div}\left(g\frac{\nabla\Phi}{|\nabla\Phi|}\right)\Psi\,dx \\
&- \int_\Omega g\,\delta_\epsilon'(\Phi)|\nabla\Phi|\Psi\,dx + \int_{\partial\Omega} g\frac{\delta_\epsilon(\Phi)}{|\nabla\Phi|}\frac{\partial\Phi}{\partial\vec{n}}\Psi\,d\sigma \\
&+ \nu\int_\Omega g\,\delta_\epsilon(\Phi)\,\Psi\,dx, \\
=& -\int_\Omega \delta_\epsilon(\Phi)\,\mathrm{div}\left(g\frac{\nabla\Phi}{|\nabla\Phi|}\right)\Psi\,dx + \nu\int_\Omega g\,\delta_\epsilon(\Phi)\,\Psi\,dx \\
&+ \int_{\partial\Omega} g\frac{\delta_\epsilon(\Phi)}{|\nabla\Phi|}\frac{\partial\Phi}{\partial\vec{n}}\Psi\,d\sigma.
\end{aligned}$$

We deduce that the obtained Euler–Lagrange equation formally is

$$0 = -\delta_\epsilon(\Phi)\,\mathrm{div}\left(g\frac{\nabla\Phi}{|\nabla\Phi|}\right) + \nu\,g\,\delta_\epsilon(\Phi), \tag{9.16}$$

with boundary conditions $g\frac{\delta_\epsilon(\Phi)}{|\nabla\Phi|}\frac{\partial\Phi}{\partial\vec{n}} = 0$ on $\partial\Omega$. Local minima of this energy are obtained by driving to steady state the Euler–Lagrange equation in the unknown $\Phi(x,t)$, $t \geq 0$, $x \in \Omega$,

$$\begin{cases} \dfrac{\partial\Phi}{\partial t} = \delta_\epsilon(\Phi)\left(\mathrm{div}\left(g\dfrac{\nabla\Phi}{|\nabla\Phi|}\right) - \nu g(|\nabla I|)\right), \\ \text{in } \Omega \times [0,\infty[\\ \Phi(x,0) = \Phi_0(x). \end{cases} \tag{9.17}$$

A rescaling (see Zhao et al. [353] and Alvarez et al. [15]) can be made by replacing $\delta_\epsilon(\Phi)$ by $|\nabla\Phi|$ to apply the motion to all level sets.

It is interesting to know that models (9.3) and (9.8) are related. In [80], the equivalence using Hamiltonian mechanics is shown between the snake model with general edge detector g and without the elastic term (9.3), and the geodesic active contour model (9.8). In [29], Aubert and Blanc–Féraud have shown the equivalence using calculus and analysis in both two and three dimensions.

We conclude this section by presenting numerical results obtained using the geodesic active contour model. We start with a synthetic image representing two discs (Figure. 9.4), this example illustrating the ability of the geodesic active contour model to handle topological changes. The initialization is made with a signed distance function that encloses the two balls. The contour splits

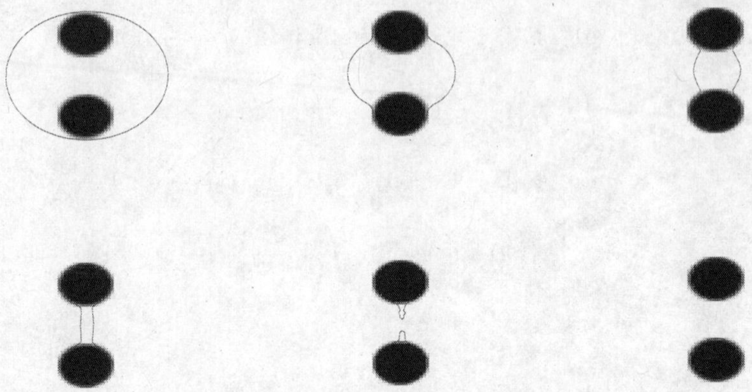

FIGURE 9.4: Segmentation of the synthetic image with two discs: the contour has split into two components. Iterations 0, 50, 100, 140, 170, 180.

into two components as expected. The second example (Figure. 9.5) is taken from [168]. One can observe that the two middle fingers touch so the evolving contour merges and a hole appears.

9.3 Alignment term

In the work of Kimmel and Bruckstein [191, 192] and Desolneux, Moisan and Morel [128], another term is proposed to enhance edge and boundary detection via curve evolution and energy minimization. It is called the alignment functional and it is defined as

$$E_A(C) = -\int_0^L \langle \nabla I(C(s)), n(s) \rangle ds, \tag{9.18}$$

or in robust form

$$E_{AR}(C) = -\int_0^L |\langle \nabla I(C(s)), n(s) \rangle| ds, \tag{9.19}$$

where L is the length of the curve, n is the normal to the curve C and C is assumed to be parameterized by its arc length. The negative of the inner product is small if the curve normal aligns with the image gradient direction.

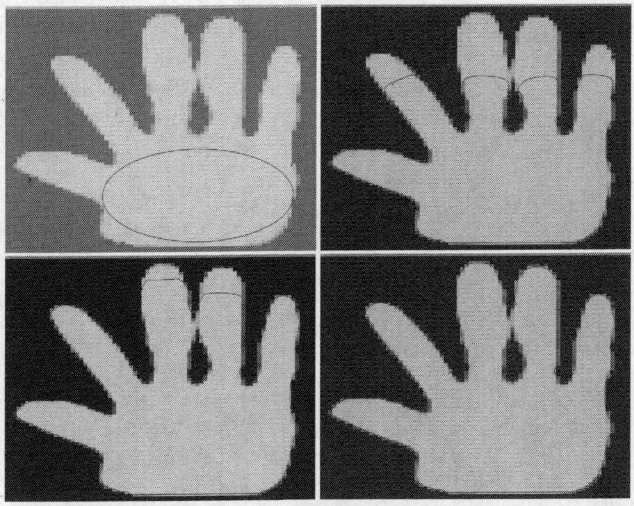

FIGURE 9.5: Steps of the segmentation. Iterations 0, 500, 780, 890.

9.4 Topology-preserving snakes model

The implicit framework of the level set method has several advantages when tracking propagating fronts. Indeed, the evolving contour is embedded in a higher dimensional level set function and its evolution can be phrased in terms of a Eulerian formulation. The ability of this intrinsic method to handle topological changes (merging and breaking) makes it useful in a wide range of applications (fluid mechanics, computer vision) and particularly in image segmentation.

The level set formulation can be straightforwardly derived using the theory of curve evolution (see [254], [256], [285], [286] for instance). The implicit representation of the curve *via* a level set function avoids parameterization issues. The discretization of the problem is made on a fixed rectangular grid. The model is intrinsic (if we define a new parameterization, the energy is unchanged), allows for corners and the initial condition does not necessarily have the same topology as the resulting contour. The geometric properties of the contour (curvature, unit normal vector) are easily deduced from the level set function.

The topological flexibility is a great benefit since it allows for the simultaneous detection of several objects in the image, which was not possible in the case of parametric deformable models without additional modifications or additional work. However, in some applications, this flexibility is undesirable, for instance, when the final shape must be homeomorphic to the initial one, or when the topology of the final contour must be consistent with the known topology of the object. In brain imaging, for example, when located near a lobe, parts of the contour may fold up and finally have a contact point (without merging), which is not consistent with the spherical topology of the brain. The contrast of the image can also be impaired and the number of components of the object can thus be increased.

These remarks motivated the work presented next: to keep the benefits of the level set framework (parameter-free, intrinsic, easily-deduced geometrical properties of the front, etc.), we describe a non-parametric topology-constrained segmentation model based on the geodesic active contours [80], [188], introduced by Le Guyader and Vese in [208].

The model from [208] presented in this section is motivated by the works [167], [166], [168], in which Han, Xu and Prince propose a first approach to preserve the topology of the implicit contour while the embedding level set function is evolving. The key idea of the model lies in the concept of simple point from digital topology. The authors assume that the topology of the zero level set is equivalent to the topology of the boundary of the digital object it defines. The topology preservation problem is therefore simplified in the following way: the topology of the implicit contour can change only if the level set function changes sign at a grid point. This is only a necessary condition; not every change of sign of the level set function implies a topology change of the zero level set and consequently of the digital object boundary. One needs to introduce the concept of simple point to identify grid points for which, when removed or added, topology is no longer preserved. A grid point is said to be simple if it can be added or removed without changing the topology of the object. The computation of two topological numbers is sufficient to characterize simple points. Thus Han et al. introduce an algorithm that at each iteration monitors the changes of sign of the level set function and prevents the level set function from changing sign on grid points which are not simple. The procedure is therefore pixel-based. The authors also stress the arbitrariness of the result produced by the algorithm, depending on the order in which points are treated in the narrow band. Their method has been applied to the geodesic active contour model [80]. Here, we describe, in the continuum case, a geodesic active contour-based segmentation model that globally integrates a topological constraint emerging from a geometrical remark. Before depicting this model, we review some related prior works.

The interesting work of Alexandrov and Santosa [6] presents a curve evolution method based on level sets for shape optimization problems arising in material sciences. The signed distance function is used, and their method avoids overlaps of the narrow band of the evolving contour. Their method, like ours, is a variational one, but it has not been proposed for image segmentation. Only artificial tests of shape optimization illustrate their point in [6]. In the case of a shape optimization problem, their model minimizes $F_\mu(\Phi) = F(\Phi) + \mu H(\Phi)$, with $\mu << 1$, where $F(\Phi)$ is a general shape optimization functional and $H(\Phi)$ is the topology constraint term defined by

$$H(\Phi) = -\int_{\partial D} \log\Big[\Phi(x + d\nabla\Phi(x))\Big]\,ds - \int_{\partial D} \log\Big[-\Phi(x - l\nabla\Phi(x))\Big]\,ds,$$

where $D = \{x : \Phi(x) > 0\}$ and $d > 0$, $l > 0$ are given parameters. The unknown level set function Φ must be a signed distance function.

Another related and very interesting approach based on the knot energy is presented by Sundaramoorthi and Yezzi in [305], [306]. This also includes a penalty in the segmentation functional in the form of a double integral, that could prevent a curve C from changing topology. The relevant term that is minimized inspired from the knot energy is

$$E(C) = \frac{1}{2}\int\int_{C\times C} \frac{dpdp'}{\|C(p) - C(p')\|^{\gamma}},$$

where $\|\cdot\|$ is the usual Euclidean norm and $\gamma > 0$. The topology depends on the choice of the parameter γ. This term is also coupled with the method from [167], [168] to insure that the topology is always preserved. Experimental results similar to ours are presented in [305] and [306].

The related work of Rochery, Jermyn and Zerubia [269], [270], [271], although devoted to a different problem, uses a similar idea to prevent pieces of the same curve from coming into contact to merge or break. The goal is to track thin long objects that evolve, with applications to the automatic extraction of road networks in remote sensing images. In [269], [270], [271], the authors have proposed interesting nonlocal regularization terms on a curve C parameterized by $p \in [0,1]$. In [269], these are of the form $E(C) = -\int_0^1\int_0^1 \vec{t}(p)\cdot\vec{t}(p')\Psi(\|C(p) - C(p')\|)dpdp'$, where $\vec{t}(p)$, $\vec{t}(p')$ are the tangent vectors to the curve at points $C(p), C(p')$ and $\|C(p) - C(p')\|$ is again the Euclidean distance between the curve points $C(p)$ and $C(p')$. The function Ψ is chosen to be $\Psi(l) = \sinh^{-1}(1/l) + l - \sqrt{1+l^2}$, thus decreasing on $[0,\infty[$. Other nonlocal forms are considered as well, and geometric motions of thin long objects are obtained. As in [208], the implicit representation by level sets is used for implementation in [269], [271]. In [270], the authors carry on their ideas, but this time in a phase-field approach for the representation of the motion. The work [269] is continued with more details in [271].

The Ph.D. thesis of Cecil [81] is also devoted to this problem, with satisfactory results based on the binary active contour model without edges presented in Chapter 8. The penalty is defined in a completely different way, but for more details we refer the reader to [81].

9.4.1 Description of self-repelling snakes model with topology constraint

Let Ω be a bounded open subset of $\mathbb{R}^n$, $\partial\Omega$ its boundary and let I be a given bounded image function defined by $I : \bar{\Omega} \to \mathbb{R}$. For the purpose of illustration, we consider $n = 2$.

The segmentation method introduced in [208] is based on the geodesic active contour model and includes a topological constraint based on a geometrical observation. In the following, we assume that Φ is a signed-distance function to $\mathcal{C}$

$$\Phi(x) = \begin{cases} dist(x, \mathcal{C}) & \text{if } x \in \Omega - \bar{\omega} \\ -dist(x, \mathcal{C}) & \text{if } x \in \omega \ . \end{cases} \tag{9.20}$$

In that case, Φ satisfies $|\nabla\Phi| = 1$. The gradient $\nabla\Phi$ is perpendicular to the isocontours of Φ. It points in the direction of increasing Φ, so the unit outward normal vector to the zero level line at point x is merely $\nabla\Phi(x)$ (more generally, the unit outward normal vector to the k level line at point x is $\nabla\Phi(x)$). Also, l denoting a real number, the set $\{x \in \Omega \mid \Phi(x) = l\}$ is obtained by shifting the points from the zero level line $\mathcal{C}$ by a quantity l in the outward normal to $\mathcal{C}$, that is

$$\{x \in \Omega \mid \Phi(x) = l\} = \{x + l\nabla\Phi(x), x \in \mathcal{C}\}.$$

Let us now consider two points $(x, y) \in \Omega \times \Omega$ belonging to the zero level line of Φ, close enough to each other, and let $\nabla\Phi(x)$ and $\nabla\Phi(y)$ be the unit outward normal vectors to the contour at these points. As shown in Figure 9.6, when the contour is about to merge, split or have a contact point (that is when the topology of the evolving contour is to change), then $\langle\nabla\Phi(x), \nabla\Phi(y)\rangle \simeq -1$, $\langle\cdot,\cdot\rangle$ denoting the Euclidean scalar product in $\mathbb{R}^2$.

FIGURE 9.6: Geometric characterization of points in a zone where the curve is to split, merge, or have a contact point.

Instead of working with only the points of the zero level line, this approach considers the points contained in a narrow band around the zero level line, more precisely, on the set of points $\{x \in \Omega \mid -l \leq \Phi(x) \leq l\}$, l being a level parameter. It is proposed to minimize the following functional

$$F(\Phi) + \mu E(\Phi), \tag{9.21}$$

F being the functional of the geodesic active contour model (9.15) and E the functional defined by

$$E(\Phi) \tag{9.22}$$
$$= -\int_\Omega\int_\Omega \Big[e^{-\frac{\|x-y\|_2^2}{d^2}} \langle\nabla\Phi(x),\nabla\Phi(y)\rangle H(\Phi(x)+l)H(l-\Phi(x))$$
$$H(\Phi(y)+l)H(l-\Phi(y))\Big]dxdy,$$
$$=\int_\Omega\int_\Omega \Big[G(\|x-y\|_2^2)g(\nabla\Phi(x),\nabla\Phi(y))\cdot h(\Phi(x))h(\Phi(y))\Big]dxdy.$$

The potential G such that $G(\|x-y\|_2^2) = exp(-\frac{\|x-y\|_2^2}{d^2})$ measures the nearness of the two points x and y. Here H denotes the one-dimensional Heaviside function and therefore $h(\Phi(x)) = H(\Phi(x)+l)H(l-\Phi(x)) = \chi_{\{z\in\Omega,\ -l\leq\Phi(z)\leq l\}}(x)$. In the calculations, the Heaviside function is not differentiable and is replaced by a smooth approximation denoted by H_ϵ. At last, setting $g(\nabla\Phi(x),\nabla\Phi(y)) = -\langle\nabla\Phi(x),\nabla\Phi(y)\rangle$, if the unit outward normal vectors to the level lines passing through x and y have opposite directions, the functional is not minimal. In the following, we will use the notations $g = g(z_1,z_2,z_3,z_4)$ and we will denote by $\nabla_{(z_1,z_2)}g = (g_{z_1}, g_{z_2})^T$ and by $\nabla_{(z_3,z_4)}g = (g_{z_3}, g_{z_4})^T$. Here $\mu > 0$ is a tuning parameter (note that we use the same notation $g(|\nabla I|)$ for the edge function in (9.15), and also for the function $g(z_1,z_2,z_3,z_4)$ in (9.22), but we think that there will be no confusion).

9.4.2 Evolution problem

We derive here the Euler–Lagrange equation associated with the minimization problem $\min_\Phi F(\Phi)+\mu E(\Phi)$, $\mu>0$. It can be formally obtained as follows. Let us set $\Upsilon(\nu) = f(\nu)+\mu e(\nu)$, with $f(\nu) = F(\Phi+\nu\Psi)$ and $e(\nu) = E(\Phi+\nu\Psi)$, ν being a small parameter and Ψ a function like Φ. The minimizer Φ will satisfy $\Upsilon'(0) = 0 = f'(0)+\mu e'(0)$ for all functions Ψ. One has

$$e(\nu) = \int_\Omega\int_\Omega\Big[G(\|x-y\|_2^2)g(\nabla\Phi(x)+\nu\nabla\Psi(x),\nabla\Phi(y)+\nu\nabla\Psi(y))$$
$$h(\Phi(x)+\nu\Psi(x))h(\Phi(y)+\nu\Psi(y))\Big]dxdy,$$

and a simple calculus gives

$$
\begin{aligned}
e'(\nu) = \int_\Omega\int_\Omega \Big[& G(||x-y||_2^2) \\
& \cdot g(\nabla\Phi(x) + \nu\nabla\Psi(x), \nabla\Phi(y) + \nu\nabla\Psi(y)) \\
& \cdot \Big(h'(\Phi(x) + \nu\Psi(x)) h(\Phi(y) + \nu\Psi(y))\Psi(x) \\
& + h(\Phi(x) + \nu\Psi(x)) h'(\Phi(y) + \nu\Psi(y))\Psi(y)\Big)\Big] dxdy \\
& + \int_\Omega\int_\Omega \Big[G(||x-y||_2^2) h(\Phi(x) + \nu\Psi(x)) \\
& \cdot h(\Phi(y) + \nu\Psi(y)) \Big(\langle \nabla_{(z_1,z_2)} g(\nabla\Phi(x) \\
& + \nu\nabla\Psi(x), \nabla\Phi(y) + \nu\nabla\Psi(y)), \nabla\Psi(x)\rangle \\
& + \langle \nabla_{(z_3,z_4)} g(\nabla\Phi(x) + \nu\nabla\Psi(x), \nabla\Phi(y) \\
& + \nu\nabla\Psi(y)), \nabla\Psi(y)\rangle \Big)\Big] dxdy.
\end{aligned}
$$

Then

$$
\begin{aligned}
e'(0) = \int_\Omega\int_\Omega \Big[& G(||x-y||_2^2) g(\nabla\Phi(x), \nabla\Phi(y)) \\
& \Big(h'(\Phi(x)) h(\Phi(y))\Psi(x) \\
& + h(\Phi(x)) h'(\Phi(y))\Psi(y)\Big)\Big] dxdy \\
& + \int_\Omega\int_\Omega \Big[G(||x-y||_2^2) h(\Phi(x)) h(\Phi(y)) \\
& \Big(\langle \nabla_{(z_1,z_2)} g(\nabla\Phi(x), \nabla\Phi(y)), \nabla\Psi(x)\rangle \\
& + \langle \nabla_{(z_3,z_4)} g(\nabla\Phi(x), \nabla\Phi(y)), \nabla\Psi(y)\rangle \Big)\Big] dxdy.
\end{aligned}
$$

We can switch x and y in the first integral and we get as first component

$$
\begin{aligned}
& \int_\Omega\int_\Omega \Big[G(||x-y||_2^2) g(\nabla\Phi(x), \nabla\Phi(y)) \cdot h'(\Phi(x)) h(\Phi(y))\Psi(x)\Big] dxdy \\
& + \int_\Omega\int_\Omega \Big[G(||x-y||_2^2) g(\nabla\Phi(y), \nabla\Phi(x)) \cdot h'(\Phi(x)) h(\Phi(y))\Psi(x)\Big] dydx.
\end{aligned}
$$

Assuming that we can interchange the order of integration, we get as first component

$$
\begin{aligned}
& \int_\Omega \Big\{ \int_\Omega \Big[G(||x-y||_2^2) g(\nabla\Phi(x), \nabla\Phi(y)) \cdot h'(\Phi(x)) h(\Phi(y))\Big] dy \Big\} \Psi(x) dx \\
& + \int_\Omega \Big\{ \int_\Omega \Big[G(||x-y||_2^2) g(\nabla\Phi(y), \nabla\Phi(x)) \cdot h'(\Phi(x)) h(\Phi(y))\Big] dy \Big\} \Psi(x) dx.
\end{aligned}
$$

For the second component the same reasoning is applied. We switch x and y in the second part of the component. Assuming once again that we can interchange the order of integration, we have

$$\int_\Omega \left\{ \int_\Omega \Big[G(||x-y||_2^2) h(\Phi(x)) h(\Phi(y)) \right.$$
$$\left. \cdot \langle \nabla_{(z_1,z_2)} g(\nabla\Phi(x), \nabla\Phi(y)), \nabla\Psi(x) \rangle \Big] dx \right\} dy$$
$$+ \int_\Omega \left\{ \int_\Omega \Big[G(||x-y||_2^2) h(\Phi(x)) h(\Phi(y)) \right.$$
$$\left. \cdot \langle \nabla_{(z_3,z_4)} g(\nabla\Phi(y), \nabla\Phi(x)), \nabla\Psi(x) \rangle \Big] dx \right\} dy.$$

Integrating by parts with respect to x, interchanging the order of integration and setting the necessary boundary conditions to zero, we obtain

$$e'(0) = \int_\Omega \left\{ \int_\Omega G(||x-y||_2^2) g(\nabla\Phi(x), \nabla\Phi(y)) \cdot h'(\Phi(x)) h(\Phi(y)) dy \right\} \Psi(x) dx$$
$$+ \int_\Omega \left\{ \int_\Omega G(||x-y||_2^2) g(\nabla\Phi(y), \nabla\Phi(x)) \cdot h'(\Phi(x)) h(\Phi(y)) dy \right\} \Psi(x) dx$$
$$- \int_\Omega \left\{ \int_\Omega \mathrm{div} \Big(G(||x-y||_2^2) h(\Phi(x)) h(\Phi(y)) \right.$$
$$\left. \cdot \nabla_{(z_1,z_2)} g(\nabla\Phi(x), \nabla\Phi(y)) \Big) dy \right\} \Psi(x) dx$$
$$- \int_\Omega \left\{ \int_\Omega \mathrm{div} \Big(G(||x-y||_2^2) h(\Phi(x)) h(\Phi(y)) \right.$$
$$\left. \cdot \nabla_{(z_3,z_4)} g(\nabla\Phi(y), \nabla\Phi(x)) \Big) dy \right\} \Psi(x) dx,$$

for all Ψ.

Using (9.16), we deduce that the obtained Euler–Lagrange equation formally is

$$- \delta_\epsilon(\Phi) \mathrm{div}\Big(g(|\nabla I|) \frac{\nabla\Phi}{|\nabla\Phi|} \Big) + \mu \left\{ \int_\Omega \Big[G(||x-y||_2^2) \Big(g(\nabla\Phi(x), \nabla\Phi(y)) \right.$$
$$+ g(\nabla\Phi(y), \nabla\Phi(x)) \Big) h'(\Phi(x)) h(\Phi(y)) \Big] dy - \int_\Omega \Big[\mathrm{div} \Big(G(||x-y||_2^2) h(\Phi(x))$$
$$h(\Phi(y)) \cdot \nabla_{(z_1,z_2)} g(\nabla\Phi(x), \nabla\Phi(y)) \Big) \Big] dy - \int_\Omega \Big[\mathrm{div} \Big(G(||x-y||_2^2) h(\Phi(x))$$
$$\left. h(\Phi(y)) \cdot \nabla_{(z_3,z_4)} g(\nabla\Phi(y), \nabla\Phi(x)) \Big) \Big] dy \right\} = 0,$$

with associated boundary conditions.

After some intermediate computations, we eventually obtain

$$2H_\epsilon(\Phi(x)+l)H_\epsilon(l-\Phi(x))\int_\Omega\Big[-2\frac{\mu}{d^2}\cdot\Big((x_1-y_1)\frac{\partial\Phi}{\partial y_1}(y)+(x_2-y_2)\frac{\partial\Phi}{\partial y_2}(y)\Big)$$
$$\cdot\, e^{-\frac{||x-y||_2^2}{d^2}}H_\epsilon(\Phi(y)+l)H_\epsilon(l-\Phi(y))\Big]dy-\delta_\epsilon(\Phi)\mathrm{div}\Big(g(|\nabla I|)\frac{\nabla\Phi}{|\nabla\Phi|}\Big)=0.$$

Therefore, parameterizing the gradient descent direction by an artificial time $t\geq 0$, the Euler–Lagrange equation in $\Phi(x,t)$ with $\Phi(x,0)=\Phi_0(x)$ (signed-distance function to the initial contour) and $x=(x_1,x_2)\in\Omega$, $y=(y_1,y_2)\in\Omega$ is:

$$\begin{aligned}\frac{\partial\Phi}{\partial t}=&\delta_\epsilon(\Phi)\mathrm{div}\big(g(|\nabla I|)\frac{\nabla\Phi}{|\nabla\Phi|}\big)+4\frac{\mu}{d^2}H_\epsilon(\Phi(x)+l)H_\epsilon(l-\Phi(x))\\ &\cdot\int_\Omega\Big[\Big((x_1-y_1)\frac{\partial\Phi}{\partial y_1}(y)+(x_2-y_2)\frac{\partial\Phi}{\partial y_2}(y)\Big)\\ &\cdot\, e^{-\frac{||x-y||_2^2}{d^2}}H_\epsilon(\Phi(y)+l)H_\epsilon(l-\Phi(y))\Big]dy.\end{aligned}$$

We replace $\delta_\epsilon(\Phi)$ by $|\nabla\Phi|$ (rescaling stage, see Zhao et al. [353] and Alvarez et al. [15]) and consider Neumann homogeneous boundary conditions: $\frac{\partial\Phi}{\partial\vec{\nu}}=0$ on $\partial\Omega$ with $\vec{\nu}$ denoting the unit outward normal to the boundary of Ω. The speed of convergence can be increased by adding the component $kg(|\nabla I|)|\nabla\Phi|$ in the evolution equation, k being a constant. Thus we consider the following evolution problem

$$\begin{cases}\dfrac{\partial\Phi}{\partial t}=|\nabla\Phi|\Big[\mathrm{div}\big(g(|\nabla I|)\dfrac{\nabla\Phi}{|\nabla\Phi|}\big)+kg(|\nabla I|)\Big]\\ +4\dfrac{\mu}{d^2}H_\epsilon(\Phi(x)+l)H_\epsilon(l-\Phi(x))\\ \cdot\displaystyle\int_\Omega\Big[\Big((x_1-y_1)\frac{\partial\Phi}{\partial y_1}(y)+(x_2-y_2)\frac{\partial\Phi}{\partial y_2}(y)\Big)\\ \cdot e^{-\frac{||x-y||_2^2}{d^2}}H_\epsilon(\Phi(y)+l)H_\epsilon(l-\Phi(y))\Big]dy,\\ \Phi(x,0)=\Phi_0(x),\\ \dfrac{\partial\Phi}{\partial\vec{\nu}}=0,\text{ on }\partial\Omega.\end{cases}$$

9.4.2.1 Role of self-reppeling term $E(\Phi)$

Let us consider the zero level line $\mathcal{C}$. After calculation, $\mathcal{C}$ moves according to

$$\begin{aligned}\frac{\partial\mathcal{C}}{\partial t}=&g(|\nabla I|)\kappa\vec{n}-\langle\nabla g(|\nabla I|),\vec{n}\rangle\vec{n}+kg(|\nabla I|)\vec{n}\\ &+\frac{\mu'}{|\nabla\Phi|}\Big[\int_\Omega\Big(\langle x-y,\nabla\Phi(y)\rangle e^{-\frac{||x-y||_2^2}{d^2}}\\ &\cdot H_\epsilon(\Phi(y)+l)H_\epsilon(l-\Phi(y))\Big)dy\Big]\vec{n},\end{aligned}$$

$\vec{n}$ being the unit inward normal vector, $x = \mathcal{C}(t, q)$, κ the curvature and $\mu' > 0$ a constant. Consider now two points x and y close enough to each other, x belonging to $\mathcal{C}$ and y belonging to the narrow band of width $2l$. In Figure 9.7, the contour $\mathcal{C}$ is about to merge or to have a contact point. If we compute $\langle x - y, \nabla\Phi(y)\rangle$, we get a positive quantity.

If we now take two points x and y close enough to each other that both belong to $\mathcal{C}$ (Figure 9.8), the contribution $\langle x - y, \nabla\Phi(y)\rangle$ is almost zero. Lastly, in the case of Figure 9.9, $\langle x - y_1, \nabla\Phi(y_1)\rangle$ is positive while $\langle x - y_2, \nabla\Phi(y_2)\rangle$ is negative. Globally, the contribution is zero in this part of the narrow band. Thus in critical cases (when the curve is about to change topology), we can expect to have a positive quantity of the component $\int_\Omega \langle x - y, \nabla\Phi(y)\rangle e^{-\frac{||x-y||_2^2}{d^2}} H_\epsilon(\Phi(y) + l)H_\epsilon(l - \Phi(y))dy$, and then the following term $\left[\int_\Omega \langle x - y, \nabla\Phi(y)\rangle e^{-\frac{||x-y||_2^2}{d^2}} H_\epsilon(\Phi(y) + l)H_\epsilon(l - \Phi(y))dy\right] \vec{n}$ can be interpreted as a repelling force, which ensures topology preservation.

9.4.2.2 Short time existence and uniqueness

The obtained evolution problem is hard to handle from a theoretical view. A suitable setting would be the viscosity solution theory (owing to the nonlinearity induced by the modified mean curvature term), but the dependency of the nonlocal term on the gradient $\nabla\Phi(y)$ and the failure to fulfill the monotony property in Φ make it difficult. For this reason, Forcadel and Le Guyader ([148]) considered a slightly modified problem: they assume that the topological constraint is only applied to the zero level line. Assuming that Φ is a signed distance function, the topological constraint E is then rephrased as

$$E(\Phi) = -\int_\Omega\int_\Omega [\exp\left(-\frac{||x-y||_2^2}{d^2}\right)\langle\nabla\Phi(x), \nabla\Phi(y)\rangle\delta(\Phi(x))\delta(\Phi(y))dx\,dy,$$

with δ the Dirac measure.

Computing the Euler–Lagrange equation, then applying a gradient descent method and doing an integration by parts and a rescaling by replacing $\delta(\Phi)$ by $|\nabla\Phi|$, yields the evolution problem (defined on $\mathbb{R}^2$ for the space coordinates for the sake of simplicity, and given $T > 0$):

$$\begin{aligned} &\frac{\partial\Phi}{\partial t} = |\nabla\Phi|\left\{\operatorname{div}\left(g(|\nabla I|)\frac{\nabla\Phi}{|\nabla\Phi|}\right) + c_0 * [\Phi(\cdot, t)]\right\} \text{ in } \mathbb{R}^2 \times (0, T), \\ &\Phi(x, 0) = \Phi_0(x) \text{ in } \mathbb{R}^2, \end{aligned} \tag{9.23}$$

with $[\Phi(\cdot, t)]$ the characteristic function of the set $\{\Phi(\cdot, t) > 0\}$ and

$$c_0 : \left\{ \begin{array}{l} \mathbb{R}^2 \to \mathbb{R} \\ x \mapsto \frac{4\mu}{d^2}\left(2 - \frac{2}{d^2}||x||_2^2\right)\exp\left(-\frac{||x||_2^2}{d^2}\right) \end{array} \right. .$$

The authors of [148] assume that

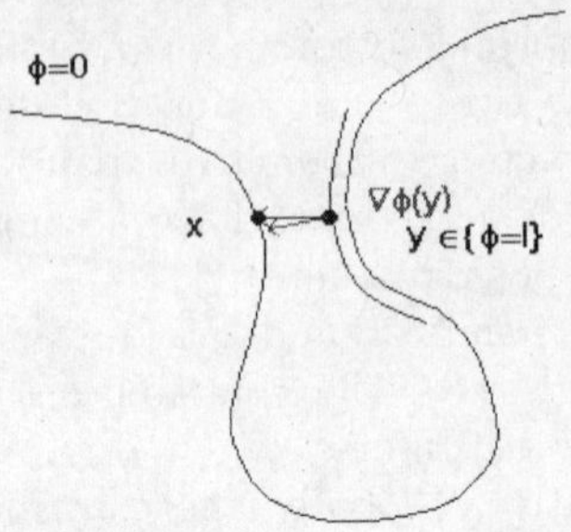

FIGURE 9.7: Case where x belongs to $\mathcal{C}$ and y belongs to the l level line, x and y close enough.

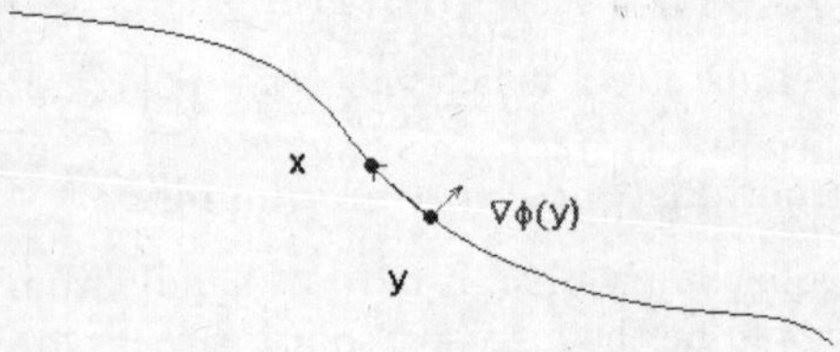

FIGURE 9.8: Case where the two points are close enough and both belong to $\mathcal{C}$.

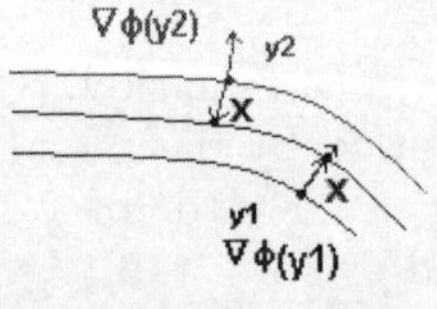

FIGURE 9.9: Case where the two points are close enough, x belonging to $\mathcal{C}$, and y belonging to the l level line or $-l$ level line.

(1) Φ_0 is such that $\nabla\Phi_0 \in W^{1,\infty}(\mathbb{R}^2)$ (we denote by B_0 its Lipschitz constant). In particular, Φ_0 is $L^\infty_{loc}(\mathbb{R}^2)$.

(2) $\exists\zeta > 0$, $\forall x \in \mathbb{R}^2$, $\zeta < g(x) \leq 1$.

(3) $g, g^{\frac{1}{2}}$ and ∇g are bounded and Lipschitz continuous on $\mathbb{R}^2$ with Lipschitz

constant κ_g, $\kappa_{g^{\frac{1}{2}}}$ and $\kappa_{\nabla g}$ respectively. For simplicity of notation, we set $L_g = \max\left(\kappa_g, \kappa_{\nabla g}, \kappa_{g^{\frac{1}{2}}}\right)$.

Equipped with these hypotheses, Forcadel and Le Guyader prove a short time existence and uniqueness result for this equation, stated as follows.

Theorem 33 *(short time existence and uniqueness)*
Assume that (1)-(3) hold. Let $\Phi_0 : \mathbb{R}^2 \to \mathbb{R}$ be such that $\nabla\Phi_0 \in W^{1,\infty}(\mathbb{R}^2)$ and

$$|\nabla\Phi_0| < B_0 \ \text{in } \mathbb{R}^2 \ \text{and } \frac{\partial\Phi_0}{\partial x_2} > b_0 > 0 \ \text{in } \mathbb{R}^2.$$

Then there exists $T^* > 0$, depending only on b_0, B_0, c_0 and g such that there exists a unique viscosity solution of the considered problem in $\mathbb{R}^2 \times [0, T^*)$. Moreover, the solution is Lipschitz continuous in space and time.

The strategy of the proof is the same as the one applied in [16] and [147], i.e., based on the use of a fixed point method by freezing the nonlocal term. More precisely, the authors apply a fixed point method on a functional space E whose definition lies, in particular, in estimations on the gradients. The key point is thus to get estimates on the Lipschitz constant in space and time of the solution as well as a bound from below on the gradient in space. The main difficulties come from the fact that the mean curvature term is balanced by a function of the space variable x and so, to obtain the estimate from below on the gradient, the mean curvature term has to be bounded. This is done using the Lipschitz regularity of the solution. For more details, we refer the reader to [148].

9.4.3 Description of numerical algorithm

For the discretization stage, we use an additive operator splitting scheme (AOS), and we follow Weickert et al. (see [341] for more details), which allows for the decomposition of the initial two-dimensional problem into two one-dimensional subproblems. This splitting-based scheme is easy to implement and requires a linear computational cost at each step.

To simplify the notation, we use a vectorial representation of the function Φ *via* a concatenation of the rows (respectively, columns) of the associated matrix. Thus $\Phi \in \mathbb{R}^{N\times M}$, where N is the number of lines and M the number of columns. The center of gravity of a pixel i coincides with the node of the mesh of coordinates x_i. Let us denote by h the space step (assumed to be equal to 1) and τ the time step. Let Φ_i^n be an approximation of $\Phi(x_i, t_n) = \Phi(x_i, n\tau)$.

The considered discretization of the problem is as follows

$$\Phi^{n+1} = \frac{1}{2} \sum_{l \in \{x,y\}} (Id - 2\tau A_l(\Phi^n))^{-1} \Big[\Phi^n$$
$$+ 4\frac{\mu\tau}{d^2} H_\epsilon(\Phi^n + l) H_\epsilon(l - \Phi^n)$$
$$\cdot \int_\Omega \Big(\langle x - y, \nabla\Phi^n(y) \rangle e^{-\frac{||x-y||_2^2}{d^2}} H_\epsilon(\Phi^n(y) + l)$$
$$H_\epsilon(l - \Phi^n(y)) \Big) dy + \tau k g(|\nabla I|) |\nabla \Phi^n| \Big],$$

with the entries of $A_l(\Phi^n)$ defined by

$$a_{ij_l} = \begin{cases} |\nabla \Phi_i^n| \dfrac{2}{\left(\dfrac{|\nabla\Phi|}{g(|\nabla I|)}\right)_i^n + \left(\dfrac{|\nabla\Phi|}{g(|\nabla I|)}\right)_j^n} \\ \text{if } j \in \bigwedge_l(i), \\ -|\nabla \Phi_i^n| \displaystyle\sum_{m \in \bigwedge_l(i)} \dfrac{2}{\left(\dfrac{|\nabla\Phi|}{g(|\nabla I|)}\right)_i^n + \left(\dfrac{|\nabla\Phi|}{g(|\nabla I|)}\right)_m^n} \\ \text{if } i = j, \\ 0 \text{ otherwise,} \end{cases}$$

and where $\bigwedge_l(i)$ represents the neighboring pixels of i with respect to direction l.

The AOS scheme can be rewritten by means of the matrices $B_l(\Phi^n) = Id - 2\tau A_l(\Phi^n)$, $l \in \{x, y\}$. We obtain two disconnected linear systems that are both tridiagonal if one reorders the nodes of the mesh in a different way for the second system. The matrices B_l are tridiagonal, strictly diagonally dominant (which ensures uniqueness of the solution of both linear systems thanks to Gershgorin's theorem for instance) and lead to systems that can be solved computing Thomas algorithm with linear complexity.

The study of the stability of the proposed scheme requires some preliminary results. We first recall the definition of a monotone matrix as well as an equivalent characterization.

Definition 21 *A real matrix $A \in \mathbb{R}^{N\times N}$ is said to be monotone if it is invertible and if its inverse entries are all non-negative.*

One has the following characterization of monotone matrices.

Lemma 3 *A real matrix $A \in \mathbb{R}^{N\times N}$ is monotone if and only if, for all vectors $x \in \mathbb{R}^N$, one has*

$$(Ax \geq 0 \;\Rightarrow\; x \geq 0),$$

$x \geq 0$ meaning that all entries of x are non-negative.

Proposition 6 *The matrices $B_l = B_l(\Phi^n)$, $l \in \{x, y\}$ involved in the linear sub-systems to be solved are monotone.*

Proof. For the sake of simplicity, we assume that $B_l \in \mathbb{R}^{n\times n}$. Let $v \in \mathbb{R}^n$ be such that $B_l v \geq 0$. Let us denote by i_0 an index such that $v_{i_0} = \min_{i\in\{1,\cdots,n\}} v_i$. We treat the general case $i_0 \in \{2, \cdots, n-1\}$, the reasoning being similar for $i_0 = 1$ or $i_0 = n$). Let us express that the component of index i_0 of $B_l v$, denoted by $(B_l v)_{i_0}$, is non-negative. After intermediate computations,

$$\frac{4\tau|\nabla\Phi^n_{i_0}|}{\left(\dfrac{|\nabla\Phi|}{g(|\nabla I|)}\right)^n_{i_0} + \left(\dfrac{|\nabla\Phi|}{g(|\nabla I|)}\right)^n_{i_0-1}} (v_{i_0} - v_{i_0-1})$$
$$+\frac{4\tau|\nabla\Phi^n_{i_0}|}{\left(\dfrac{|\nabla\Phi|}{g(|\nabla I|)}\right)^n_{i_0} + \left(\dfrac{|\nabla\Phi|}{g(|\nabla I|)}\right)^n_{i_0+1}} (v_{i_0} - v_{i_0+1}) + v_{i_0} \geq 0.$$

Due to the definition of v_{i_0}, necessarily $v_{i_0} \geq 0$, and the conclusion is straightforward. □

The coefficients of matrices $B_l^{-1}(\Phi^n)$ are thus non-negative and the sum of the elements of each row of $B_l^{-1}(\Phi^n)$ is equal to one. Thus if we neglect the topological constraints and the component $kg(|\nabla I|)|\nabla\Phi|$, it implies that each element of Φ^{n+1} is computed from a convex combination of components of Φ^n. The discrete minimum-maximum principle applies and stability is therefore guaranteed for any time step.

Regarding the constant motion $kg(|\nabla I|)|\nabla\Phi|$, the function $g(|\nabla I|)$ controls this term and annihilates the action of the force when the curve is approaching edge points. A classical discretization of the gradient can lead to the creation of loops. The entropy condition ([285],[286]) prevents the curve from propagating where it has already been and thus avoids corner formation. The resulting discretization of the gradient comes from the hyperbolic conservation laws and can be found in [285] and [286]:

$$(|\nabla\Phi|^n_i)^2 \simeq$$

$$\Big[max(D^{-x}\Phi^n_i, 0)^2 + min(D^{+x}\Phi^n_i, 0)^2$$
$$+max(D^{-y}\Phi^n_i, 0)^2 + min(D^{+y}\Phi^n_i, 0)^2\Big]$$
$$\text{if } k < 0,$$

$$\Big[max(D^{+x}\Phi^n_i, 0)^2 + min(D^{-x}\Phi^n_i, 0)^2$$
$$+max(D^{+y}\Phi^n_i, 0)^2 + min(D^{-y}\Phi^n_i, 0)^2\Big]$$
$$\text{if } k > 0,$$

using the classical finite difference operators $D^{+x}\Phi_i = \Phi_{i+1} - \Phi_i$ and $D^{-x}\Phi_i = \Phi_i - \Phi_{i-1}$ in one dimension.

As for the topological constraint, we have used the $\mathcal{C}^\infty$ regularization of the Heaviside function defined by

$$H_\epsilon(z) = \frac{1}{2}(1 + \frac{2}{\pi}\arctan(\frac{z}{\epsilon})).$$

The partial derivatives are approximated using central finite difference schemes. While the topological constraint introduced in the modeling enables us to use the level set framework to track interfaces whose topology must be preserved, the main drawback is computational cost due to the nonlocal term appearing in the evolution equation. This limitation can be overcome by slightly simplifying the issue: instead of computing the nonlocal term over the whole grid, we compute it over a square centered on the current point. The window size depends on the level parameter l. This simplification looks realistic since we are mainly interested in the behavior of Φ in a neighborhood of the current point.

Lastly, we have assumed that Φ is a signed distance function. Thus we need to periodically apply a reinitialization process. We have used the scheme obtained by Russo and Smereka in [278], defined by

$$\Phi_{i,j}^{n+1} =$$

$$\begin{cases} \Phi_{i,j}^n - \dfrac{\Delta t}{\Delta x}\big(sgn(\Phi_{i,j}^0)|\Phi_{i,j}^n| - D_{i,j}\big) \\ \text{if } \Phi_{i,j}^0\Phi_{i-1,j}^0 < 0 \text{ or } \Phi_{i,j}^0\Phi_{i+1,j}^0 < 0 \\ \text{ or } \Phi_{i,j}^0\Phi_{i,j-1}^0 < 0 \text{ or } \Phi_{i,j}^0\Phi_{i,j+1}^0 < 0, \\ \\ \Phi_{i,j}^n - \Delta t sgn(\Phi_{i,j}^0)G(\Phi)_{i,j} \text{ otherwise,} \end{cases}$$

with D_{ij}, the distance of node (i,j) from the interface and where

$$\begin{aligned} &G(\Phi)_{i,j} = \\ &\Big[max\Big(|max(D_x^-\Phi_{i,j},0)|^2, |min(D_x^+\Phi_{i,j},0)|^2\Big) \\ &+ max\Big(|max(D_y^-\Phi_{i,j},0)|^2, |min(D_y^+\Phi_{i,j},0)|^2\Big) \\ &- 1\Big]^{1/2} \text{ if } \Phi_{i,j}^0 > 0, \end{aligned}$$

and

$$\begin{aligned} &G(\Phi)_{i,j} \\ &= \Big[max\Big(|min(D_x^-\Phi_{i,j},0)|^2, |max(D_x^+\Phi_{i,j},0)|^2\Big) \\ &+ max\Big(|min(D_y^-\Phi_{i,j},0)|^2, |max(D_y^+\Phi_{i,j},0)|^2\Big) \\ &- 1\Big]^{1/2} \text{ if } \Phi_{i,j}^0 < 0. \end{aligned}$$

9.4.4 Experimental results

We conclude this section by presenting several numerical results using the self-reppeling snakes model and comparisons on both synthetic and real images in two and three dimensions. The experimental tests and images are similar to those from the related prior work by Han et al. [167], [168] and by Sundaramoorthi and Yezzi [305], [306] and can thus be considered comparisons. We can conclude that the self-reppeling snakes model qualitatively performs in a similar way, but by a different approach. Parameters are given together with a verification of the topologies of the final contours (surfaces). We also show comparisons with the geodesic active contour model from Section 9.2 (without topology constraint), and we will see that the use of the penalty term improves the accuracy of the segmentation. Many of the numerical results are applied to medical images; however, we do not use anatomical or physiological information. By the illustrated results, we only wish to demonstrate how the method performs in practice; for more accurate segmentations in specialized medical applications, the methods and the parameters must be modified and adapted to such applications. For any of the tests, the parameter μ can be simply set to zero in order not to constrain the topology.

We start with a synthetic image representing two discs in Figure 9.10 (similar to tests performed in related works by Han et al. [167], [168] and by Sundaramoorthi and Yezzi [305], [306]). We aim at segmenting these two discs while maintaining the same topology throughout the process, which means that we expect to get one path-connected component. The initialization is made with a signed distance function $\Phi^0 = \Phi_0$ that encloses the two balls and which is the cone defined by: $\Phi^0(x, y) = \sqrt{(x-48h)^2 + (y-50h)^2} - 26$. A comparison is made (Figure 9.11 shows the same results as Figure 9.4) with the results obtained in the case where no topological constraint is enforced.

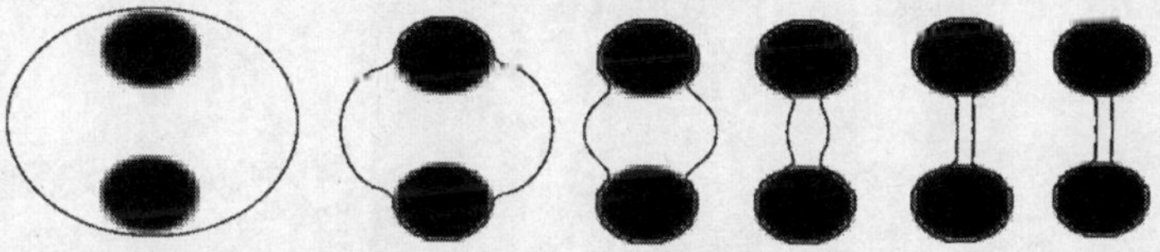

FIGURE 9.10: Segmentation of the synthetic image with two disks when topological constraints are applied. The parameters are: $h = 1$, $\tau = 0.5$, $l = 1$, size of the window: 5, $k = 0.3$, $\mu = 0.4$. Iterations 0, 50, 100, 150, 180, 210.

The second example in Figure 9.12 was taken from [167]. We can thus compare the described method with the result obtained in [167] on the same hand image. One can observe that for this test, the two middle fingers touch and with the use of classical geodesic active contours, the evolving contour is going to merge and a hole will then appear (see Figure 9.13, same results as in Figure 9.5), which is undesirable. With the described method, the repelling forces prevent the curve from merging, as in [167] and [168].

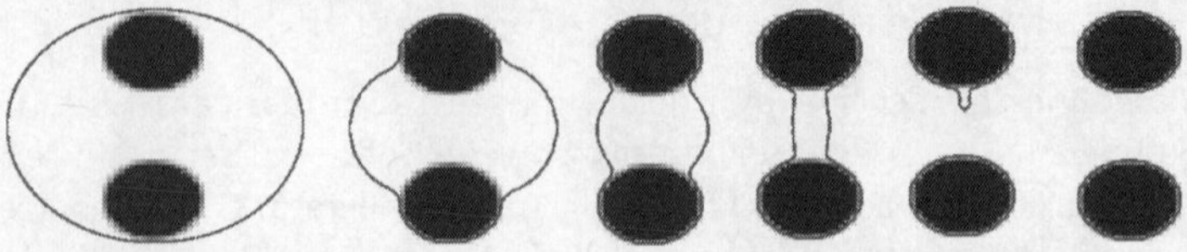

FIGURE 9.11: Segmentation of the synthetic image with two disks when no topological constraint is enforced with the same parameters as above. The contour has split into two components. Iterations 0, 50, 100, 140, 190, 210.

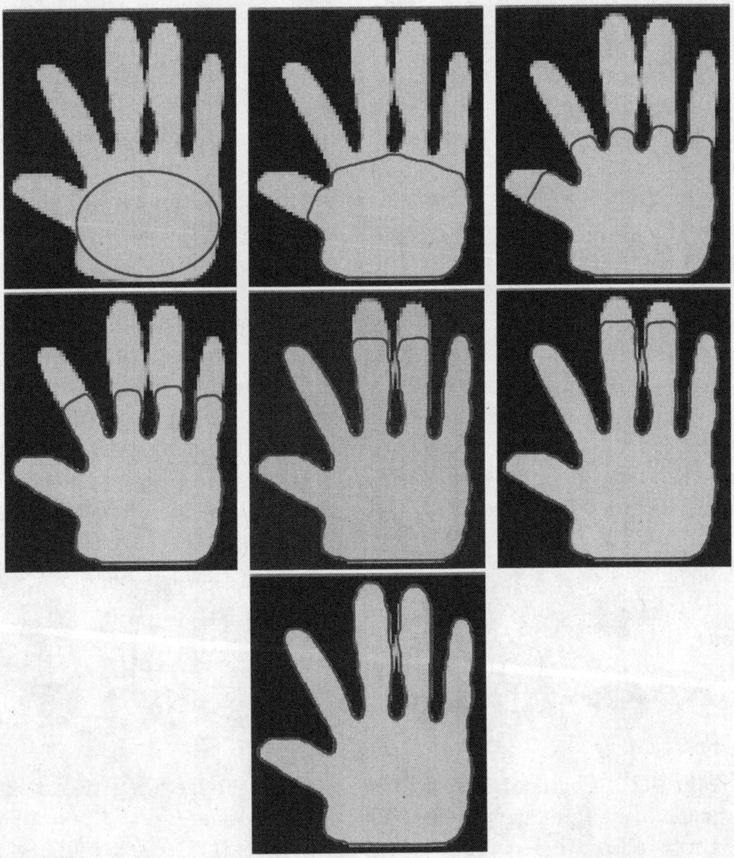

FIGURE 9.12: Segmentation of the hand image taken from [167] by the described method with the topology constraint. The result is obtained with the following parameters: $h = 1$, $\tau = 0.5$, $l = 1$, size of the window: 7, $k = -0.2$, $\mu = 0.2$. Iterations 0, 50, 150, 250, 450, 500, and 600.

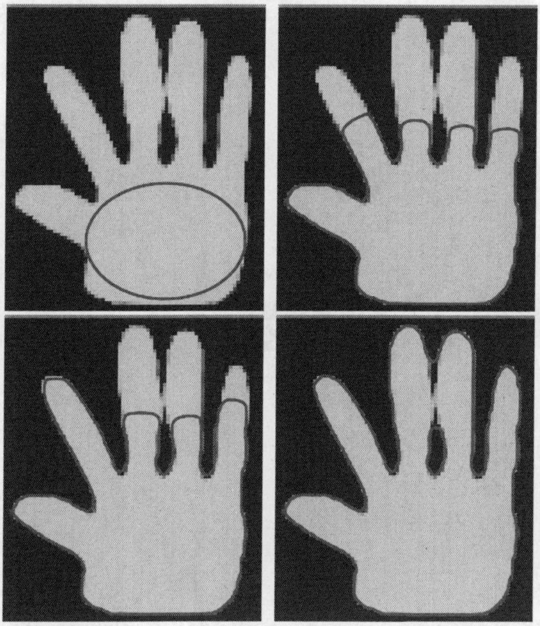

FIGURE 9.13: Steps of the process with the same parameters when no topological constraint is enforced.

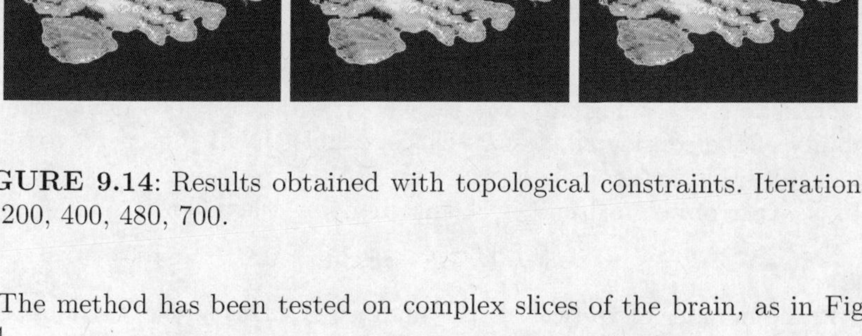

FIGURE 9.14: Results obtained with topological constraints. Iterations 0, 50, 200, 400, 480, 700.

The method has been tested on complex slices of the brain, as in Figure 9.14.

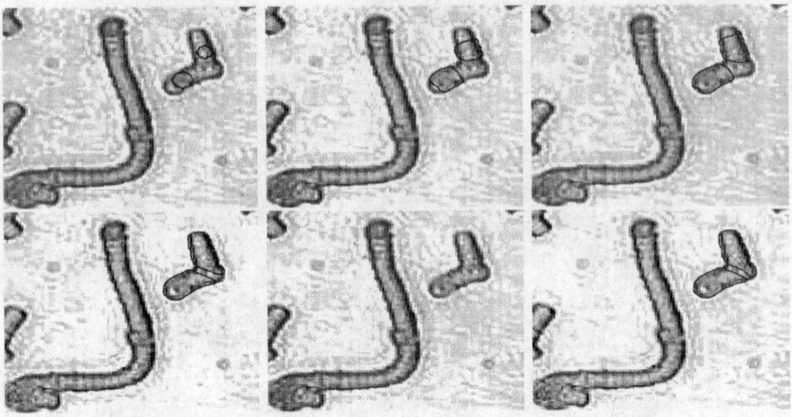

FIGURE 9.15: Segmentation of two blood cells that are very close to each other. Iterations 0, 50, 100, and 360.

Figure 9.15 illustrates the initial condition made of two disjoint closed curves. We expect to have both curves evolving without merging. This test is similar to tests presented in related work by Han et al. [167], [168] and by Sundaramoorthi and Yezzi [305], [306].

We conclude this section by presenting three-dimensional experiments in Figure 9.16. The proposed image consists of slices of the brain. We start with an initial condition such that the zero level set encloses the brain. Owing to data noise and quality impairment of the image when no topological constraint is enforced, contact points may appear in the gray matter, which is not consistent with the known spherical topology. When the constraint is included in the model, the details of the lobes are clearly extracted without contact points; there are no handles or anatomically incorrect structures. The spherical topology is kept through the process.

For the two-dimensional experiments, only by visual inspection can we see that the topology constraint has been correctly imposed. However, for the three-dimensional results, visual inspection only is not sufficient to verify that the correct desired topology has been achieved.

To insure algorithm efficiency, we computed for the three-dimensional example, the Euler characteristic χ from the obtained closed surface. The Euler characteristic is a topological invariant which, with the knowledge of the orientability of the considered closed surface, enables us to (uniquely) topologically classify this surface.

In the case of polyhedra, it is classically defined as follows:

$$\chi = V - E + F,$$

with V the number of vertices, E the number of edges and F the number of faces.

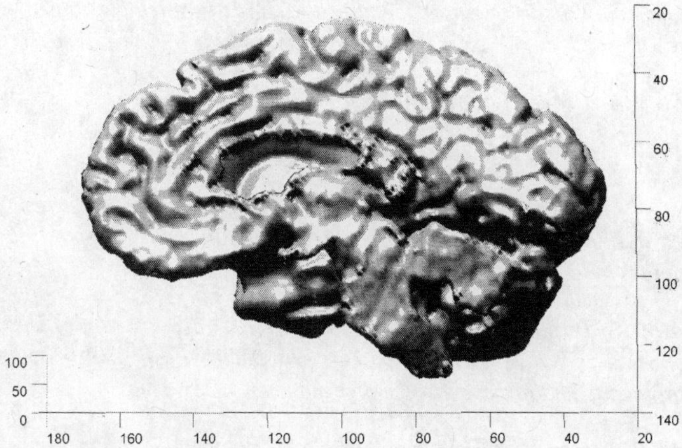

FIGURE 9.16: Top view of the final result.

It is known that for any polyhedron homeomorphic to a sphere, the Euler characteristic is equal to 2. Thus, as we aim at making surfaces topologically equivalent to a sphere, we can expect to get a Euler characteristic equal to 2 in our experiments. In the case of the brain (Figure 9.16), we have obtained $\begin{cases} V = 63626 \\ E = 190872 \\ F = 127248 \end{cases}$, that is $\chi = 2$.

9.5 Exercises

Exercise 9.1 *Prove the validity of the Euler–Lagrange equations (9.2).*

Exercise 9.2 *Using the notations from Section 9.1.1, derive the Euler–Lagrange equations given in (9.7) associated with the minimization of (9.6). Discretize the system of equations using finite differences and apply the obtained algorithm to a given image I, with an input data f defined from I as suggested in Section 9.1.1. Repeat the discretization and implementation when the Laplacian is substituted by the normalized infinity Laplacian.*

Exercise 9.3 *Prove that the geodesic energy*

$$J(C) = \int_a^b |C'(q)| g(|\nabla I(C(q))|) dq$$

is an intrinsic quantity to the curve. In other words, prove that if we make the change of variable $q = \psi(r)$, $\psi : [c,d] \mapsto [a,b]$ *differentiable and* $\psi' > 0$, *and let* $\bar{C}(r) = C(q) = C(\psi(r))$ *be a new parametrization of the curve,* $J(C) = J(\bar{C})$.

Exercise 9.4 *Prove that the flow given by*

$$\frac{\partial C}{\partial t} = g\kappa\vec{N} - (\nabla g \cdot \vec{N})\vec{N},$$

decreases the geodesic energy (9.8). Comment on each component of this evolution equation.

Exercise 9.5 *Implement the geodesic active contours model in the level set formulation given in (9.17). You can choose to keep* $\delta_\epsilon(\Phi)$ *or replace it by* $|\nabla\Phi|$ *(rescaling). Compare the two implementations.*

Exercise 9.6 *Consider a given image* $I : \Omega \to \mathbb{R}$ *in two dimensions and a vector field* $V(x) = (u(x), v(x))$. *Consider an arbitrary differentiable parametrization of the closed curve* C, $C(q)$, $0 \leq q \leq 1$. *Define a general alignment term (Subsection 9.3)*

$$E_A(C) = \int_0^1 \langle V, \vec{n} \rangle dq,$$

where $\vec{n} = \frac{\left(-\frac{\partial C_2}{\partial q}, \frac{\partial C_1}{\partial q}\right)}{|\frac{\partial C}{\partial q}|}$ *is the normal to the curve. Derive the time-dependent flow that most decreases* E_A. *What will this flow become if* $V = -\nabla I$ *?*

Exercise 9.7 *Prove the validity of (9.23) when applying the topological constraint only to the zero level line of* Φ.

Exercise 9.8 *Prove Lemma 3.*

Part III

Applications

Purpose of Part III

We have seen several applications peppered throughout the book but now we are going to examine those that require all or most of the techniques we have built so far.

Chapter 10

Nonlocal Mumford–Shah and Ambrosio–Tortorelli Variational Models for Color Image Restoration

Some of the restoration methods discussed in the previous chapters can be easily extended to the restoration of color images and other vector-valued data. However, there are several problems that have to be considered. We mention some of them. First, the notion of image edges identified using the norm of the gradient for scalar images has several different extensions in the vectorial case. The gradient is now a matrix and we can define different gradient norms $|\nabla \boldsymbol{u}|$ of the Jacobian matrix $\nabla \boldsymbol{u}$ when $\boldsymbol{u}$ is a vector-valued function; such gradient norms could define edges in different ways. Second, care has to be taken when processing color images; new colors should not be introduced in the restored image by the diffusion process.

We would like to refer the reader to three PhD thesis manuscripts on color image processing by Blomgren [53], by Kang [184] and by Tschumperlé [329] and to other work by Sapiro and Ringach [281], Blomgren and Chan [54], Deriche and Tschumperlé [331].

We present in this chapter several models for the restoration of color images and experimental results, obtained by combining techniques and ideas presented in Chapters 3, 4, 6, and 7.

Following Jung et al. [183, 180], we present nonlocal Mumford–Shah (MS) and Ambrosio–Tortorelli regularizers for solving several inverse problems in image restoration. We apply these nonlocal MS regularizers to color image denoising, color image deblurring in the presence of Gaussian or impulse noise, color image inpainting, color image super-resolution, and to color filter array demosaicing by incorporating proper fidelity terms.

Let us consider the degradation model of a color image

$$f = Ku + n, \quad (f^i = K^i u^i + n^i, \, i = R, G, B) \tag{10.1}$$

where K is a linear operator accounting for blur, sub-sampling, or missing pixels, and n is random noise. Given f and the degradation operator K, solving for u from (10.1) is a highly ill-posed problem. To overcome this difficulty, we follow again the idea from Tikhonov regularization (Chapter 2, Section 2.1). We thus formulate the restoration problem within the variational framework as

$$\inf_u \Big\{\Phi(f - Ku) + \Psi(u)\Big\},$$

where $u \mapsto \Phi(f - Ku)$ defines a data fidelity term (e.g., depending on the statistics of the noise n), while $u \mapsto \Psi(u)$ defines the regularization that enforces a prior constraint on u. The regularization term alleviates the ill-posedness of the inverse problem by reflecting some a priori properties. The properties that we wish to impose on u are that u is formed of homogeneous regions (smooth in intensity or made of uniform texture), and separated by sharp edges. To achieve these desirable properties on u, we are led to consider the Mumford and Shah regularizers presented in Chapter 6, together with their approximations presented in Chapter 7, and combined with the notion of nonlocal gradient for texture preservation (Chapter 4). This method is applied to the restoration of color images in the RGB mode.

We first extend the nonlocal gradient presented in Chapter 4 to color RGB images. Let Ω be the planar image domain (an open, bounded and connected subset of $\mathbb{R}^2$), and $u = (u^R, u^G, u^B) : \Omega \to \mathbb{R}^3$ be the unknown image to be restored. We define the nonlocal gradient norm $\|\nabla_w u\| : \Omega \to \mathbb{R}$ of the color image u by

$$\begin{aligned}\|\nabla_w u\|(x) \quad &:= \quad \sqrt{\sum_{i=R,G,B} |\nabla_w u^i|^2(x)} \\ &:= \quad \sqrt{\sum_{i=R,G,B} \int_\Omega (u^i(y) - u^i(x))^2 w(x,y)dy},\end{aligned}$$

with the weight function w at $(x, y) \in \Omega \times \Omega$ depending on an image $q : \Omega \to \mathbb{R}$ (to be specified later for each application) given by

$$w(x,y) = e^{\left(-\frac{d_a(q(x),q(y))}{h^2}\right)},$$

and the patch distance d_a given by

$$d_a(q(x), q(y)) \quad = \quad \int_{\mathbb{R}^2} G_a(t)\, |q(x+t) - q(y+t)\,|^2 dt,$$

where $q(x)$ and $q(y)$ are the pixel intensities at x and y, G_a is the Gaussian kernel with standard deviation a determining the patch size, and $h > 0$ is a parameter function of the noise level. To reduce the computational cost in practice, we use for a given pixel $x \in \Omega$ a search window $S(x) = \{y \in \Omega : |x - y| \leq r\}$, instead of Ω, to compute $w(x, y)$ for all $y \in S(x)$.

The nonlocal MSH^1 regularizer in the weak form can (at least formally) be expressed as

$$\Psi^{NL/MSH^1}(u) = \beta \int_{\Omega \setminus S_u} \|\nabla_w u\|^2(x)dx + \alpha \mathcal{H}^1(S_u),$$

where S_u is a closed subset of Ω playing the role of an edge set in the color space (roughly speaking, points x with high nonlocal gradient norm, but we do not have a clear definition of S_u in this case). Again, α and β are positive constants. Since this formulation may not be well defined or may not be easily minimized, we prefer to work with the Ambrosio–Tortorelli and Shah approximations from Chapter 7 to the Mumford–Shah-H^1 and Mumford–Shah-TV regularizers, to define the following nonlocal Mumford–Shah and Ambrosio–Tortorelli regularizers as $\epsilon \to 0$:

$$\Psi^{NL/MS}(u,v) = \beta \int_\Omega v^2 \phi(\|\nabla_w u\|^2) dx + \alpha \int_\Omega \left(\epsilon |\nabla v|^2 + \frac{(v-1)^2}{4\epsilon}\right) dx.$$

Here, $v : \Omega \to [0,1]$ is the edge function of the RGB image u. If we take $\phi(s) = s$, we obtain a nonlocal version of the Ambrosio–Tortorelli elliptic approximations,

$$\Psi^{NL/MSH^1}(u,v) = \beta \int_\Omega v^2 \|\nabla_w u\|^2 dx + \alpha \int_\Omega \left(\epsilon |\nabla v|^2 + \frac{(v-1)^2}{4\epsilon}\right) dx,$$

while taking $\phi(s) = \sqrt{s}$ gives a nonlocal version of the Shah approximation,

$$\Psi^{NL/MSTV}(u,v) = \beta \int_\Omega v^2 \|\nabla_w u\| dx + \alpha \int_\Omega \left(\epsilon |\nabla v|^2 + \frac{(v-1)^2}{4\epsilon}\right) dx.$$

10.1 Well-definedness of nonlocal regularizers

In this section we will focus on the well-definedness of the component

$$\int_\Omega v^2 \|\nabla_w u\|^2 \, dx = \int_\Omega v^2 \left(\sum_{i=R,G,B} \int_\Omega \left(u^i(y) - u^i(x)\right)^2 w(x,y) \, dy \right) dx$$

appearing in the definition of Ψ^{NL/MSH^1}.

Let us assume that w can be written $w(x,y) = w_\varepsilon(x,y) = \frac{\rho_\varepsilon(x-y)}{|x-y|^2}$ with (ρ_ε) a family of mollifiers satisfying

$$\rho_\varepsilon \geq 0, \quad \int_{\mathbb{R}^2} \rho_\varepsilon(x) \, dx = 1,$$

$$\lim_{\varepsilon \to 0} \int_{|x|>\delta} \rho_\varepsilon(x) \, dx = 0 \text{ for all } \delta > 0,$$

$$\rho_\varepsilon \text{ is radial, that is, } \rho_\varepsilon(x) = \hat{\rho}_\varepsilon(|x|),\ x \in \mathbb{R}^2.$$

In [263], Ponce establishes the following theorem:

Theorem 34 *[263, Theorem 1] With $\Omega \subset \mathbb{R}^2$ an open set such that $\partial\Omega$ is compact and Lipschitz, if $f \in W^{1,p}(\Omega)$, $p \geq 1$, then there exists $C > 0$ such that*

$$\int_\Omega \int_\Omega \frac{|f(x) - f(y)|^p}{|x-y|^p} \rho_\varepsilon(x-y)\, dx\, dy \leq C \ \ \forall \varepsilon > 0.$$

Moreover,

$$\lim_{j \to +\infty} \int_\Omega \int_\Omega \frac{|f(x) - f(y)|^p}{|x-y|^p} \rho_{\varepsilon_j}(x-y)\, dx\, dy = \int_\Omega \left(\int_{S^1} |Df \cdot \sigma|^p \, d\mu(\sigma) \right) = K_{p,2} \int_\Omega |Df|^2,$$

since ρ_ε is radial and with (ε_j) a suitable subsequence that converges to 0, μ a particular Radon measure on S^1 (see [263, Subsection 1.1] for further details) and $K_{p,2} = \fint_{S^1} |e_1 \cdot \sigma| \, d\mathcal{H}^1$.

With the above hypotheses on w and owing to the definition of v ($v : \Omega \to [0,1]$), it follows that if $u \in W^{1,2}(\Omega, \mathbb{R}^3)$, the component $\int_\Omega v^2 \|\nabla_w u\|^2 \, dx$ is well-defined.

We are now going to adapt the arguments of Ponce ([263]) to prove that

$$\lim_{j \to +\infty} \int_\Omega v^2(x) \int_\Omega \frac{|u^i(x) - u^i(y)|^2}{|x-y|^2} \rho_{\varepsilon_j}(x-y)\, dy\, dx = C \int_\Omega v^2 \, |Du^i|^2 \, dx,$$

with $i = R, G, B$ and we need the following preliminary results.

Lemma 4 *(close to [263, Lemma 1.]) Assume $w : [0,\infty) \to [0,\infty)$ is convex. If $f \in W^{1,1}_{loc}(\mathbb{R}^2)$, then (assuming that v is compactly supported and extended to 0 on $\mathbb{R}^2$)*

$$\int_{\mathbb{R}^2} v^2(x) \int_{\mathbb{R}^2} w\left(\frac{|f(x) - f(y)|}{|x-y|} \right) \rho_\varepsilon(x-y)\, dy\, dx \leq \int_{\mathbb{R}^2} \int_{\mathbb{R}^2} w\left(\left| Df(x) \cdot \frac{h}{|h|} \right| \right) \rho_\varepsilon(h)\, dh\, dx \ \ \forall \varepsilon > 0.$$

Proof. The proof is similar to that in [263] remarking again that $v : \Omega \to [0,1]$. □

Proposition 7 *(regular case) Assume $\Omega \subset \mathbb{R}^2$ is bounded and convex, and let $w : [0,\infty) \to [0,\infty)$ be a continuous function.*
If $f \in \mathcal{C}^2(\bar{\Omega})$, then

$$\lim_{j \to +\infty} \int_\Omega v^2(x) \int_\Omega w\left(\frac{|f(x) - f(y)|}{|x-y|} \right) \rho_{\varepsilon_j}(x-y)\, dy\, dx = \int_\Omega v^2 \int_{S^1} w(|Df \cdot \sigma|) \, d\mu(\sigma)\, dx.$$

Proof. We follow the same arguments as Ponce ([263]). We set $M_f = \|Df\|_{L^\infty}$, f being $\mathcal{C}^2(\bar{\Omega})$. Since w is continuous on $[0, M_f]$, it is uniformly continuous on $[0, M_f]$ and for any given $\delta > 0$, there exists $C_\delta > 0$ (that may change from line to line) such that

$$|w(s) - w(t)| \leq C_\delta \, |s - t| + \delta \quad \forall s, t \in [0, M_f].$$

It yields, owing to the triangle inequality and Taylor's formula, to

$$\begin{aligned}
&\left| w\left(\frac{|f(x) - f(y)|}{|x - y|}\right) - w\left(\left|Df(x) \cdot \frac{x - y}{|x - y|}\right|\right)\right| \\
&\leq C_\delta \, \frac{|f(x) - f(y) - Df(x) \cdot (x - y)|}{|x - y|} + \delta, \\
&\leq C_\delta \, |x - y| + \delta, \quad x \neq y.
\end{aligned}$$

Therefore,

$$\begin{aligned}
&\int_\Omega v^2(x) \int_\Omega \left| w\left(\frac{|f(x) - f(y)|}{|x - y|}\right) - w\left(\left|Df(x) \cdot \frac{x - y}{|x - y|}\right|\right)\right| \rho_\varepsilon(x - y) \, dy \, dx \\
&\leq \int_\Omega \int_{\{y \in \Omega \,||x - y| \leq 1\}} (C_\delta |x - y| + \delta) \; \rho_\varepsilon(x - y) \, dy \, dx \\
&+ 2 \max_{[0, M_f]} w \int_\Omega \int_{\{y \in \Omega \,||x - y| > 1\}} \rho_\varepsilon(x - y) \, dy \, dx,
\end{aligned}$$

still using the properties of v and the continuity of w. Making the change of variable $h = x - y$, it follows that

$$\begin{aligned}
&\int_\Omega v^2(x) \int_\Omega \left| w\left(\frac{|f(x) - f(y)|}{|x - y|}\right) - w\left(\left|Df(x) \cdot \frac{x - y}{|x - y|}\right|\right)\right| \rho_\varepsilon(x - y) \, dy \, dx \\
&\leq |\Omega| \, \delta \int_{\mathbb{R}^2} \rho_\varepsilon(h) \, dh + |\Omega| \, C_\delta \int_{\{|h| \leq 1\}} |h| \, \rho_\varepsilon(h) \, dh \\
&+ 2 \max_{[0, M_f]} w \, |\Omega| \int_{\{|h| > 1\}} \rho_\varepsilon(h) \, dh.
\end{aligned}$$

Now using the property of ρ_ε that states that $\int_{\mathbb{R}^2} \rho_\varepsilon(x) \, dx = 1$, one obtains

$$\begin{aligned}
&\int_\Omega v^2(x) \int_\Omega \left| w\left(\frac{|f(x) - f(y)|}{|x - y|}\right) - w\left(\left|Df(x) \cdot \frac{x - y}{|x - y|}\right|\right)\right| \rho_\varepsilon(x - y) \, dy \, dx \\
&\leq |\Omega| \, \delta + |\Omega| \, C_\delta \int_{\{|h| \leq 1\}} |h| \, \rho_\varepsilon(h) \, dh + 2 \max_{[0, M_f]} w \, |\Omega| \int_{\{|h| > 1\}} \rho_\varepsilon(h) \, dh.
\end{aligned}$$

As $\displaystyle\lim_{\varepsilon \to 0} \int_{|x| > 1} \rho_\varepsilon(x) \, dx = 0$, the last term in the right-hand side tends to zero as ε goes to 0. Also, it can be proved (see [263], [213]) that

$\lim_{\varepsilon \to 0} \int_{\{|h| \le 1\}} |h|\, \rho_\varepsilon(h)\, dh = 0$, so that the second term of the right-hand side also tends to zero when ε goes to 0, for every fixed $\delta > 0$. By passing to the limit when $\delta \to 0$ in the resulting expression, we have

$$\lim_{\varepsilon \to 0} \int_\Omega v^2(x) \int_\Omega \left| w\left(\frac{|f(x) - f(y)|}{|x-y|} \right) - w\left(\left| Df(x) \cdot \frac{x-y}{|x-y|} \right| \right) \right| \rho_\varepsilon(x-y)\, dy\, dx$$
$$= 0,$$

so that it suffices to prove that

$$\lim_{j \to +\infty} \int_\Omega v^2(x) \int_\Omega w\left(\left| Df(x) \cdot \frac{x-y}{|x-y|} \right| \right) \rho_{\varepsilon_j}(x-y)\, dy\, dx$$
$$= \int_\Omega v^2 \int_{S^1} w\left(|Df \cdot \sigma| \right) d\mu(\sigma)\, dx.$$

The end of the proof is similar to that of [263, Proposition 1] (still using the properties of v) and is left to the reader. □

Equipped with these two preliminary results, we are now able to establish the following theorem.

Theorem 35 *With the given hypotheses on Ω and ρ_ε, and i denoting R, G or B, if $u^i \in W^{1,2}(\Omega)$, then*

$$\lim_{j \to +\infty} \int_\Omega v^2(x) \int_\Omega \frac{|u^i(x) - u^i(y)|^2}{|x-y|^2} \rho_{\varepsilon_j}(x-y)\, dy\, dx = K_{2,2} \int_\Omega v^2 |Du^i|^2\, dx.$$

Before giving the sketch of the proof, we need to introduce the following notations. Let us fix $\Psi \in \mathcal{C}_0^\infty(\mathbb{R}^2)$ such that $\int_{\mathbb{R}^2} \Psi\, dx = 1$ and $supp\, \Psi \subset B(0,1)$. Given an open set $\Omega \subset \mathbb{R}^2$, $\delta > 0$ and a function $f \in L^1_{loc}(\Omega)$, for every $x \in \Omega$ such that $dist(x, \partial\Omega) > \delta$, we define the mollification of f by

$$f_\delta(x) = \frac{1}{\delta^2} \int_\Omega \Psi\left(\frac{x-y}{\delta} \right) f(y)\, dy.$$

Proof. We follow the arguments of [263] and consider an extension $\bar{u^i} \in W^{1,2}(\mathbb{R}^2)$ of u^i. For any $h \in \mathcal{C}_0^\infty(\mathbb{R}^2)$, using the triangle inequality in Banach spaces,

$$\left| \left(\int_\Omega \int_\Omega v^2(x) \frac{|u^i(x) - u^i(y)|^2}{|x-y|^2} \rho_{\varepsilon_j}(x-y)\, dy\, dx \right)^{\frac{1}{2}} \right.$$
$$\left. - \left(\int_\Omega \int_\Omega v^2(x) \frac{|h(x) - h(y)|^2}{|x-y|^2} \rho_{\varepsilon_j}(x-y)\, dy\, dx \right)^{\frac{1}{2}} \right|$$
$$\le \left(\int_\Omega \int_\Omega v^2(x) \frac{|(u^i - h)(x) - (u^i - h)(y)|^2}{|x-y|^2} \rho_{\varepsilon_j}(x-y)\, dy\, dx \right)^{\frac{1}{2}}.$$

Lemma 4 and Fubini's theorem enable us to conclude that

$$\left| \left(\int_\Omega \int_\Omega v^2(x) \frac{|u^i(x) - u^i(y)|^2}{|x-y|^2} \rho_{\varepsilon_j}(x-y)\, dy\, dx \right)^{\frac{1}{2}} - \left(\int_\Omega \int_\Omega v^2(x) \frac{|h(x) - h(y)|^2}{|x-y|^2} \rho_{\varepsilon_j}(x-y)\, dy\, dx \right)^{\frac{1}{2}} \right| \leq \left(\int_{\mathbb{R}^2} |D\bar{u^i} - Dh|^2\, dx \right)^{\frac{1}{2}}.$$

We then let j tend to zero. Using the density of $\mathcal{C}_0^\infty(\mathbb{R}^2)$ in $W^{1,2}(\mathbb{R}^2)$ as well as a variant of Proposition 7 for $\mathcal{C}_0^\infty$ functions, the conclusion is straightforward. □

We refer the reader to [213] for theoretical results related to the well-definedness of terms of the form $\int_\Omega \left(\int_\Omega \left(\frac{|f(x)-f(y)|}{|x-y|} \right)^2 \rho_\varepsilon(x-y)\, dy \right)^{\frac{1}{2}} dx$.

10.2 Characterization of minimizers u

We follow here techniques for characterization of minimizers presented in Chapters 3–4.

As in Chapter 4, in order to obtain such characterization of minimizers u (for fixed v only) for some of the models presented here, we first need to show that the nonlocal regularizing terms in u satisfy the properties of a seminorm on the vector space of RGB functions on Ω:

- $|u| \geq 0$ for all functions u
- $|\lambda u| = |\lambda||u|$ for all functions u and $\lambda \in \mathbb{R}$
- Triangle inequality: $|u + U| \leq |u| + |U|$ for all functions u and U

We assume that $u, U : \Omega \to \mathbb{R}^3$, $v : \Omega \to \mathbb{R}$ is fixed, and $w : \Omega \times \Omega \to \mathbb{R}$ is a given nonnegative and symmetric weight function. We define (for fixed v)

$$|u|_{NL/MSTV,v} = \int_\Omega v^2(x) \left(\sqrt{\sum_i \int_\Omega (u^i(y) - u^i(x))^2 w(x,y) dy} \right) dx,$$

and

$$|u|_{NL/MSH^1,v} = \sqrt{\int_\Omega v^2(x) \left(\sum_i \int_\Omega (u^i(y) - u^i(x))^2 w(x,y) dy \right) dx}.$$

Note that $|u|_{NL/MSTV,v}$ is directly obtained from $\Psi^{NL/MSTV}$, but we modified

Ψ^{NL/MSH^1} to have the required homogeneity; this is a minor modification which will not be used in the numerical results, since it is easier to use in practice Ψ^{NL/MSH^1} as initially defined.

We define

$$|u| = \int_\Omega g(x) \left(\sqrt{\sum_i \int_\Omega (u^i(y) - u^i(x))^2 w(x,y) dy} \right) dx$$

for any nonnegative function $g : \Omega \to \mathbb{R}$. Obviously, it is sufficient to check triangle inequality: $|u + U| \leq |u| + |U|$ for all $u, U : \Omega \to \mathbb{R}^3$.

First, we have the equality

$$\begin{aligned}
&\sum_i \int_\Omega ((u^i + U^i)(y) - (u^i + U^i)(x))^2 w(x,y) dy \\
&= \sum_i \int_\Omega (u^i(y) - u^i(x))^2 w(x,y) dy + \sum_i \int_\Omega (U^i(y) - U^i(x))^2 w(x,y) dy \\
&+ 2 \sum_i \int_\Omega (u^i(y) - u^i(x))(U^i(y) - U^i(x)) w(x,y) dy.
\end{aligned}$$

Using Cauchy–Schwarz inequality, we have

$$\begin{aligned}
&\int_\Omega (u^i(y) - u^i(x))(U^i(y) - U^i(x)) w(x,y) dy \\
&\leq \left(\int_\Omega \left((u^i(y) - u^i(x)) \sqrt{w(x,y)} \right)^2 dy \right)^{1/2} \\
&\cdot \left(\int_\Omega \left((U^i(y) - U^i(x)) \sqrt{w(x,y)} \right)^2 dy \right)^{1/2}.
\end{aligned}$$

We denote

$$a^i = \int_\Omega (u^i(y) - u^i(x))^2 w(x,y) dy, \quad b^i = \int_\Omega (U^i(y) - U^i(x))^2 w(x,y) dy.$$

Using $\sum_i \sqrt{a^i}\sqrt{b^i} \leq \sqrt{\sum_i a^i}\sqrt{\sum_i b^i}$, we obtain

$$\begin{aligned}
&\sum_i \int_\Omega ((u^i + U^i)(y) - (u^i + U^i)(x))^2 w(x,y) dy \\
&\leq \sum_i (a^i + b^i + 2\sqrt{a^i}\sqrt{b^i}), \\
&= \sum_i a^i + \sum_i b^i + 2 \sum_i \sqrt{a^i}\sqrt{b^i}, \\
&\leq \sum_i a^i + \sum_i b^i + 2 \sqrt{\sum_i a^i} \sqrt{\sum_i b^i}, \\
&= \left(\sqrt{\sum_i a^i} + \sqrt{\sum_i b^i} \right)^2,
\end{aligned}$$

which finally leads to

$$\sqrt{\sum_i \int_\Omega ((u^i+U^i)(y)-(u^i+U^i)(x))^2 w(x,y)dy}$$
$$\le \sqrt{\sum_i \int_\Omega (u^i(y)-u^i(x))^2 w(x,y)dy}$$
$$+\sqrt{\sum_i \int_\Omega (U^i(y)-U^i(x))^2 w(x,y)dy}.$$

Multiplying by $g(x)$ and integrating both sides w.r.t x, we obtain

$$\int_\Omega g(x)\sqrt{\sum_i \int_\Omega ((u^i+U^i)(y)-(u^i+U^i)(x))^2 w(x,y)dydx}$$
$$\le \int_\Omega g(x)\sqrt{\sum_i \int_\Omega (u^i(y)-u^i(x))^2 w(x,y)dydx}$$
$$+\int_\Omega g(x)\sqrt{\sum_i \int_\Omega (U^i(y)-U^i(x))^2 w(x,y)dydx}.$$

Thus $|u|$ satisfies triangle inequality, so we conclude that $|u|$ is a semi-norm. In particular, taking $g(x)=v^2(x)$, we deduce that $|u|_{NL/MSTV,v}$ is a semi-norm.

Similarly, we can also show that $|u|_{NL/MSH^1,v}$ is a semi-norm using Cauchy–Schwarz inequality and the inequality $\sum_i \sqrt{a^i}\sqrt{b^i} \le \sqrt{\sum_i a^i}\sqrt{\sum_i b^i}$:

$$\begin{aligned}
& \sum_i \int_\Omega\int_\Omega v^2(x)((u^i+U^i)(y)-(u^i+U^i)(x))^2 w(x,y)dydx \\
= \; & \sum_i \int_\Omega\int_\Omega v^2(x)(u^i(y)-u^i(x))^2 w(x,y)dydx \\
+ \; & \sum_i \int_\Omega\int_\Omega v^2(x)(U^i(y)-U^i(x))^2 w(x,y)dydx \\
+ \; & 2\sum_i \int_\Omega\int_\Omega v^2(x)(u^i(y)-u^i(x))(U^i(y)-U^i(x))w(x,y)dydx.
\end{aligned}$$

Using Cauchy–Schwarz inequality, we obtain

$$\int_\Omega\int_\Omega v^2(x)(u^i(y)-u^i(x))(U^i(y)-U^i(x))w(x,y)dydx$$
$$\le \left(\int_\Omega\int_\Omega (v(x)(u^i(y)-u^i(x))\sqrt{w(x,y)})^2 dydx\right)^{1/2}$$
$$\left(\int_\Omega\int_\Omega (v(x)(U^i(y)-U^i(x))\sqrt{w(x,y)})^2 dydx\right)^{1/2},$$

and the inequality $\sum_i \sqrt{a^i}\sqrt{b^i} \leq \sqrt{\sum_i a^i}\sqrt{\sum_i b^i}$ yields

$$\sqrt{\sum_i \int_\Omega \int_\Omega v^2(x)((u^i+U^i)(y)-(u^i+U^i)(x))^2 w(x,y)dydx}$$
$$\leq \sqrt{\sum_i \int_\Omega \int_\Omega v^2(x)(u^i(y)-u^i(x))^2 w(x,y)dydx}$$
$$+\sqrt{\sum_i \int_\Omega \int_\Omega v^2(x)(U^i(y)-U^i(x))^2 w(x,y)dydx}.$$

Hence, $|u|_{NL/MSH^1,v}$ satisfies triangle inequality as well, so $|u|_{NL/MSH^1,v}$ is a semi-norm.

As in Chapters 3–4, we now want to characterize the minimizers u (for fixed v) of the functionals formulated with nonlocal MS regularizers, in the case of impulse noise with regularized L^1 data fidelity terms (since the quadratic data fidelity term for additive Gaussian noise has been considered in Chapters 3–4). We also refer the reader to [236, 309].

Let $\langle \cdot,\cdot \rangle$ denote the $(L^2(\Omega))^3$ inner product. Assuming that a functional $\|\cdot\|$ on a subspace X of $(L^2(\Omega))^3$ is a semi-norm, we can define the dual norm of $f \in (L^2(\Omega))^3 \subset (L^1(\Omega))^3$ as $\|f\|_* := \sup_{\varphi \in X, \|\varphi\| \neq 0} \frac{\langle f,\varphi\rangle}{\|\varphi\|} \leq +\infty$, so that the usual duality $\langle f,\varphi\rangle \leq \|\varphi\|\|f\|_*$ holds for $\|\varphi\| \neq 0$.

For the case of impulse noise, we define the functionals (here $Ku := k * u$),

$$\begin{aligned} E(u,v) &= \int_\Omega \sqrt{\sum_i |f^i - Ku^i|^2 + \eta^2}dx + \beta |u|_{NL/MS} \\ &+\alpha \int_\Omega \Big(\epsilon|\nabla v|^2 + \frac{(v-1)^2}{4\epsilon}\Big)dx, \\ E'(u,v) &= \sum_i \int_\Omega \sqrt{|f^i - Ku^i|^2 + \eta^2}dx + \beta |u|_{NL/MS} \\ &+\alpha \int_\Omega \Big(\epsilon|\nabla v|^2 + \frac{(v-1)^2}{4\epsilon}\Big)dx, \end{aligned}$$

where $\alpha > 0$, $\beta > 0$, $\epsilon > 0$ and $\eta > 0$, and $|u|_{NL/MS} \in \{|u|_{NL/MSH^1,v}, |u|_{NL/MSTV,v}\}$ with

$$|u|_{NL/MSH^1,v} = \sqrt{\int_\Omega v^2(x)\|\nabla_w u\|^2(x)dx},$$
$$|u|_{NL/MSTV,v} = \int_\Omega v^2(x)\|\nabla_w u\|(x)dx.$$

As we have shown, the regularizers $|u|_{NL/MSH^1,v}$ and $|u|_{NL/MSTV,v}$ are semi-norms. In addition, as we have already mentioned, we modified the regularizing functional $|u|_{NL/MSH^1,v}$. Indeed, the square-root term replaces the original term,

$$\int_\Omega v^2(x)\|\nabla_w u\|^2(x)dx,$$

of the model. It is introduced here to enable the characterization of minimizers below, but the numerical calculations utilize the original formulations, which solve the same equivalent problems. The following characterizations of minimizers allow us to give conditions on the existence of minimizers (including the case of trivial minimizers), and allow us to associate a dual "texture" norm to the residual and to quantify its size.

Proposition 8 *Let $K : (L^2(\Omega))^3 \to (L^2(\Omega))^3$ be a linear and continuous blurring operator with adjoint K^* and let E be the associated functional. If (u, v) is a minimizer of E with $v \in [0, 1]$, then*

$$\left\| K^* \frac{f - Ku}{\sqrt{\sum_i (f^i - Ku^i)^2 + \eta^2}} \right\|_* = \beta,$$

$$\text{and } \left\langle K^* \frac{f - Ku}{\sqrt{\sum_i (f^i - Ku^i)^2 + \eta^2}}, u \right\rangle = \beta |u|_{NL/MS},$$

where $\|\cdot\|_*$ is the corresponding dual norm of $|\cdot|_{NL/MS}$.

Proof. Let (u, v) be a minimizing pair. Considering the variation of E only with respect to u, we find that for any $\varphi \in NL/MS(\Omega) = \{u \in (L^2(\Omega))^3 : |u|_{NL/MS} < \infty\}$, we have

$$\int_\Omega \sqrt{\sum_i (f^i - Ku^i)^2 + \eta^2} dx + \beta |u|_{NL/MS}$$

$$\leq \int_\Omega \sqrt{\sum_i (f^i - K(u^i + \epsilon\varphi^i))^2 + \eta^2} dx + \beta |u + \epsilon\varphi|_{NL/MS}.$$

Let

$$g(\epsilon) := \sqrt{\sum_i (f^i - K(u^i + \epsilon\varphi^i))^2 + \eta^2}.$$

Taylor's expansion gives $g(\epsilon) = \sqrt{\sum_i (f^i - Ku^i)^2 + \eta^2} - \epsilon \frac{(f - Ku)\cdot(K\varphi)}{\sqrt{\sum_i (f^i - Ku^i)^2 + \eta^2}} +$

$\frac{\epsilon^2}{2} g''(\eta_\epsilon)$ and hence

$$\begin{aligned}
&\int_\Omega \sqrt{\sum_i (f^i - K(u^i + \epsilon\varphi^i))^2 + \eta^2 dx} \\
&\leq \int_\Omega \sqrt{\sum_i (f^i - Ku^i)^2 + \eta^2 dx} \\
&- \epsilon \left\langle \frac{f - Ku}{\sqrt{\sum_i (f^i - Ku^i)^2 + \eta^2}}, K\varphi \right\rangle + \frac{\epsilon^2}{2} |\Omega| \max_x |g''(x)|.
\end{aligned}$$

Then the first inequality implies that

$$\epsilon \left\langle \frac{f - Ku}{\sqrt{\sum_i (f^i - Ku^i)^2 + \eta^2}}, K\varphi \right\rangle \leq \epsilon\beta|\varphi|_{NL/MS} + \frac{\epsilon^2}{2}|\Omega| \max_x |g''(x)|.$$

Dividing by $\epsilon > 0$ and letting $\epsilon \downarrow 0_+$ (while noticing that $\lim_{\epsilon\to 0} \frac{\epsilon^2}{2} \max_x |g''(x)| = 0$) yields that for any $\varphi \in NL/MS(\Omega)$,

$$\left\langle K^* \frac{f - Ku}{\sqrt{\sum_i (f^i - Ku^i)^2 + \eta^2}}, \varphi \right\rangle \leq \beta|\varphi|_{NL/MS};$$

thus

$$\left\| K^* \frac{f - Ku}{\sqrt{\sum_i (f^i - Ku^i)^2 + \eta^2}} \right\|_* \leq \beta.$$

Now let $\varphi = u$. Then dividing by $\epsilon < 0$, and letting $\epsilon \uparrow 0_-$, we obtain

$$\left\langle K^* \frac{f - Ku}{\sqrt{\sum_i (f^i - Ku^i)^2 + \eta^2}}, u \right\rangle \geq \beta|u|_{NL/MS}.$$

Combining the last two inequalities concludes the proof. □

Remark 14 *If we replace E by E' in Proposition 8, we obtain a similar result: the second formula in the conclusion must be replaced by*

$$\sum_i \left\langle K^* \frac{f^i - Ku^i}{\sqrt{(f^i - Ku^i)^2 + \eta^2}}, u^i \right\rangle = \beta|u|_{NL/MS}$$

where $\langle \cdot, \cdot \rangle$ is the $L^2(\Omega)$ inner product.

10.3 Gâteaux derivative of nonlocal M–S regularizers

We show here the computation of the Gâteaux derivative for the two regularizers presented in this chapter. Note that in practice, for the NL/MSH^1 case, we work with the functional $\int_\Omega v^2(x)\,\|\nabla_w u\|^2(x)\,dx$. More specifically, the NL/MSH^1 and NL/MSTV regularizers have the following Gâteaux derivatives

$$\begin{aligned}
L^{NL/MSH^1}u &= -2\nabla_w \cdot \left(v^2(x)\nabla_w u(x)\right), \\
&= -2\int_\Omega (u(y)-u(x))w(x,y)\left[v^2(y)+v^2(x)\right]dy,
\end{aligned}$$

$$\begin{aligned}
L^{NL/MSTV}u &= -\nabla_w \cdot \left(v^2(x)\frac{\nabla_w u(x)}{|\nabla_w u(x)|}\right), \\
&= -\int_\Omega (u(y)-u(x))w(x,y)\left[\frac{v^2(y)}{|\nabla_w u|(y)}+\frac{v^2(x)}{|\nabla_w u|(x)}\right]dy.
\end{aligned}$$

The above derivatives are obtained in the following way. We compute the Gâteaux derivative of J with respect to a minimizing function $u:\Omega\to\mathbb{R}^3$, with $g:\Omega\to\mathbb{R}$ and $\phi:\mathbb{R}\to\mathbb{R}$, necessary in the computation of the Euler–Lagrange equations, where

$$J(u)=\int_\Omega g(x)\phi(\|\nabla_w u\|^2(x))dx.$$

We assume that u is a minimizer of J and define $G(\epsilon)=J(u+\epsilon h)$ for $\epsilon\in\mathbb{R}$ and a test function h. Then

$$G(\epsilon)=\int_\Omega g(x)\phi(\|\nabla_w(u+\epsilon h)\|^2(x))dx.$$

We denote here by $\langle\cdot,\cdot\rangle$ the usual inner product in $\mathbb{R}^3$. By differentiating G w.r.t ϵ, we obtain

$$\begin{aligned}
G'(\epsilon) \;=\; & 2\int_\Omega g(x)\phi'(\|\nabla_w(u+\epsilon h)\|^2(x))\int_\Omega \langle((u(y)-u(x)) \\
& +\epsilon(h(y)-h(x))),(h(y)-h(x))\rangle w(x,y)dydx.
\end{aligned}$$

Taking $\epsilon = 0$, we obtain the variation of J with respect to u, i.e.,

$$\begin{aligned}
G'(0) =& 2\int_\Omega g(x)\phi'(\|\nabla_w(u)\|^2(x)) \\
& \left[\int_\Omega \langle (u(y)-u(x)),(h(y)-h(x))\rangle w(x,y)dy\right] dx, \\
=& 2\int_\Omega \langle \int_\Omega g(x)\phi'(\|\nabla_w(u)\|^2(x))(u(y)-u(x))w(x,y)dx, h(y)\rangle dy \\
& -2\int_\Omega g(x)\phi'(\|\nabla_w(u)\|^2(x))\langle \left[\int_\Omega (u(y)-u(x))w(x,y)dy\right], h(x)\rangle dx,
\end{aligned}$$

assuming that we can interchange the order of integration. Then by switching x and y in the first component,

$$\begin{aligned}
G'(0) =& 2\int_\Omega \langle \int_\Omega g(y)\phi'(\|\nabla_w(u)\|^2(y))(u(x)-u(y))w(y,x)dy, h(x)\rangle dx \\
& -2\int_\Omega g(x)\phi'(\|\nabla_w(u)\|^2(x))\langle \left[\int_\Omega (u(y)-u(x))w(x,y)dy\right], h(x)\rangle dx, \\
=& -2\int_\Omega \langle \int_\Omega g(y)\phi'(\|\nabla_w(u)\|^2(y))(u(y)-u(x))w(x,y)dy, h(x)\rangle dx \\
& -2\int_\Omega g(x)\phi'(\|\nabla_w(u)\|^2(x))\langle \left[\int_\Omega (u(y)-u(x))w(x,y)dy\right], h(x)\rangle dx,
\end{aligned}$$

where $\phi'(s)$ is the derivative of ϕ with respect to s and $w(x,y) = w(y,x)$. Hence, we obtain

$$\begin{aligned}
Lu = -2\int_\Omega (u(y)-u(x))w(x,y)\Big[& g(y)\phi'(\|\nabla_w(u)\|^2(y)) \\
& + g(x)\phi'(\|\nabla_w(u)\|^2(x))\Big] dy,
\end{aligned}$$

where the operator L is the gradient flow corresponding to the functional J.

Specifically, by taking $g(x) = v^2(x)$ and $\phi(s) = s$ or $\phi(s) = \sqrt{s}$, we obtain two functionals and their corresponding gradient flows, respectively:

$$
\begin{aligned}
J^{NL/MSH^1}(u) &= \int_\Omega v^2(x)\|\nabla_w u\|^2(x)dx : \\
L^{NL/MSH^1}u &= -2\nabla_w \cdot \left(v^2(x)\nabla_w u(x)\right), \\
&= -2\int_\Omega (u(y)-u(x))w(x,y)\left[v^2(y)+v^2(x)\right]dy,
\end{aligned}
$$

$$
\begin{aligned}
J^{NL/MSTV}(u) &= \int_\Omega v^2(x)\|\nabla_w u\|(x)dx : \\
L^{NL/MSTV}u &= -\nabla_w \cdot \left(v^2(x)\frac{\nabla_w u(x)}{\|\nabla_w u(x)\|}\right), \\
&= -\int_\Omega (u(y)-u(x))w(x,y)\Big[\frac{v^2(y)}{\|\nabla_w u\|(y)} \\
&+ \frac{v^2(x)}{\|\nabla_w u\|(x)}\Big].
\end{aligned}
$$

10.4 Image restoration with NL/MS regularizers

We present in this section several applications of the described nonlocal Mumford–Shah regularizers to color image restoration, more precisely to denoising and deblurring, inpainting, super-resolution, and demosaicing.

10.4.1 Color image deblurring and denoising

Image blur and noise are the most common problems in photography, and can be especially significant in light limited situations, resulting in ruined photographs. Image deblurring (or deconvolution) is the process of recovering a sharp image from an input image corrupted by blur and noise, where the blurring is due to convolution with a known or unknown kernel; see [96], [170], [39], [38], [64]. Recently, new image denoising models [72], [73], [229], [308], [154], [155], [260] based on nonlocal image information have been developed to better restore texture. The standard linear degradation model for color image deblurring-denoising (or denoising) is

$$f = k * u + n \quad (f^i = k * u^i + n^i,\ i = R, G, B)$$

where k is a (known) space-invariant blurring kernel, and n is additive Gaussian noise, additive Laplace noise, or impulse noise (salt-and-pepper noise or random-valued impulse noise; in this case, the relation between $k * u$ and n is no longer of the above form).

First, in the case of Gaussian noise model, the L^2-fidelity term led by the maximum likelihood estimation is commonly used:

$$\Phi(f - k * u) = \int_\Omega \sum_i |f^i - k * u^i|^2 dx.$$

However, the quadratic data fidelity term considers the impulse noise, which might be caused by bit errors in transmissions or wrong pixels, as an outlier. So, for the impulse noise model (or the additive Laplace noise model), the L^1-fidelity term is more appropriate, due to its ability to remove outlier effects [12], [251]; moreover, we consider the case of independent channel noise [38]:

$$\Phi(f - k * u) = \int_\Omega \sum_i |f^i - k * u^i| dx.$$

Thus we present two types of total energies for color image deblurring-denoising (to be minimized with respect to u and v), depending on the type of noise, as follows (Gau, denoting Gaussian noise, Im, denoting impulse noise):

$$E^{Gau}(u, v) = \frac{1}{2} \int_\Omega \sum_i |f^i - k * u^i|^2 dx + \Psi^{NL/MS}(u, v), \tag{10.2}$$

$$E^{Im}(u, v) = \int_\Omega \sum_i |f^i - k * u^i| dx + \Psi^{NL/MS}(u, v). \tag{10.3}$$

10.4.1.1 Preprocessing step for impulse noise model

To extend the nonlocal methods to the impulse noise case, we need a preprocessing step for the weight function w since we cannot directly use the data f to compute w. In other words, in the presence of impulse noise, the noisy pixels tend to have larger weights than the other neighboring points, so it is likely to keep the noise value at such pixels. Thus we present a simple algorithm to obtain a preprocessed image $\bar{g}$, which removes the impulse noise (outliers) as well as preserves the texture as much as possible. Basically, we use the median filter known for removing impulse noise. However, for the deblurring-denoising model, if we apply one step of the median filter, the output may be too smoothed out. In order to preserve fine structures and to remove noise properly, we define a preprocessing method for the deblurring-denoising model, inspired by the Bregman iteration [63]. Thus we present the following algorithm to obtain a preprocessed image $\bar{g}$ that will be used only in the computation of the weight function w:

Initialize: $r_0^i = 0$, $g_0^i = 0$, $i = R, G, B$.

Do (iterate $m = 0, 1, 2, \ldots$)

$$g_{m+1} = \text{median}(f + r_m, [s\ s])$$

$$r_{m+1} = r_m + f - k * g_{m+1}$$

$$\text{while } \sum_i \|f^i - k * g^i_m\|_1 > \sum_i \|f^i - k * g^i_{m+1}\|_1,$$

where f is the given noisy-blurry data, and median$(f, [s\ s])$ is the median filter of size $s \times s$ with input f. The residual energy $\sum_i \|f^i - k * g^i_m\|_1$ has a minimum value at the l-th iteration; thus we obtain a preprocessed image $\bar{g} = g_l$. We show in Figure 10.1 the residual norm $\sum_i \|f^i - k * g^i_m\|_1$ versus steps m. The preprocessed image $\bar{g}$ is a deblurred and denoised version of f, but it still includes some irregularities (or remaining impulse noise) that will be handled by constructing binary weights (values of 0 or 1) proposed for detecting and removing irregularities from texture in [155].

The image $\bar{g}$ will be used only in the computation of the weights w, while keeping f in the data fidelity term; thus artifacts are not introduced by the median filter. The iterative algorithm for the computation of the weights is simple, fast, and satisfactorily constructs the weights.

10.4.2 Color image inpainting

Image inpainting, also known as image interpolation, is the process of reconstructing lost or corrupted parts of an image. This is an interesting and important inverse problem with many applications such as removal of scratches in old photographs, removal of overlaid text or graphics, and filling in missing blocks in unreliably transmitted images. Non-texture image inpainting has received considerable interest since the pioneering paper by Masnou and Morel [233], [232], who proposed variational principles for image disocclusion. A recent wave of interest in inpainting started from Bertalmio et al. [47], whose applications to the movie industry, video and art restoration were unified. These authors proposed nonlinear partial differential equations for non-texture inpainting. Moreover, many contributed works have been proposed for the solution of this interpolation task based on (a) diffusion and transport PDE and variational principles [37], [45], [89], [291], [138], [290], [261], [35], (b) exemplar region fill-in [135], [340], [48], [123], [113], and (c) compressive sensing [137]. Inpainting corresponds to the operation K in (10.1) of losing pixels from an image, i.e., the observed data f is given by

$$f = u \quad \text{on} \quad \Omega - D$$

where $D = D^0$ is the region where the input data u has been damaged. Thus inspired by [89], we present the total energy functional for color image inpainting, introduced by Jung et al. in [180], and defined by

$$E^{Inp}(u, v) = \frac{\lambda}{2} \int_\Omega \chi_{\Omega - D}(x) \sum_i |f^i - u^i|^2 dx + \Psi^{NL/MS}(u, v), \qquad (10.4)$$

where $\chi_{\Omega-D}$ is the characteristic function of $\Omega - D$ (i.e. $\chi_{\Omega-D}(x) = 1$ if $x \in \Omega - D$, 0 otherwise), and $\lambda > 0$ is a parameter. In addition, we update the weights w only in the damaged region D at every mth iteration for u, using the patch distance

$$d_a^R(u(x), u(y)) = \int_{\mathbb{R}^2} \chi_{\Omega-R}(x+t) G_a(t) \|u(x+t) - u(y+t)\|^2 dt,$$

where $\chi_{\Omega-R}$ is a characteristic function on Ω defined above, and $R \subset D$ is an un-recovered region (still missing region). Therefore, the missing region $D = D^0$ is recovered by the following iterative algorithm, producing the un-recovered regions D^i, $i = 0, 1, 2, ...,$ with $D^0 \supset D^1 \supset D^2 \supset \cdots$:

(1) Compute weights w for $x \in \Omega$ s.t. $P(x) \bigcap (\Omega - D^0) \neq \emptyset$ using $d_a^{D^0}(u^0(x), u^0(y))$ with $u^0 = f$ in $\Omega - D^0$ and ∞ in D^0, a patch $P(x)$ centered at x, and $y \in S(x) \cap (\Omega - D^0)$, $S(x)$ being the search window.

(2) Iterate $n = 1, 2, ...$ to obtain a minimizer (u, v) starting with $u = u^0$:

   (a) For fixed u, update v in Ω to obtain v^n.

   (b) For fixed v, update u in Ω to obtain u^n with a recovered region $\Omega - D^n \supset \Omega - D^0$: at every mth iteration, update weights w only in $x \in D^0$ s.t. $P(x) \bigcap (\Omega - D^{n,m}) \neq \emptyset$ with

$$d_a^{D^{n,m}}(u(x), u(y))$$

   where $y \in S(x) \cap (\Omega - D^{n,m})$, $D^{n,m}$ is an un-recovered region in D^0 until the mth iteration with $D^{n,m} \supset D^{n,2m} \supset \cdots \supset D^{n,n} = D^n$.

10.4.3 Color image super-resolution

Super-resolution corresponds to the recovery of a high resolution image from a filtered and down-sampled image. It is usually applied to a sequence of images in video; see [230], [149], [142], [119], [265]. We consider here a simpler problem of increasing the resolution of a single still image, and the observed data f is given by

$$f^i = D_k(h * u^i), \qquad i \in R, G, B$$

where h is a low-pass filter and $D_k : \mathbb{R}^{n \times n} \to \mathbb{R}^{p \times p}$ (with $p = [\frac{n}{k}]$ where $[q]$ is the integer part of q) is the down-sampling operator by factor k along each axis. We want to recover a high resolution image $u \in (\mathbb{R}^{n \times n})^3$ by minimizing

$$E^{Sup}(u, v) = \frac{1}{2} \int_\Omega \sum_i |f^i - D_k(h * u^i)|^2 dx + \Psi^{NL/MS}(u, v). \qquad (10.5)$$

In addition, we use a super-resolved image $\bar{g} \in (\mathbb{R}^{n \times n})^3$ obtained by bicubic interpolation of $f \in (\mathbb{R}^{p \times p})^3$ only for the computation of the weights w.

10.4.4 Color filter array demosaicing

In a demosaicing algorithm we have to reconstruct a full color image from the incomplete color samples output from an image sensor overlaid with a color filter array (CFA). A color filter array is a mosaic of color filters in front of the image sensor, and we use the Bayer filter [43] that utilizes alternating green (G) and red (R) filters for odd rows and alternating blue (B) and green (G) filters for even rows. Since each pixel of the sensor is behind a color filter, the output is an array of pixel values, each indicating a raw intensity of one of the three filter colors. Thus an algorithm is needed to estimate for each pixel the color levels for all color components, rather than a single component; see [4], [190], [162], [163], [224], [174], [73]. In this variational framework, we consider the observed data f as

$$f^i = H^i \cdot u^i, \quad i \in \{R, G, B\}$$

where $\cdot$ is the pointwise product, and H^i is the down-sampling operator; H^G has alternating 1 and 0 values for odd rows and alternating 0 and 1 values for even rows, H^R has alternating 0 and 1 values for odd rows and only 0 values for even rows and H^B has only 0 values for odd rows and alternating 1 and 0 values for even rows. We describe the following minimization problem to recover a full color image u

$$E^{Demo}(u, v) = \frac{1}{2} \int_\Omega \sum_i |f^i - H^i \cdot u^i|^2 dx + \Psi^{NL/MS}(u, v). \tag{10.6}$$

Moreover, we use the interpolated image obtained by applying the Hamilton–Adams algorithm [4] to the green channel and bilinear interpolation for $R-G$ and $B-G$, to compute the initial weight function w. In the Hamilton–Adams method, the evaluation of the gradient at the missing green pixel is corrected by the second order derivatives of the red or blue channels. In addition, as in [73], in order to gradually correct the erroneous structures and artifacts of the initial color image u_0, we also proceed by an iterative strategy, refining at each step the similarity search by reducing the value of parameter h in the weights as:

(1) Initialize $u = u_0$ with an interpolated image with Hamilton–Adams algorithm.

(2) Iterate for h (e.g. $h = \{16, 8, 4\}$):

   (a) Construct the weight function $w = w_u$ using the image u.

   (b) Compute a minimizer (u, v) by minimizing the functional $E^{Demo}(u, v)$.

Finally, we refer the reader to [73] which inspired the work on demosaicing presented in this chapter.

10.5 Numerical discretizations

Minimization of the presented functionals (10.2)-(10.6): E^{Gau}, E^{Im}, E^{Inp}, E^{Sup}, E^{Demo} in u and v is carried out using the Euler–Lagrange equations

$$\begin{aligned}
\frac{\partial E^{Gau,Im,Inp,Sup,Demo}}{\partial v} =& 2\beta v\phi(\|\nabla_w u\|^2) - 2\epsilon\alpha\triangle v + \alpha\left(\frac{v-1}{2\epsilon}\right) = 0,\\
\frac{\partial E^{Gau}}{\partial u} =& \tilde{k} * (k * u - f) + L^{NL/MS}u = 0,\\
\frac{\partial E^{Im}}{\partial u} =& \tilde{k} * \operatorname{sign}(k * u - f) + L^{NL/MS}u = 0,\\
\frac{\partial E^{Inp}}{\partial u} =& \lambda\,\chi_{\Omega-D}(u - f) + L^{NL/MS}u = 0,\\
\frac{\partial E^{Sup}}{\partial u} =& \tilde{h} * (D_k^T(D_k(h * u) - f)) + L^{NL/MS}u = 0,\\
\frac{\partial E^{Demo}}{\partial u} =& H \cdot (H \cdot u - f) + L^{NL/MS}u = 0,
\end{aligned}$$

where $\tilde{k}(x) = k(-x)$, $\tilde{h}(x) = h(-x)$, $D_k^T : (\mathbb{R}^{p\times p})^3 \to (\mathbb{R}^{n\times n})^3$ is the transpose of D_k i.e. the up-sampling operator, and

$$\begin{aligned}
L^{NL/MS}u =& -2\int_\Omega \Big\{(u(y) - u(x))w(x,y)\\
&\cdot\left[v^2(y)\phi'(\|\nabla_w u\|^2(y)) + v^2(x)\phi'(\|\nabla_w u\|^2(x))\right]\Big\}dy.
\end{aligned}$$

To solve two Euler–Lagrange equations simultaneously, the alternate minimization approach is applied. Note that since the energy functionals are not convex in the joint variable (u, v), we may compute only a local minimizer. In the algorithm, we define the initial guess u^0 to be the data f (except for the super-resolution problem). We also tested other initial conditions (u^0 a constant, a random image, or the data f perturbed by a random component; v^0 constant equal to 0 or 1); we observed that the final steady state result does not change; only the number of iterations needed to reach the same result changes; we noticed that, if the initial image u^0 is the input data f, fewer iterations are needed to reach the steady state. Due to its simplicity, we use Gauss–Seidel scheme for v, and an explicit scheme for u using the gradient descent method, leading to the following iterative algorithm:

(1) Initialization: $u^0 = f$, $v^0 = 1$.

(2) Iterate $n = 0, 1, 2, \ldots$, until $(\|u^{n+1} - u^n\|_2 < \eta\|u^n\|_2)$.

(a) Solve the equation for v^{n+1} using Gauss–Seidel scheme:

$$\left(2\beta\phi(\|\nabla_w u^n\|^2) + \frac{\alpha}{2\epsilon} - 2\epsilon\alpha\triangle\right)v^{n+1} = \frac{\alpha}{2\epsilon}.$$

(b) Set $u^{n+1,0} = u^n$ and solve for u^{n+1} by iterating on l:

$$u^{n+1,l+1} = u^{n+1,l} - dt \cdot \frac{\partial E}{\partial u}(u^{n+1,l}, v^{n+1}).$$

Here η is a small positive constant. The basic discretizations are explained next. Let u_k^i denote the value of a pixel k in the image $(1 \le k \le N)$ with channel i (i.e., the discretized version of $u^i(x)$ defined on Ω), and let $p_{k,l}^i$ be the discretized version of $p^i(x,y)$ with $x, y \in \Omega$. Also, $w_{k,l}$ is the sparsely discrete version of $w = w(x,y) : \Omega \times \Omega \to \mathbb{R}$. We use the neighbor set $l \in N_k$ defined by $l \in N_k := \{l : w_{k,l} > 0\}$. Then we have ∇_{wd} and div_{wd}, the discretizations of ∇_w and div_w, respectively, given by [155]:

$$\begin{aligned} \nabla_{wd}(u_k^i) &:= (u_l^i - u_k^i)\sqrt{w_{k,l}}, \quad l \in N_k, \\ div_{wd}(p_{k,l}^i) &:= \sum_{l \in N_k} (p_{k,l}^i - p_{l,k}^i)\sqrt{w_{k,l}}. \end{aligned}$$

Moreover, the magnitude of $p_{k,l}^i$ at k is $|p^i|_k = \sqrt{\sum_l (p_{k,l}^i)^2}$ and thus the discretization of $\|\nabla_w u\|^2(x)$ is

$$\|\nabla u_{wd}\|_k^2 = \sum_{i \in \{R,G,B\}} |\nabla_{wd} u^i|_k^2 = \sum_i \sum_l (u_l^i - u_k^i)^2 w_{k,l}.$$

Basically, we construct the weight function $w_{k,l}$, following the algorithm in [154]: for each pixel k, (1) take a patch P_k around a pixel k, compute the distances $(d_a)_{k,l}$ (a discretization of d_a) to all the patches P_l in the search window $l \in S(k)$, and construct the neighbor set N_k by taking the m most similar and the four nearest neighbors of the pixel k; (2) compute the weights $w_{k,l}$, $l \in N_k$ and set to zero for all the other connections ($w_{k,l} = 0$, $l \notin N_k$); (3) set $w_{k,l} = w_{l,k}$, $l \in N_k$. For deblurring in the presence of impulse noise, we used $m = 5$, so a maximum of up to $2m + 4$ neighbors for each pixel is allowed, and 5×5 pixel patches with $a = 10$, a search window of size 11×11. The complexity of computing the weights using this algorithm is $N \times Window_{size} \times (Patch_{size} \times Channel_{size} + \log m)$. Thus, in this case, we need $121 \times (25 \times 3 + 2.5) \approx 9619$ operations per pixel. Note that when we use a preprocessed image $\bar{g}$ to compute w in the impulse noise model, we construct the weight function w with the binary values of 0 or 1 [155]: in step (2) above, for $l \in N_k$, we assign the value 1 to $w_{k,l}$ and $w_{l,k}$.

10.6 Experimental results and comparisons

The nonlocal MS regularizers presented here, NL/MSH[1] and NL/MSTV, were tested on several color images corrupted by different blur kernels or different noise types, on color images with missing regions, on sub-sampled color

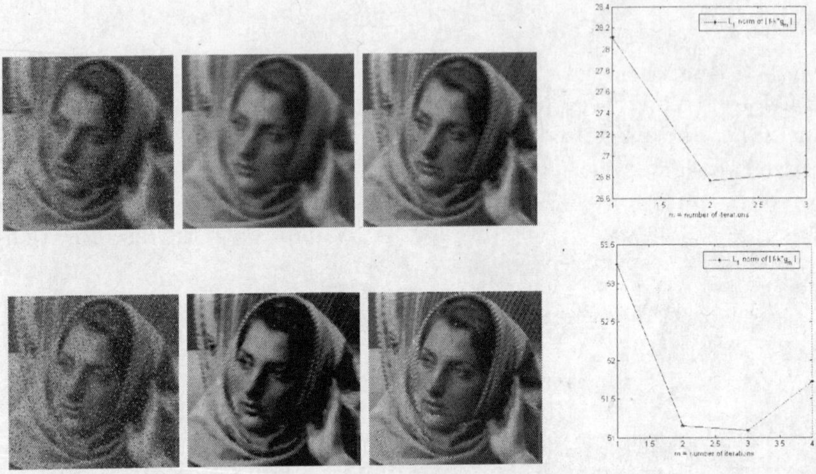

FIGURE 10.1: Preprocessed images $\bar{g}$ in the presence of random-valued impulse noise. First through fourth columns: data f, g_1, $\bar{g}$, $\|f - k * g_m\|_1$ versus m. Data f: (top) motion blur kernel with length of 8 and orientation 0, noise density $d = 0.1$, (bottom) motion blur kernel with length of 4 and orientation 0, noise density $d = 0.2$. Preprocessed images: $\bar{g} = g_2$ (top), $\bar{g} = g_3$ (bottom) with 3×3 median filter.

TABLE 10.1: Peak signal-to-noise ratio (PSNR) values of recovered Barbara and girl images with random-valued impulse noise. l stands for the length of the motion blur kernel, d stands for density of noise and r stands for the radius of the out of focus blur kernel.

Image	Fig. 10.2		Fig. 10.4	
Data	$l = 8$ $d = 0.1$	$l = 4$ $d = 0.2$	$r = 5$ $d = 0.3$	$r = 3$ $d = 0.4$
data f	16.9528	14.9649	12.4583	11.2622
preprocessed $\bar{g}$	21.7047	22.3935	27.9720	25.7044
TV	25.2780	23.9149	30.7566	29.5530
MSH[1]	25.2128	23.9964	30.9531	27,6346
MSTV	25.4572	24.1592	32.4722	29.8059
NL/TV	26.0389	24.1542	32.6614	29.9388
NL/MSH[1]	25.7009	24.2128	31.4532	28.1454
NL/MSTV	**26.2093**	**24.2962**	**33.0390**	**30.1133**

FIGURE 10.2: Deblurring in the presence of random-valued impulse noise. Top two rows: results related to data f given in Figure 10.1 top row. Bottom two rows: results related to data f given in Figure 10.1 bottom row. First column: (top) data f, (bottom) preprocessed image $\bar{g}$. Second through fourth columns: recovered images using (top) local regularizers (MSH^1, MSTV, TV) and (bottom) nonlocal regularizers (NL/MSH^1, NL/MSTV, NL/TV).

FIGURE 10.3: Edge set v obtained during the restoration process using MSH^1, NL/MSH^1, MSTV, NL/MSTV in Figure 10.2, bottom two rows.

(a) (b) (c)

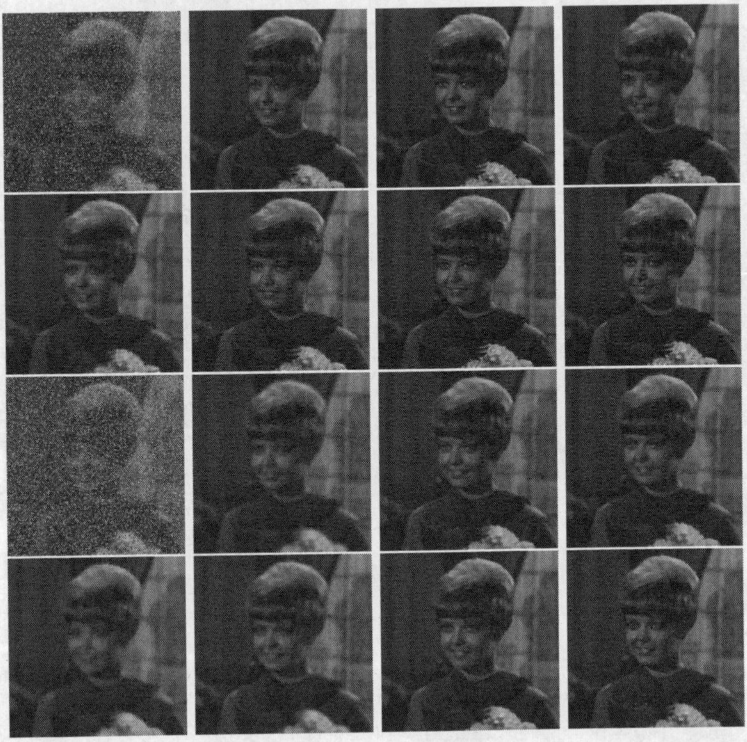

FIGURE 10.4: Deblurring in the presence of random-valued impulse noise. (a) in top row shows original image. Data f: second and third rows show results related to original image corrupted by out of focus blur kernel with radius $r = 5$ (top row, (b)) and noise density $d = 0.3$; fourth and fifth rows show results related to out of focus blur kernel with radius $r = 3$ (top row, (c)) and noise density $d = 0.4$. First column: data f (top), preprocessed image $\bar{g}$ (bottom) with 9×9 (second and third rows) and 11×11 (fourth and fifth rows) median filters. Second through fourth columns: recovered images using local regularizers (MSH^1, MSTV and TV, top) and nonlocal regularizers (NL/MSH^1, NL/MSTV and NL/TV, bottom).

TABLE 10.2: Parameter selections (λ, β) for Figures 10.2 and 10.4

	TV	NLTV	MSH^1	NL/MSH^1	MSTV	NL/MSTV
Barbara	17.5	25	0.4	0.09	0.065	0.04
	3	6	2	0.6	0.3	0.2
Girl	14	33	0.3	0.12	0.05	0.03
	2.5	5	5	2	0.4	0.18

FIGURE 10.5: Edge set v obtained during the restoration process using (left to right) MSH^1, NL/MSH^1, MSTV, NL/MSTV shown in second and third rows of Figure 10.4.

images, as well as on incomplete color sample outputs. We compared them with their local versions from Chapter 7 and Bar et al. [38]. For deblurring in the presence of impulse noise, we also present results with total variation regularization (TV) [64] and nonlocal total variation regularization (NL/TV) from [223] and Chapter 4.

First, in Figures 10.1 and 10.2, we recover blurred images contaminated by random-valued impulse noise with noise density $d = 0.1$ or $d = 0.2$. Figure 10.1 presents the preprocessed images $\bar{g}$ obtained by iterative median filter of size 3×3, and the corresponding residual energies $\|f - k * g_m\|_1$ versus m. As m increases, the image g_m gets deblurred to some extent where the residual energy has a minimum. Thus the image having the minimum energy value is chosen as a preprocessed image $\bar{g}$, but this still contains some impulse noise or artifacts. However, the recovered images using nonlocal regularizers in Figure 10.2 show that the artifacts on the preprocessed images are well handled when constructing weights and thus do not influence the quality of the final recovered images. Hence, by computing the weight function w based on the preprocessed images $\bar{g}$, all nonlocal regularizers recover texture better and reduce the artifacts due to the blurring kernel (especially on the face and hand), providing cleaner images as well as higher PSNR values (as seen in Table 10.1). Figure 10.3 provides the edge set v of local or nonlocal MS regularizers, concurrently obtained during the restoration process shown in Figure 10.2.

Additionally, in Figure 10.4, we use the "girl" image corrupted by random-valued impulse noise in two cases: strong blur and some noise with $d = 0.3$ (second and third rows); and weak blur and stronger noise with $d = 0.4$ (fourth

FIGURE 10.6: Deblurring in the presence of Gaussian noise. (a) original image, (b) blurry-noisy data $f = k * u + n$ with Gaussian blur kernel k with $\sigma_b = 1.2$ and noise variance $\sigma_n = 0.01$, recovered images using (c) MSTV and (d) NL/MSTV. Peak signal-to-noise ratio (PSNR): f: 17.2534, MSTV: 19.4849, NL/MSTV: 19.8815. β: MSTV: 0.06, NL/MSTV: 0.02.

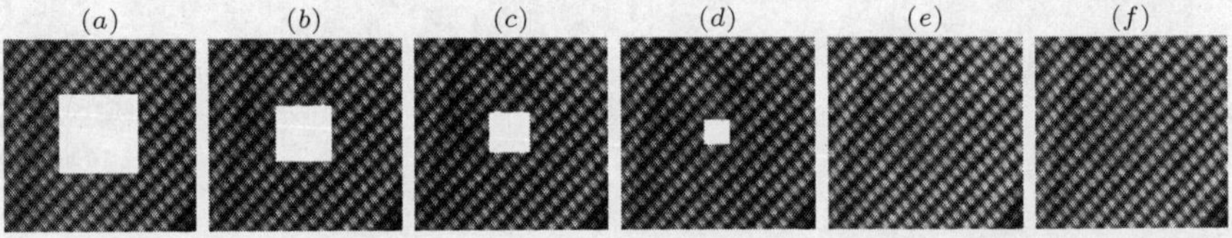

FIGURE 10.7: Inpainting of 100×100 size image with 40×40 missing part. (a) data f, (b) through (d): intermediate steps of the inpainting process with NL/MSH^1, recovered images using (e) NL/MSH^1: PSNR = 35.6704 and (f) NL/MSTV: PSNR = 35.8024, with 41×41 search window and 9×9 patch.

and fifth rows). In the first case, NL/MSH^1 reduces the ringing effect (especially seen on the cloth part of the result obtained using local MSH^1 model), providing cleaner image, and NL/TV gives a much better restored image than TV, leading to much higher PSNR (as seen in Table 10.1). Even though MSTV gives a desired recovered image, NL/MSTV additionally reduces the staircase effect (seen on the image obtained with MSTV), resulting in a more realistic image and higher PSNR. Both NL/MSTV and NL/TV give very good recovered images visually and according to PSNR values. In the second case with less blur but more noise, NL/MSTV and NL/TV give sharper images than those obtained by the local regularizing models.

In Figure 10.6, we only test the MSTV and NL/MSTV models for the Gaussian noise model. For the noisy-blurry castle image, NL/MSTV gives cleaner and sharper restored images, leading to higher PSNR values.

In Figures 10.7 and 10.8, we use the NL/MS regularizers to recover images with texture and large missing regions. In Figure 10.7, we present the process of inpainting, and final recovered images using NL/MS regularizers. We can easily see that both nonlocal regularizers recover the missing regions very well, and moreover NL/MSTV gives slightly better results than NL/MSH^1 according to PSNR even though these visually seem to produce very similar results.

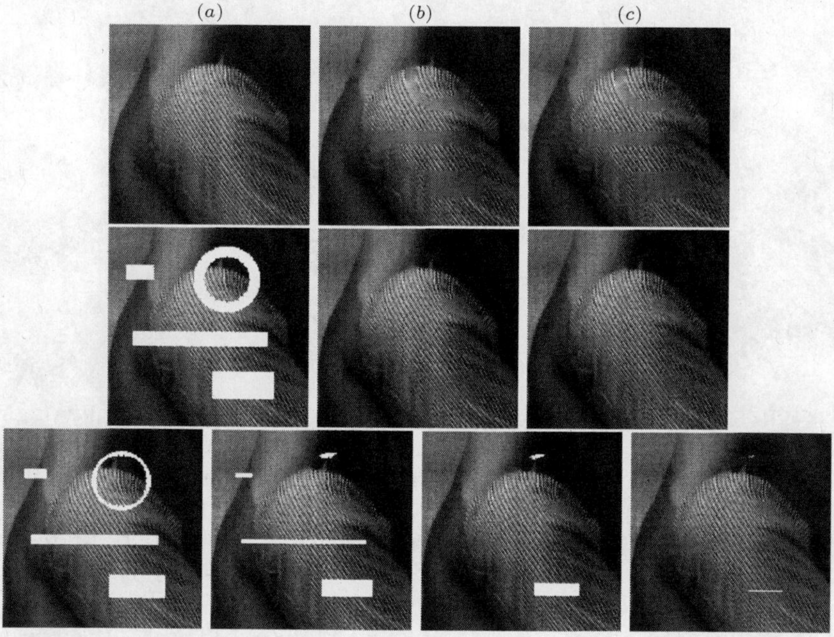

FIGURE 10.8: Inpainting of 150×150 size image. First row, (a): original; second row, (a): data f. First row, (b) through (c): recovered images using MSH^1: PSNR = 29.2797 and MSTV: PSNR = 29.4205. Second row, (b) through (c): recovered images using NL/MSH^1: PSNR = 34.4953 and NL/MSTV: PSNR = 34.2406, with 51×51 search window and 9×9 patch. Bottom row: intermediate steps of the inpainting process with NL/MSH^1 in 50th, 100th, 200th, 350th iterations.

However, in Figure 10.8 with a real image, NL/MSH^1 gives slightly higher PSNR values, especially recovering better the part damaged by the rectangle in the bottom. Both NL/MS regularizers gradually recover the missing regions, as seen in Figure 10.8, bottom row, while local MS regularizers fail to recover them.

In Figures 10.9 and 10.10, we recover an image filtered with a low-pass filter and then sub-sampled, using MSTV and NL/MSTV. As seen in both recovered images and edge sets v, NL/MSTV (incorporating weights based on the interpolated image $\bar{g}$) provides clearer edges, leading to much higher PSNR, while MSTV produces some artifacts especially on the edges. For comparison, we applied the Lucy–Richardson deconvolution iterative algorithm [225, 268] to the blurry image $\bar{g}$, but this produced severe artifacts and provided an even lower PSNR than for $\bar{g}$ itself and lower than the values of recovered images using MSTV and NL/MSTV.

In Figures 10.11 through 10.13, we reconstruct full color images from the

FIGURE 10.9: Super-resolution of still image. Top row, left to right: original image of size 272×272; down-sampled data $f = D_k(h * u)$ of size 68×68 with Gaussian blur kernel h with $\sigma_b = 2$ and $k = 4$; preprocessed (up-sampled) image $\bar{g}$ using bicubic interpolation; deblurred image from applying Lucy–Richardson deconvolution algorithm to $\bar{g}$ at 10th iteration. Bottom row, left to right: recovered images using MSTV with corresponding edge set v and NL/MSTV with corresponding edge set v. $h * u$: PSNR = 21.9423, $\bar{g}$: PSNR = 18.3970, MSTV: PSNR = 23.4551, and NL/MSTV: PSNR = 24.3336.

FIGURE 10.10: Another super-resolution of still image. Top row: original image of size 272×272, blurred image $h * u$ with out-of-focus blur kernel h with radius $r = 3$, down-sampled data $f = D_k(h * u)$ of size 68×68 with $k = 4$, preprocessed (up-sampled) image $\bar{g}$ using bicubic interpolation. Bottom row: recovered images using (left) MSTV, (right) NL/MSTV. $h * u$: PSNR = 21.7459, $\bar{g}$: PSNR = 19.9989, MSTV: PSNR = 21.8869 and NL/MSTV: PSNR = 22.1116.

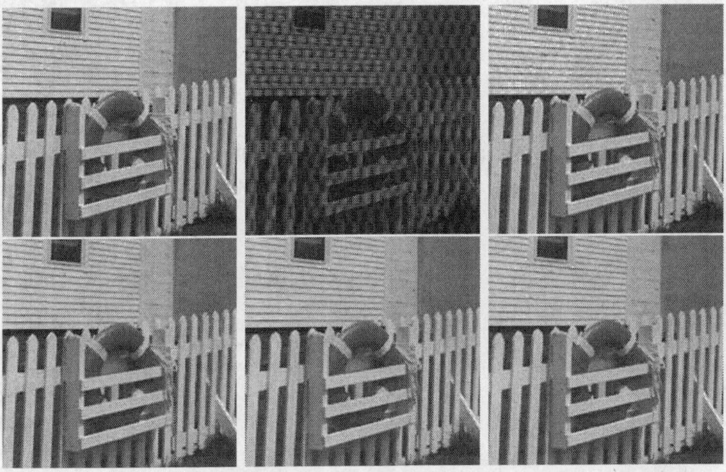

FIGURE 10.11: Demosaicing using NL/MSTV with iterative algorithm. Top row, left to right: original image; data f; interpolated image (PSNR = 27.3076) using Hamilton–Adams for green and bilinear interpolation for $R-G$ and $B-G$. Bottom row, left to right: demosaiced images with decreasing sequence of $h = \{16, 8, 4\}$ and corresponding PSNR values: first = 29.5101, second = 29.7029, third = 29.7128.

FIGURE 10.12: Demosaicing using NL/MSTV with $h = \{16, 8, 2\}$. PSNR values: Hamilton–Adams (top right) = 26.5008, NL/MSTV (bottom row, left to right): 28.1480, 28.2220, and 28.2615.

FIGURE 10.13: Demosaiced images using NL/MS regularizers with $h = \{16\}$, and corresponding residuals with the original image. Top row, left to right: original; data f; interpolated image with Hamilton–Adams (PSNR = 36.5672); residual. Bottom row, left to right: recovered images using NL/MSH^1 (37.3598) and NL/MSTV (37.3606).

FIGURE 10.14: Edge sets v of NL/MSH^1 (first and third) and NL/MSTV (second and fourth) obtained in Fig. 10.11 and 10.13.

incomplete color samples outputs by the Bayer color filter, using NL/MSTV regularizer with the decreasing sequences of h. In all the examples, we use the initial color image u_0 obtained by applying the Hamilton–Adams algorithm to the green channel (G) and bilinear interpolation to both $R - G$ and $B - G$. Figures 10.11 through 10.12 show that, with NL/MSTV regularizer, the color artifacts of u_0 are much reduced even after one iteration, and these are gradually further reduced by using decreasing sequences $h = \{16, 8, 4\}$ or $h = \{16, 8, 2\}$. Even in the case with the well-interpolated initial u_0 as in Figure 10.13, NL/MSTV regularizer improves u_0 in one iteration, by recovering the window parts, which can be seen in the residual with the original image. Note that the NL/MSTV regularizer produces better reconstructed images than using NL/MSH^1 or local MS regularizers.

Finally, we note that the parameters α, β and ϵ were selected manually to provide the best PSNR results. The smoothness parameter β increases with the noise level, while the other parameters α, ϵ are approximately fixed, $\alpha = 0.001$, $\epsilon = 0.00000001$ for deblurring-denoising, inpainting, super-resolution, demosaicing, and $\alpha = 0.1$, $\epsilon = 0.001$ for denoising (although in theory $\epsilon \to 0$, it is common in practice to work with a small fixed ϵ). For the weights w, we used 11×11 search window with a 5×5 patch for the deblurring-denoising model (or denoising), a 21×21 search window with a 9×9 patch for super-resolution, and a 15×15 search window with a 3×3 patch for demosaicing, while larger search windows are needed for inpainting.

10.7 Exercises

Exercise 10.1 *State and prove the corresponding result with Proposition 8 for the functional $E'(u, v)$ given in Section 10.2 (see Remark 14).*

Exercise 10.2 *Derive the system of Euler–Lagrange equations associated with the minimization of*

$$E^{Gau}(u, v) = \frac{1}{2}\int_\Omega \sum_i |f^i - k * u|^2 dx + \Psi^{NL/MSH^1}(u, v).$$

Exercise 10.3 *Derive the system of Euler–Lagrange equations associated with the minimization of*

$$E^{Gau}(u, v) = \frac{1}{2}\int_\Omega \sum_i |f^i - k * u|^2 dx + \Psi^{NL/MSTV}(u, v).$$

Exercise 10.4 *Derive the system of Euler–Lagrange equations associated with the minimization of*

$$\begin{aligned} E(u, v) &= \int_\Omega \sqrt{\sum_i |f^i - Ku^i|^2 + \eta^2} dx + \beta \sqrt{\int_\Omega v^2(x) \|\nabla_w u\|^2(x) dx} \\ &\quad + \alpha \int_\Omega \Big(\epsilon|\nabla v|^2 + \frac{(v-1)^2}{4\epsilon}\Big) dx. \end{aligned}$$

Exercise 10.5 *Derive the system of Euler–Lagrange equations associated with the minimization of*

$$\begin{aligned} E(u, v) &= \int_\Omega \sqrt{\sum_i |f^i - Ku^i|^2 + \eta^2} dx + \beta \int_\Omega v^2(x) \|\nabla_w u\|(x) dx \\ &\quad + \alpha \int_\Omega \Big(\epsilon|\nabla v|^2 + \frac{(v-1)^2}{4\epsilon}\Big) dx. \end{aligned}$$

Exercise 10.6 *Implement the preprocessing step given in Subsection 10.4.1.1 and apply it to a blurry image contaminated by impulse noise.*

Chapter 11

A Combined Segmentation and Registration Variational Model

11.1 Introduction

Image segmentation and image registration are challenging issues that are encountered in a wide range of fields such as medical imaging (object detection and shape tracking, comparison of images taken at different instants, data fusion from images that have not necessarily been acquired with the same modality, comparison of data to a common reference frame) and pattern recognition or geophysics.

Considering two images called template and reference, image registration consists in finding an optimal diffeomorphic transformation to map the template into the reference, while segmentation aims at detecting and visualizing the contours of the objects contained in a given image. For images of the same modality, a well-registered template has geometric features and intensity distribution matched with the reference. For images produced by different mechanisms and possessing distinct modalities, the goal of registration is to correlate the images while maintaining the modality of the template.

We present in this chapter a segmentation model based on the binary active contour model without edges from Chapter 8 that no longer has as unknown a level set function. The model is now solved using registration techniques. Therefore, a displacement field models the deformation of the initial curve into the final segmented boundary via registration. Thus, following the work of Le Guyader and Vese [209, 210], the binary segmentation functional from Chapter 8

$$F(c_1, c_2, \phi) = \int_\Omega \Big[\nu_1 |R - c_1|^2 H(\phi) + \nu_2 |R - c_2|^2 (1 - H(\phi))\Big]\, d\mathbf{x} \\ + \mu \int_\Omega |DH(\phi)|$$

(R is the given image, ϕ is a level set function describing the unknown contour, H is the one-dimensional Heaviside function) can be reformulated as a warping problem between the binary image defining the initial contour and the (unknown) binary segmented image. Or the described model can also be used for registration between two (binary-like) images: having a segmentation of one of the images defined via a displacement field, this is used as initial guess in the "registration segmentation" model, to segment and/or register the second image. The main ingredients of the described minimization model are thus the binary segmentation model from Chapter 8 and registration via a nonlinear elasticity smoother solved in a simplified way. The unknown transformation is substituted for the unknown level set function ϕ, with an appropriate regularization as a substitute for the length term. Topology-preserving segmentation results can be obtained, together with applications to registration of binary-like images.

As presented in Chapter 9, in many applications, such as medical imaging, topology constraint is a desirable property in segmentation (and registration). In other words, if we know the topology of the desired contour (e.g., topological equivalent to a circle), then the initial contour (a circle for instance) should be deformed without change of topology (without merging or breaking). This is difficult to achieve in a standard level set implementation of the active contour. Topology-preserving segmentation techniques have been proposed by Han, Xu and Prince [168], Sundaramoorthi and Yezzi [306], Le Guyader and Vese [208] (see also work by Alexandrov and Santosa [6] for a material science application), and more recently by Arrate, Ratnanather and Younes [24], among other works. For more details on this subject, we refer the reader to Chapter 9.

11.1.1 Registration methods

In a survey, Sotiras et al. ([299]) provide an extensive overview of existing registration methods and analyze the three components they consider parts of a registration algorithm:

(1) a deformation model *(itself organized into three classes: geometric transformations derived from physical models — like the one considered in this chapter; geometric transformations derived from interpolation theory; knowledge-based geometric transformations)*,

(2) an objective function, i.e., a matching criterion *(geometric methods including landmarks; iconic methods; hybrid methods)*, and an

(3) optimization method *(continuous methods; discrete methods; greedy approaches)*.

An extensive review of registration techniques can also be found in the work of Modersitzki [238], [239]. These techniques can be partitioned into two classes: parametric and non-parametric. In the non-parametric methods (present framework), unlike parametric ones, the transformation is not restricted to a parameterizable set. The problem is phrased as a functional minimization whose unknown is the displacement vector field $\mathbf{u}$. Denoting by T the template and by R the reference, the introduced functional combines a distance measure component $\mathcal{D}[R, T, \mathbf{u}]$ and a smoother on the displacement vector field $\mathcal{S} = \mathcal{S}[\mathbf{u}]$ to remove the ill-posed character of the problem. Usually, the distance measure is intensity-driven and is chosen to be the L^2−norm of the difference between the deformed template and the reference (suitable when the images have been acquired through similar sensors), i.e., $\mathcal{D}[R,T,\mathbf{u}] = \frac{1}{2}\int_\Omega (T(\mathbf{x}+\mathbf{u}) - R(\mathbf{x}))^2\,d\mathbf{x}$, but one could also use correlation-based or mutual information-based techniques [238].

Several methods to regularize the displacement vector field have been investigated. One is the elastic registration introduced by Broit [67], in which the objects to be registered are considered to be the observations of an elastic body before and after being subjected to a deformation. The smoother $\mathcal{S} = \mathcal{S}[\mathbf{u}]$ is chosen to be the linearized elastic potential of the displacement vector field $\mathbf{u}$ and its expression integrates the Lamé coefficients λ, μ which reflect material properties. A drawback is that this smoother is not suitable for problems involving large deformations. To circumvent this problem, Christensen et al. [101] proposed a viscous fluid registration model (not in a variational form) in which objects are viewed as fluids evolving in accordance with the fluid-dynamic Navier–Stokes equations. More precisely, it consists in minimizing the linearized elastic potential of the velocity of the displacement vector field (nonlinearly related to the displacement vector field via the material derivative). With time, the constraint weakens as $\mathbf{u}$ reaches a steady state and large deformations are therefore possible. One drawback of this method is the computational cost. Numerically, the image-related force field is first computed at time t. Fixing the force field $\mathbf{f}$, the linear PDE satisfied by the velocity is solved by means of a successive over-relaxation (SOR) scheme. Then an explicit Euler scheme is used to advance the displacement vector field in time. In the diffusion registration model introduced by Fischer and Modersitzki [143], the smoother is based on the semi-norm of $H^1(\Omega, \mathbb{R}^n)$ of $\mathbf{u} = (u_1, \cdots, u_n)^T$, Ω being an open bounded subset of $\mathbb{R}^n$. Regularizing properties motivate this choice (it minimizes oscillations of all components of $\mathbf{u}$) rather than physical ones but here again only small deformations can be expected. In the curvature-based registration model introduced by Fischer and Modersitzki [144], [145], the $\dot{H}^2$ (biharmonic) regularization is employed. Affine linear transformations belong to the kernel of the regularizer $\mathcal{S}[\mathbf{u}]$, which is not the case in elastic, viscous fluid or diffusion registration. But here again, transformations are restricted to small deformations. To circumvent this drawback, a nonlinear elasticity-based smoother is used in this chapter that allows larger deformations.

Many improvements or alternatives of these non-parametric methods have been proposed. These include [164], [165], [349], [217], [216], [214].

In [164], Haber and Modersitzki address non-parametric image registration under volume-preserving constraints. They propose to restrict the set of feasible mappings by adding a volume-preserving constraint which forces the Jacobian of the deformation to be equal to one.

Papers [165] and [349] are dedicated to topology-preserving constraints applied to the deformation mapping to keep it diffeomorphic. These methods differ from classical regridding techniques (see [101]) in which one monitors the values of the Jacobian of the deformation, stops the process when the values drop below a threshold and reinitializes the process using the deformed template obtained at the previous step. In [165], Haber et al. propose to monitor the Jacobian of the deformation to prevent the deformation from exhibiting twists and foldings. More precisely, the authors aim at keeping the Jacobian bounded, which leads to an inequality-constrained minimization problem. An information theoretic-based approach is proposed in [349] to generate diffeomorphic mappings and to monitor the statistical distribution of the Jacobian. The authors propose to quantify the magnitude of the deformation by means of the Kullback–Leibler distance between the probability density function associated with the deformation and the identity mapping.

Also, some new frameworks have been studied. In [217], the authors propose to quantify differences between images by matching gradient fields. In [216], Liao et al. propose a level set-based framework for matching overlapping and non-overlapping shapes and open curves.

Pennec and collaborators have done extensive work on image registration for medical applications with large deformations by variational principles, fluid registration and demon algorithms. We refer the reader to [69] among many other works. Also, we refer to Ashburner [25] and Maheswaran et al. [228] for other examples of image registration techniques.

In Burger et al. [76], the authors design a hyperelastic regularizer. More precisely, they build a hyperelastic stored energy function penalizing variations of lengths and areas, and add a penalty term on the Jacobian determinant such that the energy tends to infinity as $\det \nabla\varphi$ tends to 0 and such that shrinkage and growth have the same price. The numerical implementation is based on a discretize-then-optimize strategy and the authors use a generalized Gauss-Newton scheme to compute a numerical minimizer.

Droske and Rumpf [131] propose a related work that uses nonlinear elasticity principles. The authors address the issue of non-rigid registration of multimodal image data, which means that there is no correlation of image intensities at corresponding positions. A suitable deformation is determined via the minimization of a morphological matching functional which locally measures the defect of the normal fields of the set of level lines of the template image and the deformed reference image. Their matching criterion thus already includes first order derivatives of the deformation. This matching functional is complemented by a nonlinear elastic regularization; they consider a general energy functional of the form $E_{reg}[\Phi] = \int_\Omega W(D\Phi, \mathrm{Cof}\, D\Phi, \det\, D\Phi)\, d\mu$, where $W : \mathbb{R}^{d\times d} \times \mathbb{R}^{d\times d} \times \mathbb{R} \to \mathbb{R}$ is supposed to be convex.

By comparison with some of these methods, the only input required in the method described in this chapter is a fixed level set function representing the template image, that is, partitioning the image into two regions. Also, the method jointly treats segmentation and registration: the distance measure is devised using the binary segmentation criterion from Chapter 8, while registration is jointly performed and guided by the segmentation process. The method applies to a specific class of images, since the binary criterion is used.

Before depicting the main model of this chapter, we would like to mention other relevant work for joint segmentation and registration.

11.1.2 Joint segmentation and registration methods

In [350], Yezzi et al. also suggest jointly treating segmentation and registration. Denoting by $R : \Omega \subset \mathbb{R}^2 \to \mathbb{R}$ and $T = \widehat{R} : \widehat{\Omega} \subset \mathbb{R}^2 \to \mathbb{R}$ the two images containing a common object to be registered and segmented, their goal is to find a closed curve $C \subset \Omega$ which captures the boundary of an object in image R and another closed curve $\widehat{C} \subset \widehat{\Omega}$ which captures the boundary of the corresponding object in image $\widehat{R}$. These closed curves are related through the mapping $\mathbf{g} : \mathbb{R}^2 \to \mathbb{R}^2$, $g \in G$, (G, finite dimensional group, for instance the groupe of rigid motions) such that $\widehat{C} = \mathbf{g}(C)$. If the fidelity criterion is also defined in terms of a region-based energy as in this chapter, the regularizer is different based on the mean curvature flow, in order to ensure that the evolving contour remains smooth. Also, the functional to be minimized and presented in this chapter is straightforwardly phrased in the level set framework and the class of admissible deformations (rigid, and others) is not an input in the described model. It should be pointed out that their model, first exposed in the context of rigid deformations, has been extended to non-rigid motions [345], [339], [332].

We would also like to mention the interesting work by Lord et al. [222] which uses a matching criterion based on metric structure comparison. The authors propose a unified method that simultaneously treats segmentation and registration by introducing two unknowns in the process: the deformation map and the segmenting curve. The segmentation process is guided by the registration map. The matching criterion, unlike classical registration methods, rests on the minimization of deviation from isometry. It is based on the metric structure comparison of the surfaces, more precisely on their first fundamental form, and on a homogeneity constraint as in [94]. Thus, contrary to the model described in this chapter, in which the expected curve delineates two regions with homogeneous intensity, their criterion is still based on metric structure comparisons to disconnect normal regions from abnormal ones.

We would also like to mention the related work by Vemuri et al. [334], [334]. The authors propose a coupled PDE model to perform both segmentation and registration. In the first PDE, the level sets of the source image are evolved along their normals with a speed defined as the difference between the target and the evolving source image. The second PDE allows explicit retrieval of the displacement vector field. In particular, in the work of Vemuri-Chen [333] for joint registration and segmentation, the piecewise-smooth level set segmentation model presented in Chapter 8 is combined with prior shape information through global alignment. We also refer the reader to [100] in which a geodesic active contour-based model including a shape prior is presented, and [99] in which a shape prior is incorporated in the Mumford–Shah segmentation model from Chapter 6.

Related work is presented in [133], on an atlas-based segmentation of medical images locally constrained by level sets. We wish to refer to Rouchdy et al. [274] for another segmentation method that also uses nonlinear elasticity to define the deformation of the evolving contour or surface. The segmentation criterion is based on the gradient vector flow [346], and a deformation field is computed via nonlinear elasticity using the finite element method. For completeness, we also refer the reader to [44], [237] for a variational registration method for large deformations and to [266], for a related work which also uses nonlinear elasticity regularization and is implemented using the finite element method. In [266], the authors also devise a hyperelastic material model valid for large-magnitude deformations. The proposed mathematical framework is very general to the extent that the stored energy density function is written as a function of the location in the reference configuration and the right Cauchy–Green strain tensor, without additional details. This yields a general variational problem which is solved using a three-dimensional finite element approach.

Finally, we would like to mention the work of Droske et al. [132, 130], where segmentation and registration criteria are also combined as in the present chapter. The general Mumford and Shah functional from Chapter 6 is used in the minimization combined with registration of the unknown edge sets. The regularization on the deformation is imposed via variational nonlinear elasticity principles in [130], and via the above-mentioned morphological matching functional in [132], as in [131]. Moreover, in [132] and [130], the deformation and the edge sets are considered separate variables. For the numerical implementation, [132] and [130] use finite elements.

In the method described in this chapter, we view the shapes to be matched as Ciarlet–Geymonat materials, and we minimize a segmentation-like functional that is solved using registration techniques. Its main characteristics are:

(1) only the deformation is the explicit unknown, and not the edge set;

(2) for the numerical implementation, we use finite differences and thus a fixed regular grid (also, a finite differences algorithm on a regular grid can be much more easily extended to three dimensions, by comparison with a finite element discretization);

(3) we also propose a different method for the minimization, based on the introduction of an auxiliary variable and the use of an augmented Lagrangian method; and

(4) we present a complete theoretical analysis of the minimization problem, proving the existence of minimizers.

11.2 Description of model

The scope of the proposed method is twofold:

(1) Devise a theoretically well-motivated model in which (binary) segmentation and registration are jointly performed.

(2) Large and smooth deformations must be authorized while keeping the deformation map topology-preserving.

We see in this section how these criteria are fulfilled.

Let us emphasize that the focus of this chapter is on the mathematical presentation of a nonlinear elasticity-based combined segmentation and registration method. Hence, the computational results are still restricted to two dimensions. However, as will be seen in Section 11.3, the proposed algorithm can be easily adapted to the three-dimensional case.

Distance measure criterion

There are forward and backward registrations. In this chapter, we adopt the Eulerian framework to find a backward transformation φ such that the grid points $\mathbf{y}$ in the deformed image template originate from non-grid points $\mathbf{x} = \varphi(\mathbf{y})$.

Let Ω be a connected bounded open subset of $\mathbb{R}^3$ with Lipschitz boundary $\partial\Omega$. Let us denote by $R : \bar{\Omega} \to \mathbb{R}$ the reference image to be segmented (later we will discuss how the proposed method can be used for registration between a template image $T : \bar{\Omega} \to \mathbb{R}$ and the reference image R; initially, the method is defined as a segmentation method based on Chapter 8. The shape from the template image is supposed to be modelled via a Lipschitz function Φ_0 whose zero-level line is the shape boundary. Denoting by $\mathcal{C}$ the zero-level set of Φ_0 and $w \subset \Omega$ the open set it delineates, Φ_0 is such that

$$\begin{cases} \mathcal{C} = \{\mathbf{x} \in \Omega \mid \Phi_0(\mathbf{x}) = 0\} \\ w = \{\mathbf{x} \in \Omega \mid \Phi_0(\mathbf{x}) > 0\} \\ \Omega - \bar{w} = \{\mathbf{x} \in \Omega \mid \Phi_0(\mathbf{x}) < 0\} . \end{cases}$$

For theoretical and numerical purposes, we may consider a linear extension operator (see [65, p. 158]) $P : W^{1,\infty}(\Omega) \to W^{1,\infty}(\mathbb{R}^3)$ such that for all $\Phi \in W^{1,\infty}(\Omega)$ (i) $P\Phi_{|\Omega} = \Phi$, (ii) $\|P\Phi\|_{L^\infty(\mathbb{R}^3)} \leq C\,\|\Phi\|_{L^\infty(\Omega)}$ and (iii) $\|P\Phi\|_{W^{1,\infty}(\mathbb{R}^3)} \leq C\,\|\Phi\|_{W^{1,\infty}(\Omega)}$, with C depending only on Ω. By this extension process, we consider that $\Phi_0 \in W^{1,\infty}(\mathbb{R}^3)$ to ensure that $\Phi_0 \circ \varphi$ with φ introduced later is always defined. The deformation of the evolving curve is made to satisfy a segmentation criterion. Indeed, the distance measure we employ is related to the fitting term of the active contours without edges model from Chapter 8. In this way, registration and segmentation are correlated and we expect at the end of the process to obtain the segmentation of the reference image as well as a smooth deformation map. It results in a region-based intensity approach and no longer in a pointwise process as usually done.

Let $\varphi : \bar{\Omega} \to \mathbb{R}^3$ be the deformation. A deformation is a smooth enough mapping that is orientation-preserving and injective except possibly on $\partial\Omega$. As stressed by Ciarlet ([106, p. 26]), the reason a deformation may lose its injectivity on the boundary of Ω is that self-contact must be allowed. We also denote by $\mathbf{u}$ the associated displacement such that $\varphi = \mathrm{id} + \mathbf{u}$. The deformation gradient is $\nabla\varphi = \mathrm{I} + \nabla\mathbf{u}$, $\bar{\Omega} \to M_3(\mathbb{R})$, $M_3(\mathbb{R})$ being the set of all real square matrices of order 3. Thus the idea is to find a smooth vector field $\mathbf{u}$ defined above, $\mathbf{u} = (u_1, u_2, u_3) : \Omega \to \mathbb{R}^3$, $\mathbf{x} \mapsto (u_1(\mathbf{x}), u_2(\mathbf{x}), u_3(\mathbf{x}))$, for each $\mathbf{x} \in \Omega$, such that the zero-level line of Φ defined by $\Phi(\mathbf{x}) = \Phi_0(\mathbf{x} + \mathbf{u}(\mathbf{x}))$ fits the boundary of the object to be warped in the given reference image. We emphasize that Φ_0 is not an unknown in the segmentation-registration problem. Denoting by H the one-dimensional Heaviside function, by $\nu_1, \nu_2 > 0$ two fixed parameters and c_1 and c_2 being two unknown constants depending on Φ_0 (fixed), given image R and unknown displacement $\mathbf{u}$, the distance measure functional F_d (the segmentation criterion) is defined by

$$\begin{aligned} F_d(c_1, c_2, \mathbf{u}) &= \nu_1 \int_\Omega |R(\mathbf{x}) - c_1|^2 H\left(\Phi_0(\mathbf{x} + \mathbf{u}(\mathbf{x}))\right) d\mathbf{x} \\ &+ \nu_2 \int_\Omega |R(\mathbf{x}) - c_2|^2 \left(1 - H\left(\Phi_0(\mathbf{x} + \mathbf{u}(\mathbf{x}))\right)\right) d\mathbf{x}. \quad (11.1) \end{aligned}$$

In the calculations, the Heaviside function being not differentiable, it is replaced by a $\mathcal{C}^\infty$ regularization defined by $H_\epsilon(z) = \dfrac{1}{2}\left(1 + \dfrac{2}{\pi}\arctan\dfrac{z}{\epsilon}\right)$ as in Chapters 2 and 8, so we set

$$\begin{aligned} F_d(c_1, c_2, \mathbf{u}) &= \nu_1 \int_\Omega |R(\mathbf{x}) - c_1|^2 H_\epsilon\left(\Phi_0(\mathbf{x} + \mathbf{u}(\mathbf{x}))\right) d\mathbf{x} \\ &+ \nu_2 \int_\Omega |R(\mathbf{x}) - c_2|^2 \left(1 - H_\epsilon\left(\Phi_0(\mathbf{x} + \mathbf{u}(\mathbf{x}))\right)\right) d\mathbf{x}. \quad (11.2) \end{aligned}$$

We need to add a regularization term of the form $F_{reg}(\mathbf{u})$ to (11.2), which is a substitute for the length term of the evolving curve, and therefore the unknown $\Phi(\mathbf{x})$ from Chapter 8 is replaced by $\Phi_0(\mathbf{x} + \mathbf{u}(\mathbf{x}))$, with Φ_0 fixed. Thus we obtain a binary segmentation method that can also be used for registration. The segmentation problem impacts the registration result and conversely; therefore, these problems cannot be considered independent in the model.

Introduction of a nonlinear elasticity-based regularizer

A regularizing term F_{reg} is presented to ensure the smoothness of the displacement vector field $\mathbf{u}$. To allow large displacements, we use a nonlinear elasticity-based smoother. As stressed by Fischer and Modersitzki ([145]), the smoother depends on the specific properties required for the displacement vector field and is related to the physics of the object under consideration. As mentioned in [266], the theory of linear elasticity is not suitable for the present work because it assumes small strains and validity of Hooke's law, both of which are invalid for large-magnitude material deformations. It is thus necessary to turn to the nonlinear deformation theory. We view the shapes to be registered as hyperelastic materials.

We recall that the right Cauchy–Green strain tensor is defined by $\mathbf{C} = \nabla\varphi^T\nabla\varphi = \mathbf{F}^T\mathbf{F} \in \mathcal{S}^3$ with $\mathcal{S}^3 = \{\mathbf{A} \in M_3 = M_3(\mathbb{R}),\ \mathbf{A} = \mathbf{A}^T\}$, the set of all symmetric matrices of order 3. Physically, the right Cauchy–Green tensor can be regarded as a quantifier of the square of local change in distances due to deformation.

The Green–Saint-Venant strain tensor is defined by $\mathbf{E} = \frac{1}{2}\left(\nabla\mathbf{u} + \nabla\mathbf{u}^T + \nabla\mathbf{u}^T\nabla\mathbf{u}\right)$. Associated with a given deformation φ, it is a measure of the deviation between φ and a rigid deformation. We also need the following notations:

- $M_3^+(\mathbb{R}) = \{\mathbf{A} \in M_3(\mathbb{R}),\ \det(\mathbf{A}) > 0\}$
- $\mathbf{A} : \mathbf{B} = \mathrm{tr}\mathbf{A}^T\mathbf{B}$, the matrix inner product in $\mathbb{R}^n$
- $||\mathbf{A}|| = \sqrt{\mathbf{A} : \mathbf{A}}$, matrix norm in $\mathbb{R}^n$
- $\mathrm{Cof}\,\mathbf{A}$, cofactor matrix of A

The stored energy of an isotropic, homogeneous, hyperelastic material, if the reference configuration is a natural state, is of the form

$$W(\mathbf{F}) = \widehat{W}(\mathbf{E}) = \frac{\lambda}{2}(\mathrm{tr}\ \mathbf{E})^2 + \mu\mathrm{tr}\ \mathbf{E}^2 + o(||\mathbf{E}||^2),\ \ \mathbf{F}^T\mathbf{F} = \mathbf{I} + 2\mathbf{E}. \tag{11.3}$$

The stored energy function of a Saint-Venant–Kirchhoff material is defined by

$$W(\mathbf{F}) = \widehat{W}(\mathbf{E}) = \frac{\lambda}{2}(\mathrm{tr}\ \mathbf{E})^2 + \mu\mathrm{tr}\ \mathbf{E}^2. \tag{11.4}$$

The Saint-Venant–Kirchhoff material is thus the simplest one that agrees with the expansion (11.3) and its stored energy can be written

$$W(\mathbf{F}) = -\frac{3\lambda + 2\mu}{4}\mathrm{tr}\ \mathbf{C} + \frac{\lambda + 2\mu}{8}\mathrm{tr}\ \mathbf{C}^2 + \frac{\lambda}{4}\mathrm{tr}\ \mathrm{Cof}\ \mathbf{C} + \frac{6\mu + 9\lambda}{8}. \tag{11.5}$$

Raoult ([267]) proved that the stored energy function of a Saint-Venant–Kirchhoff material is not polyconvex. It is in fact not rank-1 convex and consequently not quasiconvex, which raises a drawback of theoretical nature since we cannot obtain the weak lower semi-continuity of the introduced functional. Also, as stressed by Ciarlet in [105], the stored energy lacks a term preventing $\det(\nabla\varphi)$ to approach zero. Despite these theoretical drawbacks, the simpler stored energy of a Saint-Venant–Kirchhoff material will be used in the numerical implementations providing satisfactory experimental results.

In [107], Ciarlet and Geymonat build a family of Ogden materials whose polyconvex stored energy is derived to comply with the expansion of the stored energy function of a homogeneous, isotropic, hyperelastic material near a natural state, defined by (11.3).

Given $\lambda > 0$ and $\mu > 0$ Lamé coefficients, we consider the Ciarlet–Geymonat material with stored energy

$$W_{CG} : \mathbf{F} \in M_3^+(\mathbb{R}) \mapsto W_{CG}(\mathbf{F}) = a_1 \|\mathbf{F}\|^2 + a_2 \|\mathbf{F}^T\mathbf{F}\|^2 + b \|\text{Cof } \mathbf{F}\|^2 + \Gamma(\det(\mathbf{F})) + e, \tag{11.6}$$

which can be rewritten, with $\mathbf{C} = \mathbf{F}^T\mathbf{F}$ as

$$W_{CG} : \mathbf{F} \in M_3^+(\mathbb{R}) \mapsto W_{CG}(\mathbf{F}) = a_1 \text{tr } \mathbf{C} + a_2 \text{tr } \mathbf{C}^2 + b \text{tr Cof } \mathbf{C} + \Gamma(\sqrt{\det (\mathbf{C})}) + e,$$

with $a_1, a_2 > 0$, $b > 0$, $\Gamma(\delta) = c\,\delta^2 - d \ln \delta$, $c > 0$, $d > 0$, $e \in \mathbb{R}$ and Γ convex, twice differentiable at the point 1. One has: $\lim_{\delta \to 0^+} \Gamma(\delta) = +\infty$ and $\lim W_{CG}(\mathbf{F}) = +\infty$ when $\det \mathbf{F} \to 0^+$. Thus the expression of the stored energy function is reminiscent of the one of the Saint-Venant–Kirchhoff material. The norm of the Jacobian of the deformation $\|\mathbf{F}\|$ controls the isotropically averaged change of length under the deformation. The norm of the matrix of cofactors of the Jacobian of the deformation $\|\text{Cof}\,\mathbf{F}\|$ is the proper measure for the averaged change of area. Finally, the local volume transformation under a deformation φ is represented by $\det(\nabla\varphi) = \det(\mathbf{F})$. The stored energy of the Ciarlet–Geymonat material will be used in the theoretical analysis of the model.

Total energy functional

Based on the above concepts, the described model uses in the theoretical analysis the Ciarlet–Geymonat stored energy function defined in (11.6) and thus the nonlinear elasticity regularizer that will be coupled with the distance measure functional F_d is defined by:

$$R_{eg}(\varphi) = \int_\Omega W_{CG}(\nabla\varphi)\, d\mathbf{x}. \tag{11.7}$$

Finally, the total energy E_{total} considered is given by

$$\begin{aligned} E_{total}(c_1, c_2, \varphi) &= F_d(c_1, c_2, \varphi) + R_{eg}(\varphi), \qquad (11.8)\\ &= \int_\Omega W_{CG}(\nabla\varphi)\, d\mathbf{x} + \nu_1 \int_\Omega (R(\mathbf{x}) - c_1)^2 H_\epsilon(\Phi_0(\varphi(\mathbf{x})))\, d\mathbf{x} \\ &\quad + \nu_2 \int_\Omega (R(\mathbf{x}) - c_2)^2 (1 - H_\epsilon(\Phi_0(\varphi(\mathbf{x}))))\, d\mathbf{x}. \end{aligned}$$

The goal is then to seek for a minimum of

$$E_{total}(\varphi) = \int_\Omega W_{CG}(\nabla\varphi)\, d\mathbf{x} + \nu_1 \int_\Omega (R(\mathbf{x}) - c_1(\varphi))^2 H_\epsilon\left(\Phi_0(\varphi(\mathbf{x}))\right)\, d\mathbf{x}$$
$$+\, \nu_2 \int_\Omega (R(\mathbf{x}) - c_2(\varphi))^2 \left(1 - H_\epsilon\left(\Phi_0(\varphi(\mathbf{x}))\right)\right)\, d\mathbf{x}, \qquad (11.9)$$

where φ belongs to a suitable space that will be defined hereafter and where

$$c_1 = \frac{\int_\Omega R(\mathbf{x}) H_\epsilon\left(\Phi_0(\varphi(\mathbf{x}))\right)\, d\mathbf{x}}{\int_\Omega H_\epsilon\left(\Phi_0(\varphi(\mathbf{x}))\right)\, d\mathbf{x}}, \qquad (11.10)$$

$$c_2 = \frac{\int_\Omega R(\mathbf{x}) \left(1 - H_\epsilon\left(\Phi_0(\varphi(\mathbf{x}))\right)\right)\, d\mathbf{x}}{\int_\Omega \left(1 - H_\epsilon\left(\Phi_0(\varphi(\mathbf{x}))\right)\right)\, d\mathbf{x}}, \qquad (11.11)$$

by fixing φ and minimizing $E_{total}(c_1, c_2, \varphi)$ with respect to c_1 and c_2.

Existence of minimizers

We start by recalling the definition of polyconvexity which is a key argument for obtaining the existence of minimizers.

Definition 22 (Definition 5.1, Chapter 5 of [114]) *A function* $f : \mathbb{R}^{N\times n} \to \mathbb{R} \cup \{+\infty\}$ *is said to be polyconvex if there exists* $F : \mathbb{R}^{\tau(n,N)} \to \mathbb{R} \cup \{+\infty\}$ *convex, such that*

$$f(\xi) = F(T(\xi)),$$

where $T : \mathbb{R}^{N\times n} \to \mathbb{R}^{\tau(n,N)}$ *is such that*

$$T(\xi) := (\xi, adj_2\,\xi, \cdots, adj_{n\wedge N}\,\xi)\,.$$

In the preceding definition, $adj_s\,\xi$ *stands for the matrix of all* $s\times s$ *minors of the matrix* $\xi \in \mathbb{R}^{N\times n}$, $2 \le s \le n \wedge N = \min\{n, N\}$ *and*

$$\tau(n,N) := \sum_{s=1}^{n\wedge N} \sigma(s), \quad \text{where } \sigma(s) := \binom{N}{s}\binom{n}{s} = \frac{N!n!}{(s!)^2(N-s)!(n-s)!}.$$

In our case, the function $g : \mathbf{H} \in M_3(\mathbb{R}) \to g(\mathbf{H}) = \mathrm{tr}\ \mathbf{H}^T\mathbf{H} = \|\mathbf{H}\|^2$ is convex, so $\mathbf{F} \in M_3(\mathbb{R}) \to \mathrm{tr}\ \mathrm{Cof}\,\mathbf{F}^T\mathrm{Cof}\,\mathbf{F}$ is polyconvex.

It is convenient to express $\|\mathbf{F}^T\mathbf{F}\|^2 = \mathrm{tr}\,(\mathbf{F}^T\mathbf{F})^2$ in terms of the singular values of $\mathbf{F}$. If v_1, v_2, v_3 are the singular values of $\mathbf{F}$, $\mathrm{tr}\,(\mathbf{F}^T\mathbf{F})^2 = v_1(\mathbf{F})^4 + v_2(\mathbf{F})^4 + v_3(\mathbf{F})^4$.

Theorem 36 *(Ball [36], Thompson and Freede [318] taken from [105], p. 131).*

Let $\Phi : [0, +\infty[^n \to \mathbb{R}$ be a convex, symmetric, and increasing function with respect to each variable. Then the function

$$W : \mathbf{F} \in M_n(\mathbb{R}) \to W(\mathbf{F}) = \Phi\left(v_1(\mathbf{F}), v_2(\mathbf{F}), \cdots, v_n(\mathbf{F})\right)$$

is convex.

The function $\mathbf{F} \mapsto v_1(\mathbf{F})^4 + v_2(\mathbf{F})^4 + v_3(\mathbf{F})^4$ is thus convex and consequently the proposed stored energy function (11.6) is polyconvex.

We now need the following preliminary results related to the weak continuity of the functions determinant and adjugate matrices.

Theorem 37 *(partially taken from Theorem 8.20, Chapter 8 of [114]).*

Let $\Omega \subset \mathbb{R}^3$ be a bounded open set, $1 < p < \infty$, and let

$$u_\nu \rightharpoonup u \quad in\ W^{1,p}(\Omega, \mathbb{R}^3).$$

If $p > 2$, then

$$adj_2\, \nabla u_\nu \rightharpoonup adj_2\, \nabla u \quad in\ L^{\frac{p}{2}}(\Omega, \mathbb{R}^9).$$

If $p \geq 3$, then

$$\det \nabla u_\nu \rightharpoonup \det \nabla u \quad in\ \mathcal{D}'(\Omega).$$

If $p > 3$, then

$$\det \nabla u_\nu \rightharpoonup \det \nabla u \quad in\ L^{\frac{p}{3}}(\Omega).$$

Theorem 38 (Theorem VI.3.3 of [207]) *Let $p \geq 2$ and $q \geq 1$ be such that $\frac{1}{p} + \frac{1}{q} = \frac{1}{s} \leq 1$. If $\Psi \in W^{1,p}(\Omega, \mathbb{R}^3)$ is such that $Cof\nabla\Psi \in L^q(\Omega, M_3)$, then $\det(\nabla\Psi) \in L^s(\Omega)$. Moreover, if $\Psi_n \rightharpoonup \Psi$ in $W^{1,p}(\Omega, \mathbb{R}^3)$, $Cof\nabla\Psi_n \rightharpoonup H$ in $L^q(\Omega, M_3)$ and $\det(\nabla\Psi_n) \rightharpoonup \delta$ in $L^r(\Omega)$ for a certain $r \geq 1$, then $H = Cof\nabla\Psi$ and $\delta = \det(\nabla\Psi)$.*

Equipped with these results, we are now able to state the main theorem of the existence of minimizers.

Theorem 39 *Let $\mathcal{W}$ be the functional space defined by*

$$\mathcal{W} = \{\varphi \in W^{1,4}(\Omega, \mathbb{R}^3) \mid Cof\nabla\varphi \in L^2(\Omega, M_3(\mathbb{R}) \sim \mathbb{R}^9),\ \det(\nabla\varphi) \in L^2(\Omega), \\ \varphi = Id \text{ on } \partial\Omega \text{ and } \det(\nabla\varphi) > 0 \text{ almost everywhere}\}.$$

Let us assume that E_{total} is proper (i.e, that there exists $\tilde{\varphi} \in \mathcal{W}$ such that $E_{total}(\tilde{\varphi}) < +\infty$). Then problem (11.9) admits at least one minimizer in $\mathcal{W}$.

Before giving the broad lines of the proof and for the sake of completeness, we introduce a useful lemma.

Lemma 5 *(generalized Poincaré's inequality, taken from [125, pp. 106-107]) Let Ω be a Lipschitz bounded domain in $\mathbb{R}^N$. Let $p \in [1, +\infty[$ and let $\mathcal{N}$ be a continuous semi-norm on $W^{1,p}(\Omega)$; that is, a norm on the constant functions. Then there exists a constant $C > 0$ depending only on Ω, N and p such that:*

$$\|u\|_{W^{1,p}(\Omega)} \leq C \left(\left(\int_\Omega |\nabla u(\mathbf{x})|^p \, d\mathbf{x} \right)^{\frac{1}{p}} + \mathcal{N}(u) \right).$$

In practice, we will take $\mathcal{N}(u) = \left(\int_{\partial\Omega} |u(\mathbf{x})|^p \, d\sigma \right)^{\frac{1}{p}}$, ($\Omega$ being of class $\mathcal{C}^1$).

Proof. For the sake of conciseness, we assume that $e = 0$.

We remark that $\forall \Psi \in \mathcal{W}$, $E_{total}(\Psi)$ is well-defined and it is an element of $\bar{\mathbb{R}}$. ($W_{CG}(\mathbf{F}) = \mathbb{W}^*(\mathbf{F}, \operatorname{Cof}\mathbf{F}, \det(\mathbf{F}))$ with $\mathbb{W}^*$ convex and continuous on $M_3 \times M_3 \times]0, +\infty[$ and $\mathbb{W}^*(F_k, H_k, \delta_k) \to +\infty$ when $F_k \to F$, $H_k \to H$ and $\delta_k \to 0^+$, so that one can set $\mathbb{W}^*(F, H, \delta) = +\infty$ when $\delta \leq 0$).

In a first step, we prove that the infimum is finite by establishing a coercivity inequality of the form

$$\begin{aligned} &E_{total}(\varphi) \qquad (11.12) \\ &\geq \alpha \left(\|\varphi\|^4_{W^{1,4}(\Omega,\mathbb{R}^3)} + \|\operatorname{Cof}\nabla\varphi\|^2_{L^2(\Omega,M_3(\mathbb{R})\sim\mathbb{R}^9)} + \|\det(\nabla\varphi)\|^2_{L^2(\Omega)} \right) + \kappa, \end{aligned}$$

with κ a real number and $\alpha > 0$.

Using the singular values of F denoted by $v_1(F)$, $v_2(F)$ and $v_3(F)$, since for all $\gamma \geq 1$ the mapping $v = (v_i) \in \mathbb{R}^3 \mapsto (v_1^\gamma + v_2^\gamma + v_3^\gamma)^{\frac{1}{\gamma}}$ is a norm on $\mathbb{R}^3$ and since all norms are equivalent in finite dimension, there exists a constant $l > 0$ such that

$$\left(v_1(F)^4 + v_2(F)^4 + v_3(F)^4\right)^{\frac{1}{4}} \geq l \left(v_1(F)^2 + v_2(F)^2 + v_3(F)^2\right)^{\frac{1}{2}},$$

so

$$\|F^T F\|^2 \geq l^4 \|F\|^4.$$

Moreover, $\Gamma(\delta) = c\delta^2 - d\ln\delta = \frac{c}{2}\delta^2 + \frac{c}{2}\delta^2 - d\ln\delta$. On $\mathbb{R}^{+*}$, the function $\eta : \delta \mapsto \frac{c}{2}\delta^2 - d\ln\delta$ reaches its minimum at $\delta = \sqrt{\frac{d}{c}}$ and $\eta\left(\sqrt{\frac{d}{c}}\right) = \frac{d}{2} - \frac{d}{2}\ln\frac{d}{c}$. Consequently, $\forall \delta \in \mathbb{R}^{+*}$,

$$\Gamma(\delta) \geq \frac{c}{2}\delta^2 + \frac{d}{2} - \frac{d}{2}\ln\frac{d}{c}$$

and

$$\begin{aligned} E_{total}(\varphi) \geq\ & a_2 l^4 \|\nabla\varphi\|^4_{L^4(\Omega,M_3)} + b\|\operatorname{Cof}\nabla\varphi\|^2_{L^2(\Omega,M_3)} + \frac{c}{2}\|\det(\nabla\varphi)\|^2_{L^2(\Omega)} \\ &+ \left(\frac{d}{2} - \frac{d}{2}\ln\frac{d}{c}\right)|\Omega|. \end{aligned}$$

The generalized Poincaré inequality enables us to obtain the desired coercivity inequality.

We now introduce a minimizing sequence (φ^k) such that $\forall k$, $\varphi^k \in \mathcal{W}$. By definition, one has

$$E_{total}(\varphi^k) \longrightarrow \inf_{\Psi \in \mathcal{W}} E_{total}(\Psi) \text{ when } k \to +\infty.$$

According to the hypotheses, there exists $\tilde{\varphi} \in \mathcal{W}$ such that $E_{total}(\tilde{\varphi}) < +\infty$. One can thus always assume that $E_{total}(\varphi^k) \leq E_{total}(\tilde{\varphi}) + 1$ so that the above coercivity inequality (11.12) indicates that φ^k is bounded in $W^{1,4}(\Omega, \mathbb{R}^3)$, $\operatorname{Cof} \nabla\varphi^k$ is bounded in $L^2(\Omega, M_3)$ and $\det(\nabla\varphi^k)$ is bounded in $L^2(\Omega)$. One can thus extract a subsequence, still denoting by (φ^k), such that

$$\begin{cases} \varphi^k \rightharpoonup \bar{\varphi} \quad \text{in} \quad W^{1,4}(\Omega, \mathbb{R}^3), \\ \operatorname{Cof} \nabla\varphi^k \rightharpoonup H \quad \text{in} \quad L^2(\Omega, M_3), \\ \det(\nabla\varphi^k) \rightharpoonup \bar{\delta} \quad \text{in} \quad L^2(\Omega). \end{cases}$$

By uniqueness of the weak limit in L^2, we deduce from Theorem 37 that $H = \operatorname{Cof} \nabla\bar{\varphi}$ and by Theorem 38 that $\bar{\delta} = \det(\nabla\bar{\varphi})$.

The last step consists in letting k tend to $+\infty$. Functional J defined by

$$J(\Psi, H, \delta) = \int_\Omega \mathbb{W}^*(\nabla\Psi(\mathbf{x}), H(\mathbf{x}), \delta(\mathbf{x}))\, d\mathbf{x}$$

with $\mathbb{W}^*(\nabla\Psi, H, \delta) = a_1\|\nabla\Psi\|^2 + a_2\, \|\nabla\Psi^T\nabla\Psi\|^2 + b\|H\|^2 + \Gamma(\delta)$ is defined on $W^{1,4}(\Omega, \mathbb{R}^3) \times L^2(\Omega, M_3) \times L^2(\Omega)$ with values in $\bar{\mathbb{R}}$. It is convex (thanks to the argument of polyconvexity) and strongly sequentially lower semi-continuous since $\mathbb{W}^*$ is convex and continuous. It is thus weakly sequentially lower semi-continuous and

$$J(\bar{\varphi}, \operatorname{Cof} \nabla\bar{\varphi}, \det(\nabla\bar{\varphi})) \leq \liminf_{k \to +\infty} J(\varphi^k, \operatorname{Cof} \nabla\varphi^k, \det(\nabla\varphi^k)).$$

The Rellich–Kondrachov embedding theorem states that $W^{1,4}(\Omega, \mathbb{R}^3)$ is compactly embedded in $\mathcal{C}^0(\bar{\Omega}, \mathbb{R}^3)$, so that weak convergence in $W^{1,4}(\Omega, \mathbb{R}^3)$ implies uniform convergence in $\mathcal{C}^0(\bar{\Omega}, \mathbb{R}^3)$. Then assuming R sufficiently smooth and with the above hypotheses on Φ_0, the dominated convergence theorem enables us to conclude that $c_1 = c_1(\varphi^k) \to c_1(\bar{\varphi})$ when k goes to $+\infty$ and $c_2 = c_2(\varphi^k) \to c_2(\bar{\varphi})$. Still using the dominated convergence theorem, we can then prove that

$$\lim_{k \to +\infty} \int_\Omega \left(R(\mathbf{x}) - c_1(\varphi^k)\right)^2 H_\epsilon(\Phi_0(\varphi^k(\mathbf{x})))\, d\mathbf{x}$$
$$= \int_\Omega \left(R(\mathbf{x}) - c_1(\bar{\varphi})\right)^2 H_\epsilon(\Phi_0(\bar{\varphi}(\mathbf{x})))\, d\mathbf{x},$$
$$\lim_{k \to +\infty} \int_\Omega \left(R(\mathbf{x}) - c_2(\varphi^k)\right)^2 (1 - H_\epsilon(\Phi_0(\varphi^k(\mathbf{x}))))\, d\mathbf{x}$$
$$= \int_\Omega \left(R(\mathbf{x}) - c_2(\bar{\varphi})\right)^2 (1 - H_\epsilon(\Phi_0(\bar{\varphi}(\mathbf{x}))))\, d\mathbf{x},$$

so that $E_{total}(\bar{\varphi}) \leq \liminf_{k \to +\infty} E_{total}(\varphi^k)$.

It remains to prove that $\bar{\varphi} \in \mathcal{W}$. By continuity of the trace, $\bar{\varphi} = Id$ on $\partial\Omega$. (Note that the set $\left\{\varphi \in W^{1,4}(\Omega, \mathbb{R}^3) \mid \varphi = Id \text{ on } \Omega\right\}$ is a strongly closed convex set). Moreover, $\bar{\varphi} \in W^{1,4}(\Omega, \mathbb{R}^3)$, $\mathrm{Cof}\, \nabla\bar{\varphi} \in L^2(\Omega, M_3)$ and $\det(\nabla\bar{\varphi}) \in L^2(\Omega)$ since $E_{total}(\bar{\varphi}) < +\infty$. At last, as

$$\mathbb{W}^* \left(\nabla\bar{\varphi}(\mathbf{x}), \mathrm{Cof}\, \nabla\bar{\varphi}(\mathbf{x}), \det(\nabla\bar{\varphi}(\mathbf{x}))\right) = +\infty \text{ when } \det(\nabla\bar{\varphi}(\mathbf{x})) \leq 0,$$

the set on which this applies must be of null measure. □

11.3 Algorithm

In the numerical discussion, we view the deformation of the initial contour into the final segmented contour as the deformation undergone by Saint-Venant–Kirchhoff materials. The stored energy of Saint-Venant–Kirchhoff materials is given by $W(\mathbf{E}) = \frac{\lambda}{2}(\mathrm{tr}\, \mathbf{E})^2 + \mu\, \mathrm{tr}\, \mathbf{E}^2$ with $\mathbf{E} = \mathbf{E}(\mathbf{u}) = \frac{1}{2}\left(\nabla\mathbf{u} + \nabla\mathbf{u}^T + \nabla\mathbf{u}^T\nabla\mathbf{u}\right)$. Thus the nonlinear elasticity regularizer that will be coupled with the distance measure functional F_d is defined by

$$F_{reg}(\mathbf{u}) = \int_\Omega W(\mathbf{E}(\mathbf{u}))\, d\mathbf{x} = \int_\Omega \left\{\frac{\lambda}{2}(\mathrm{tr}\, \mathbf{E}(\mathbf{u}))^2 + \mu \mathrm{tr}\, \mathbf{E}^2(\mathbf{u})\right\} d\mathbf{x}\ . \quad (11.13)$$

Although this functional does not satisfy known theoretical assumptions, it gives better results than those obtained with linearized models, as will be demonstrated next.

The computation of the Euler Lagrange equation satisfied by $\mathbf{u}$ is cumbersome. Inspired by [249] (which is the variable splitting method), the difficulty of minimizing the nonlinear elasticity functional is overcomed by introducing an auxiliary variable $\mathbf{V}$ for the Jacobian matrix $\nabla\mathbf{u}$. The nonlinear elasticity regularizer is thus applied to $\mathbf{V}$ and no longer to $\nabla\mathbf{u}$, that is, the nonlinearity is no longer in the derivatives of the unknown $\mathbf{u}$. More precisely, letting $\widehat{\mathbf{V}} = \frac{\mathbf{V}^\mathrm{T} + \mathbf{V} + \mathbf{V}^\mathrm{T}\mathbf{V}}{2}$, we redefine the smoothing functional $F_{reg} = F_{reg}(\nabla\mathbf{u}) = F_{reg}(\mathbf{V})$ by

$$F_{reg}(\mathbf{V}) = \int_\Omega W(\widehat{\mathbf{V}})\, d\mathbf{x}, \quad (11.14)$$

with $\mathbf{V} = \nabla\mathbf{u}$.

The total energy E_{total} considered in the remainder of this chapter (numerical part) is thus given by

$$E_{total}(c_1, c_2, \mathbf{u}, \mathbf{V}) = F_d(c_1, c_2, \mathbf{u}) + F_{reg}(\mathbf{V}), \quad (11.15)$$

and the associated constrained optimization problem is formulated by

$$\min_{c_1,c_2,\mathbf{u},\mathbf{V}} E_{total}(c_1,c_2,\mathbf{u},\mathbf{V})$$
$$\text{such that } \mathbf{V}=\nabla\mathbf{u}. \tag{11.16}$$

Problem (11.16) is solved by the augmented Lagrangian method. This algorithm reduces the possibility of ill-conditioning of the subproblems generated in this approach by introducing explicit Lagrange multiplier estimates at each step into the function to be minimized (see [252] for more details).

The augmented Lagrangian method (see also for instance [172], [264] or [272]) has been applied to nonlinear PDE's and mechanics in [157] and more recently in image processing in [311] to solve the Rudin-Osher-Fatemi restoration problem described in Chapter 3. Tai and Wu overcome the difficulty related to the non-differentiability of the considered minimization problem by also introducing an auxiliary variable. They then demonstrate that the dual methods and split Bregman iteration are different procedures to solve the same system resulting from a Lagrangian and penalty approach.

In two dimensions, the augmented Lagrangian method consists in solving problem (11.16) by

$$\begin{aligned}\min_{c_1,c_2,\mathbf{u},\mathbf{V}}\max_{\beta}\mathcal{L}_\alpha\left(c_1,c_2,\mathbf{u},\mathbf{V},\beta\right) &= E_{total}(c_1,c_2,\mathbf{u},\mathbf{V})\\ &+\frac{\alpha}{2}\int_\Omega \|\mathbf{V}-\nabla\mathbf{u}\|^2\,d\mathbf{x}+\int_\Omega \beta:(\mathbf{V}-\nabla\mathbf{u})\,d\mathbf{x},\end{aligned} \tag{11.17}$$

where $\beta=\begin{pmatrix}\beta_{11} & \beta_{12}\\ \beta_{21} & \beta_{22}\end{pmatrix}\in M_2(\mathbb{R})$ is the Lagrange multiplier, α is a positive constant, with $\mathbf{A}:\mathbf{B}=\mathrm{tr}\mathbf{A}^T\mathbf{B}$ the matrix inner product and $\|\mathbf{A}\|=\sqrt{\mathbf{A}:\mathbf{A}}$ the matrix norm.

The goal is thus to search for a saddle point of the functional $\mathcal{L}_\alpha\left(c_1,c_2,\mathbf{u},\mathbf{V},\beta\right)$. We now derive the system of optimality conditions.

We set $H'_\epsilon=\delta_\epsilon$, regularization of the Dirac measure. Fixing $\mathbf{u}$ and $\mathbf{V}$ and minimizing $E_{total}(c_1,c_2,\mathbf{u},\mathbf{V})$ with respect to c_1 and c_2 yields, as in [94]:

$$c_1=\frac{\displaystyle\int_\Omega R(\mathbf{x})H_\epsilon\left(\Phi_0\left(\mathbf{x}+\mathbf{u}(\mathbf{x})\right)\right)d\mathbf{x}}{\displaystyle\int_\Omega H_\epsilon\left(\Phi_0\left(\mathbf{x}+\mathbf{u}(\mathbf{x})\right)\right)d\mathbf{x}},\quad c_2=\frac{\displaystyle\int_\Omega R(\mathbf{x})\left(1-H_\epsilon\left(\Phi_0\left(\mathbf{x}+\mathbf{u}(\mathbf{x})\right)\right)\right)d\mathbf{x}}{\displaystyle\int_\Omega \left(1-H_\epsilon\left(\Phi_0\left(\mathbf{x}+\mathbf{u}(\mathbf{x})\right)\right)\right)d\mathbf{x}}.$$

Computing the first variation of functional $F_d(c_1,c_2,\mathbf{u})$ in (11.2) with respect to $\mathbf{u}$ gives the following gradient:

$$\begin{aligned}\partial_{\mathbf{u}}F_d(c_1,c_2,\mathbf{u}) &= \left(\nu_1(R-c_1)^2-\nu_2(R-c_2)^2\right)\delta_\epsilon\left(\Phi_0\left(\mathbf{x}+\mathbf{u}(\mathbf{x})\right)\right)\\ &\quad\nabla\mathbf{\Phi_0}\left(\mathbf{x}+\mathbf{u}(\mathbf{x})\right).\end{aligned}$$

To finish, setting $\mathbf{V} = (v_{ij})_{1\leq i,j\leq 2}$ and letting

$$c_{01} = v_{11} + v_{22} + \frac{1}{2}\left(v_{11}^2 + v_{12}^2 + v_{21}^2 + v_{22}^2\right),\ c_{02} = 2v_{11} + v_{11}^2 + v_{21}^2,$$

$$c_{03} = 2v_{22} + v_{12}^2 + v_{22}^2,\ c_{04} = v_{12} + v_{21} + v_{11}v_{12} + v_{21}v_{22},$$

we obtain

$$\begin{aligned}
\partial_{v_{11}} F_{reg}(\mathbf{V}) &= (\lambda c_{01} + \mu c_{02})(1 + v_{11}) + \mu c_{04} v_{12}.\\
\partial_{v_{12}} F_{reg}(\mathbf{V}) &= (\lambda c_{01} + \mu c_{03}) v_{12} + \mu c_{04}(1 + v_{11}).\\
\partial_{v_{21}} F_{reg}(\mathbf{V}) &= (\lambda c_{01} + \mu c_{02}) v_{21} + \mu c_{04}(1 + v_{22}).\\
\partial_{v_{22}} F_{reg}(\mathbf{V}) &= (\lambda c_{01} + \mu c_{03})(1 + v_{22}) + \mu c_{04} v_{21}.
\end{aligned} \tag{11.18}$$

The system of optimality conditions is thus given by:

$$\begin{aligned}
&\frac{\partial \mathcal{L}_\alpha}{\partial \mathbf{u}} = \partial_{\mathbf{u}} F_d(c_1, c_2, \mathbf{u}) - \alpha \boldsymbol{\Delta}\, \mathbf{u} + \alpha \begin{pmatrix} \operatorname{div} v_1 \\ \operatorname{div} v_2 \end{pmatrix} + \begin{pmatrix} \operatorname{div} \beta_1 \\ \operatorname{div} \beta_2 \end{pmatrix},\\
&\frac{\partial \mathcal{L}_\alpha}{\partial \mathbf{V}} = \partial_{\mathbf{V}} F_{reg}(\mathbf{V}) + \beta + \alpha\left(\mathbf{V} - \nabla \mathbf{u}\right),\\
&\frac{\partial \mathcal{L}_\alpha}{\partial \beta} = \mathbf{V} - \nabla \mathbf{u}
\end{aligned} \tag{11.19}$$

(where v_1 denotes the first row of $\mathbf{V}$, v_2 the second, similarly for β_1 and β_2).

The augmented Lagrangian method is based on an alternating framework to solve (11.17) as follows.

- Initialize $\beta^0 = \mathbf{0}_{\mathbf{M_2}(\mathbb{R})}$, $\mathbf{V}^0 = \mathbf{0}_{\mathbf{M_2}(\mathbb{R})}$ and $\mathbf{u^0} = \mathbf{0}_{\mathbb{R}^2}$. Select λ, μ, α and ν_1, ν_2. (These last two parameters are most often taken equal to 1.0).
- For $k = 0, 1, 2, \cdots$, compute

$$\left(\mathbf{u}^{k+1}, \mathbf{V}^{k+1}\right) = \operatorname{argmin}_{\mathbf{u},\mathbf{V}} \mathcal{L}_\alpha\left(c_1, c_2, \mathbf{u}, \mathbf{V}, \beta^{\mathbf{k}}\right). \tag{11.20}$$

Update $\beta^{\mathbf{k+1}} = \beta^{\mathbf{k}} + \alpha\left(\mathbf{V^{k+1}} - \nabla \mathbf{u^{k+1}}\right)$.

Algorithm 1: Combined segmentation and registration framework with nonlinear elasticity smoother via augmented Lagrangian method.

To solve problem (11.20), as in [311], we divide it into two subproblems:

$$\operatorname{argmin}_{\mathbf{V}} F_{reg}(\mathbf{V}) + \frac{\alpha}{2}\int_\Omega ||\mathbf{V} - \nabla \mathbf{u}||^2\, d\mathbf{x} + \int_\Omega \beta^{\mathbf{k}} : \mathbf{V}\, d\mathbf{x},$$

for fixed $\mathbf{u}$, and then

$$\operatorname{argmin}_{\mathbf{u}} F_d(c_1, c_2, \mathbf{u}) + \frac{\alpha}{2} \int_\Omega ||\mathbf{V} - \nabla \mathbf{u}||^2 \, d\mathbf{x} - \int_\Omega \beta^{\mathbf{k}} : \nabla \mathbf{u} \, d\mathbf{x},$$

for given $\mathbf{V}$ with c_1 and c_2 defined above.

Numerically, the Euler–Lagrange equations in $\mathbf{u}$ and $\mathbf{V}$ are solved using a gradient descent method, parameterizing the descent direction by an artificial time $t \geq 0$. Systems of four and two equations are obtained (solved by semi-implicit finite difference schemes), equipped with the boundary conditions $\mathbf{u} = \mathbf{0}_{\mathbb{R}^2}$ on $\partial\Omega$. Either we use a Uzawa-type approach, that is, we solve the minimization problem in $\mathbf{u}$ and $\mathbf{V}$ at each step k and then update β or follow an Arrow-Hurwicz approach, leading in this case to only one gradient descent step in the minimization of $\mathbf{u}$ and $\mathbf{V}$ and one gradient ascent step in the maximization of β.

The numerical algorithm uses a regridding technique similar to the one proposed by Christensen et al. [101], and described in more details in the following chapter. The Jacobian $\det(\nabla(\mathbf{Id} + \mathbf{u}))$ is monitored and if it drops below a defined threshold in some parts of the image, the process is reinitialized ($\mathbf{V} = \mathbf{0}_{\mathbf{M_2}(\mathbb{R})}$, $\mathbf{u} = \mathbf{0}_{\mathbb{R}^2}$ and $\beta = \mathbf{0}_{\mathbf{M_2}(\mathbb{R})}$). The only change is that instead of performing the reinitialization step with the last deformed template as done in [101], the last deformed level set function $\Phi_0\,(\cdot + \mathbf{u}(\cdot))$ is used. The overall displacement $\mathbf{u}$ is reconstructed similarly to [101].

Note that the described algorithm can be straightforwardly extended to the three-dimensional case: the expression of $F_d(c_1, c_2, \mathbf{u})$ is unchanged, $\mathbf{V}$ is now an element of $M_3(\mathbb{R})$ as well as $\nabla \mathbf{u}$ and β. The derivation of the Euler–Lagrange equations allows us to solve systems of nine and three equations, respectively and no particular difficulties are encountered.

11.4 Numerical experiments

We conclude this chapter by presenting several results on both synthetic and real images in two dimensions. In most experiments, $\nu_1 = \nu_2 = 1$ but when dealing with complex topologies involving long and thin concavities, these parameters have been increased up to 2.5, or took slightly different values. We illustrate two applications of the proposed method: image registration and topology-preserving segmentation. Since a binary criterion is used, the chosen images are well approximated by binary images. The binary criterion could be replaced by other more general terms with respect to the desired application.

Note that in the following, parameter λ (Lamé's first parameter) is always set to zero. This parameter has no physical meaning. Also, it can be related to Poisson's ratio ν via the relation $\nu = \dfrac{\lambda}{2\lambda + \mu}$. Selecting $\lambda = 0$ yields $\nu = 0$, which is not inconsistent since it is usually between -1 and 0.5. Poisson's ratio is a measure of Poisson's effect which can be regarded as the ability of a material compressed in one direction to expand in the other (two) direction(s) and conversely for stretching, so $\nu = 0$ means that if the material is compressed (its width dimininishes), its height is not altered.

11.4.1 Applications to image registration

The first experimental test shown in Figure 11.1 is similar to those performed by Modersitzki in [238] (see pages 114–115, 129–130, 150–153, 168–170 for comparisons using linear elasticity, diffusion, curvature, or the viscous fluid method), with the goal to illustrate that the model easily handles large displacements. The problem is to warp a black disk to the letter C, both defined on the same image domain. The given data are the template and reference images as well as the curve delineating the disk boundary. The method qualitatively performs in a way similar to the fluid model without requiring the expensive Navier–Stokes solver employed for numerical discretization, and provides two results: the segmentation of the reference image and a smooth displacement vector field **u**. The method allows large deformations unlike the linear elasticity model, diffusion model and curvature-based model for which the registration cannot be accomplished because the images differ too much (see pages 114–115, 150–153, 168–171 from [238]). In this example, four regridding steps were necessary; the transformation was considered admissible if the Jacobian exceeded 0.075. Note that regridding steps were also necessary with the fluid registration model. The parameters were set to $\Delta t = 0.01$ in the gradient descent method, $\lambda = 0$, $\mu = 1000$ and $\alpha = 60000$.

For the next four examples, we illustrate a medical application for image registration through the proposed segmentation-registration algorithm, which is applied to real data. As shown in Figures 11.2 through 11.6, the method is proposed for mapping a two-dimensional slice of mouse brain gene expression data (template) to its corresponding two-dimensional slice of the mouse brain atlas to facilitate the integration of anatomic, genetic, and physiologic observations from multiple subjects in a common space. Since genetic mutations and knock-out strains of mice provide critical models for a variety of human diseases, such linkage between genetic information and anatomical structure is important. The data were provided by the Center for Computational Biology of the University of California, Los Angeles. The mouse atlas database was pre-segmented. The gene expression data were manually segmented. The non-brain regions were removed and set to zero to produce better matching.

First, the described method is applied to the gene data (each top right image in Figures 11.2, 11.4, 11.5, and 11.6) using Φ_0 defining a disk, to segment it and extract a contour; then the method is applied again using as Φ_0 the new obtained contour (shown in the second rows on the left of Figures 11.2,

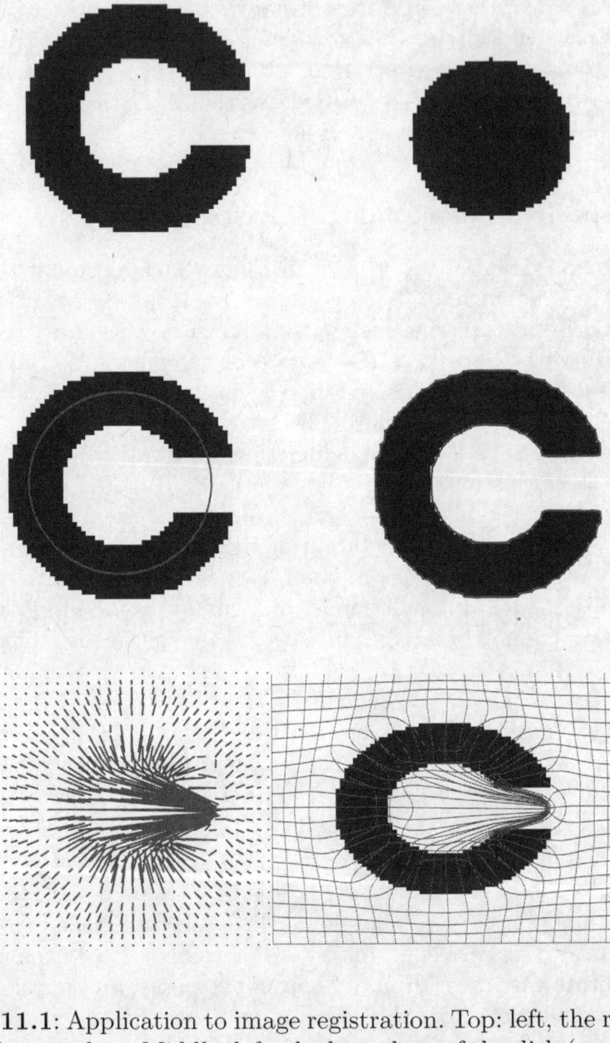

FIGURE 11.1: Application to image registration. Top: left, the reference image; right the template. Middle: left, the boundary of the disk (zero-level set of Φ_0) superimposed on the reference image; right, the segmentation of the letter C. Bottom: left, deformation field using nonlinear elasticity regularization; right, deformation grid from reference to template.

11.4, 11.5, and 11.6), to segment the atlas data (each top left image in Figures 11.2, 11.4, 11.5, and 11.6). The results of the second segmentation step are shown in the second rows, right, of Figures 11.2, 11.4, 11.5, and 11.6). In the process, we obtain a smooth deformation between the gene and the atlas

TABLE 11.1: Values obtained for registration results of mouse data.

Experiment	Regridding steps	det(Jac)≥ tol	Min(det(Jac))	Max(det(Jac))
Fig. 11.2	0	tol = 0.05	0.2813	2.0996
Fig. 11.4	1	tol = 0.05	0.036643	3.48181
Fig. 11.5	3	tol = 0.15	0.0081646	10.207976
Fig. 11.6	4	tol =0.075	0.005684	28.942492

data, as shown in Figures 11.2, 11.4, 11.5, and 11.6, third rows, together with the determinant of the Jacobian matrix of the deformation in the fourth rows. For all four results from Figures 11.2, 11.4, 11.5, and 11.6 on the mouse data, we used the parameters $\Delta t = 0.01$, $\lambda = 0$, $\mu = 3000$, $\alpha = 40000$. The number of regridding steps, together with the tolerance on the minimum of the determinant of the Jacobian and its range are summarized in Table 11.1. Regarding the experiment shown in Figure 11.2, we also display the components of $\mathbf{V}$ and $\nabla \mathbf{u}$ in Figure 11.3 to illustrate a good match between these two variables. In these numerical experiments, the segmentation was correctly achieved after 800 to 1400 iterations with a computational time of a few seconds on two-dimensional slices of the mouse brain of size 200×200.

11.4.2 Applications to topology-preserving segmentation

The following examples illustrate how the method can be used in the case of topology-preserving segmentation. We refer the reader to Chapter 9 for more details on this topic. As we have already mentioned, in many applications such as medical imaging, topology constraint is a desirable property. In other words, if we know the topology of the desired contour (e.g., equivalent with a circle), then the initial contour (a circle for instance) will be deformed without change of topology (without merging or breaking). This is difficult to achieve with a standard level set implementation of the active contour.

We begin with the example in Figure 11.7. This synthetic reference image represents two disks (similar to tests performed in related works [168], [306], [208]). The initial shape Φ_0 is a black ellipse such that when superimposed on the input image R, its boundary encloses the two disks. We aim at segmenting these two disks while maintaining the same topology throughout the process (one path-connected component) and obtaining a smooth displacement vector field $\mathbf{u}$. In this example, $\Delta t = 0.01$, $\lambda = 0$, $\mu = 500$, $\alpha = 4000$ and three regridding steps were necessary. The transformation was considered admissible if the Jacobian exceeded 0.0625. Also, min det $= 2.45\, 10^{-5}$, max det $= 39.1681$ and the total iteration number is 600 (12 main steps in k and 50 iterations in the gradient descent method to find $\mathbf{u^k}$ and $\mathbf{V^k}$).

The method has been tested on two synthetic noisy images. In both cases, the goal was to segment the boundary of the given geometrical figure (topologically equivalent to a circle). In the torus example (Figure. 11.8), $\Delta t = 0.02$, $\lambda = 0$, $\mu = 1000$, $\alpha = 30000$ and one regridding step was necessary. The transformation was considered admissible if the Jacobian exceeded 0.1. Also, min det = 0.039267, max det = 2.8981 and the iteration number is 200 (4 main steps in k and 50 iterations in the gradient descent method to find $\mathbf{u^k}$ and $\mathbf{V^k}$). A similar result is shown in Figure 11.9, where a different initialization is chosen, requiring larger smooth deformation during the process ($\triangle t = 0.02$, $\nu_1 = 1.5$, $\nu_2 = 2.5$, $\mu = 1000$, $\alpha = 30000$, $\min \det = 0.000088944$).

In the triangle example (Figure 11.10), $\Delta t = 0.02$, $\lambda = 0$, $\mu = 1000$, $\alpha = 25000$ and one regridding step was necessary. The transformation was considered admissible if the Jacobian exceeded 0.2. Also, min det = 0.16065, max det = 6.236432 and the iteration number is 250 (5 main steps in k and 50 iterations in the gradient descent method to find $\mathbf{u^k}$ and $\mathbf{V^k}$). Again, a similar result is shown in Figure 11.11, where a different initialization is chosen (less overlapping with the desired object) and required larger smooth deformation during the process ($\triangle t = 0.02$, $\nu_1 = 2.5$, $\nu_2 = 2.5$, $\mu = 1000$, $\alpha = 30000$, $\min \det = 0.0035$).

The method has also been tested on complex slices of brain data. The goal was to register a disk to the outer boundary of the brain with topology preservation. In Figure 11.12, the template image, defined on the same image domain, is made of a disk (shown superimposed on the reference). For this slice of the brain, the parameters are such that $\Delta t = 0.01$, $\lambda = 0$, $\mu = 500$, $\alpha = 54000$ and two regridding steps were necessary. The transformation was considered admissible if the Jacobian exceeded 0.075. Also, min det = 0.01419, max det = 8.369458 and the iteration number is 2500 (5 main steps in k and 500 iterations in the gradient descent method to find $\mathbf{u^k}$ and $\mathbf{V^k}$). The same problem has been investigated in Figures 11.13 and 11.14.

Finally, we show another numerical result on the hand image from Figure 11.15, illustrating topology-preserving segmentation and larger deformations. The method produced similar results with those from Chapter 9. In Figure 11.16, we show two zooms of the region between the fingers in the final segmentation, to better demonstrate that the curve did not undergo any merging or breaking.

11.5 Exercises

Exercise 11.1 *Derive the optimality condition for (11.17) with respect to* c_1 *and* c_2.

Exercise 11.2 *Derive the optimality condition for (11.17) with respect to* $\mathbf{u}$.

Exercise 11.3 *Derive the optimality condition for (11.17) with respect to* $\mathbf{V}$.

Exercise 11.4 *Derive the optimality condition for (11.17) with respect to* β.

Exercise 11.5 *Assuming linearly elastic deformation, derive the Euler–Lagrange equations associated with the minimization of*

$$\begin{aligned} E_{total}(c_1, c_2, \mathbf{u}) =& \nu_1 \int_\Omega |R(\mathbf{x}) - c_1|^2 H_\epsilon\left(\Phi_0(\mathbf{x} + \mathbf{u}(\mathbf{x}))\right) d\mathbf{x} \\ &+ \nu_2 \int_\Omega |R(\mathbf{x}) - c_2|^2 \left(1 - H_\epsilon\left(\Phi_0(\mathbf{x} + \mathbf{u}(\mathbf{x}))\right)\right) d\mathbf{x} \\ &+ \int_\Omega \left\{ \frac{\lambda}{2} (\operatorname{tr} \mathbf{E}(\mathbf{u}))^2 + \mu \operatorname{tr} \mathbf{E}^2(\mathbf{u}) \right\} dx, \end{aligned}$$

where $\mathbf{E}(\mathbf{u}) = \frac{1}{2}(\nabla \mathbf{u} + \nabla \mathbf{u}^T)$.

Exercise 11.6 *Write a numerical algorithm for minimizing the energy given in Exercise 11.5.*

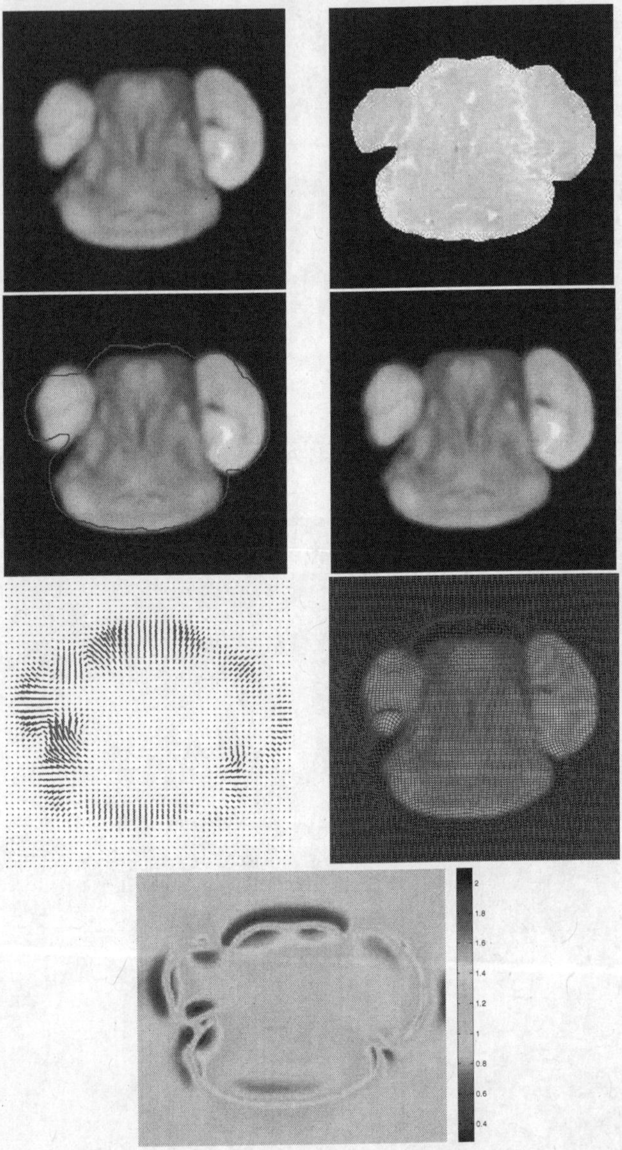

FIGURE 11.2: Application to image registration. First row: left, reference R; right, template T (mouse atlas and gene data). Second row, left to right: contour obtained by the proposed algorithm segmenting template T (starting with Φ_0 defining a disk), superimposed over the reference R; segmented reference, using as Φ_0 the output contour detected at the previous step. Third row, left to right: deformation field; final deformation grid from reference to template. Last row: $\det(\nabla\varphi)$.

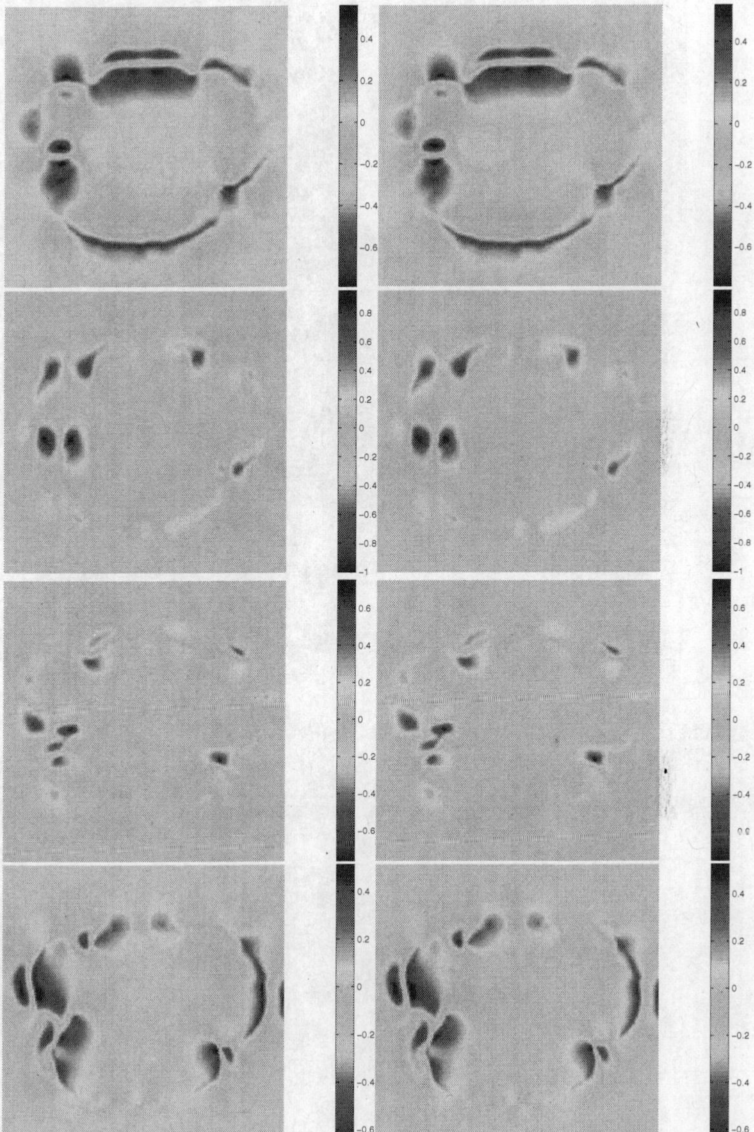

FIGURE 11.3: Components of V and ∇u illustrating good match of two variables.

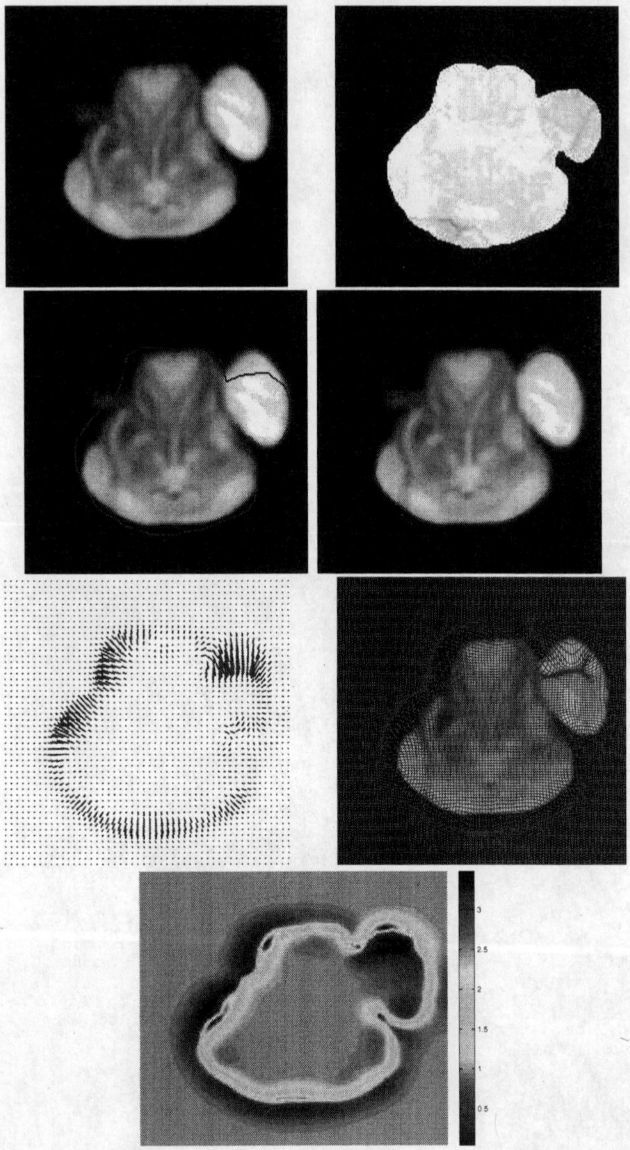

FIGURE 11.4: Application to image registration. First row: left, reference R; right, template T (mouse atlas and gene data). Second row, left to right: contour obtained by the proposed algorithm segmenting template T (starting with Φ_0 defining a disk), superimposed over the reference R; segmented reference, using as Φ_0 the output contour detected at the previous step. Third row, left to right: deformation field; final deformation grid from reference to template. Last row: $\det(\nabla\varphi)$.

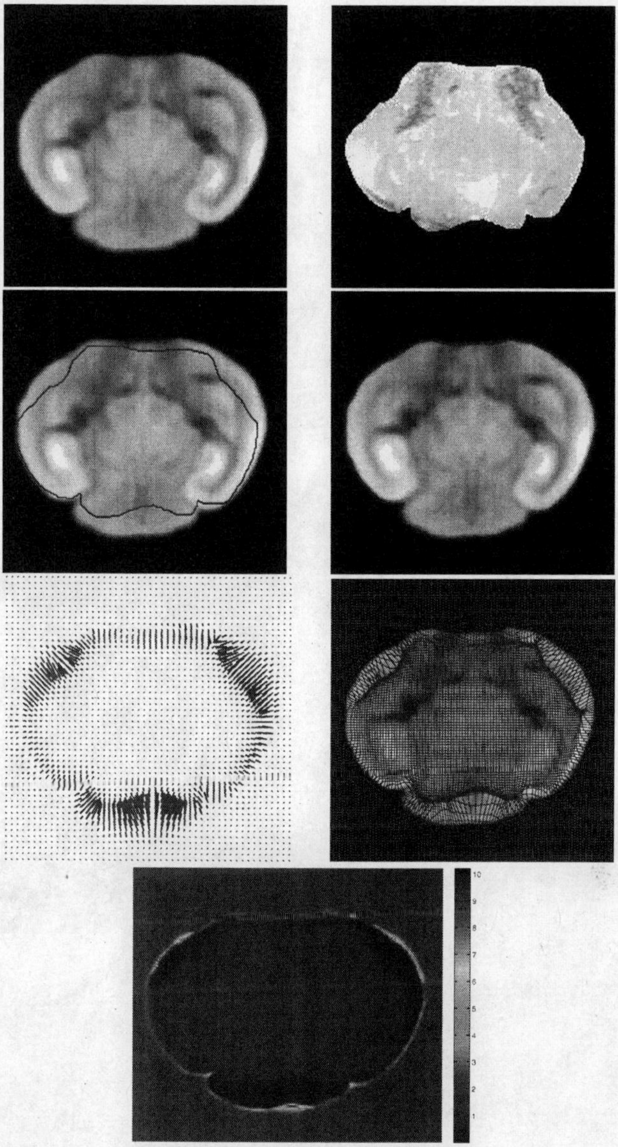

FIGURE 11.5: Application to image registration. First row: left, reference R; right, template T (mouse atlas and gene data). Second row, left to right: contour obtained by the proposed algorithm segmenting template T (starting with Φ_0 defining a disk), superimposed over the reference R; segmented reference, using as Φ_0 the output contour detected at the previous step. Third row, left to right: deformation field; final deformation grid from reference to template. Last row: det $(\nabla \varphi)$.

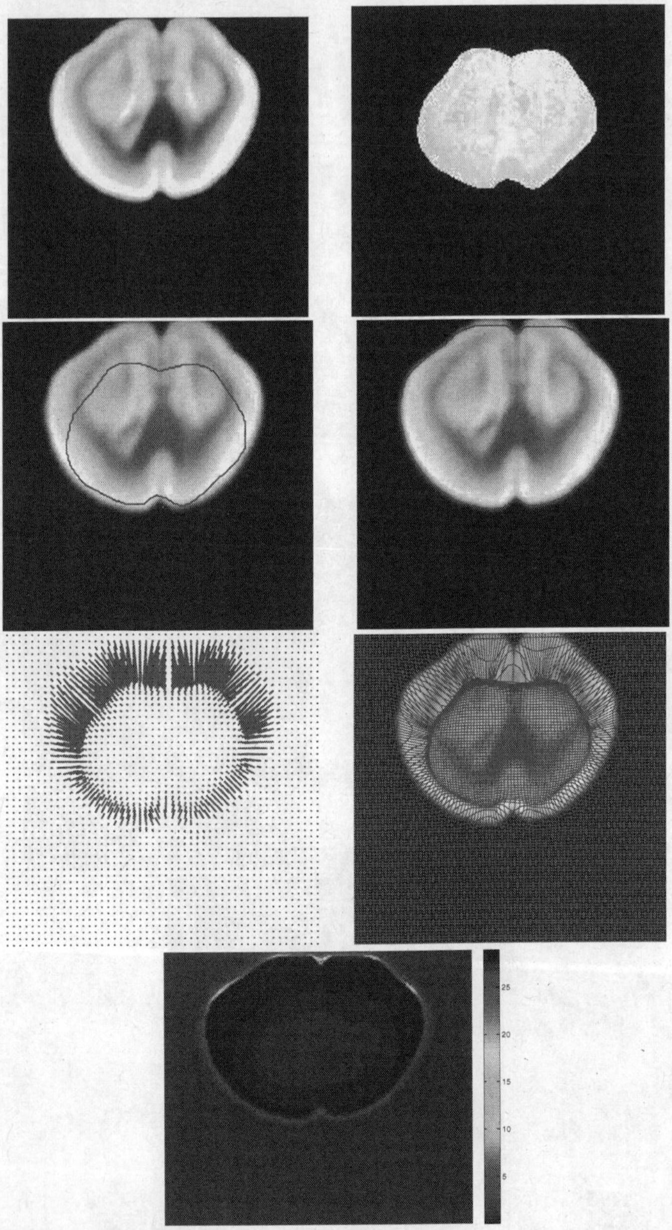

FIGURE 11.6: Application to image registration. First row: left, reference R; right, template T (mouse atlas and gene data). Second row, left to right: contour obtained by the proposed algorithm segmenting template T (starting with Φ_0 defining a disk), superimposed over the reference R; segmented reference, using as Φ_0 the output contour detected at the previous step. Third row, left to right: deformation field; final deformation grid from reference to template. Last row: $\det(\nabla\varphi)$.

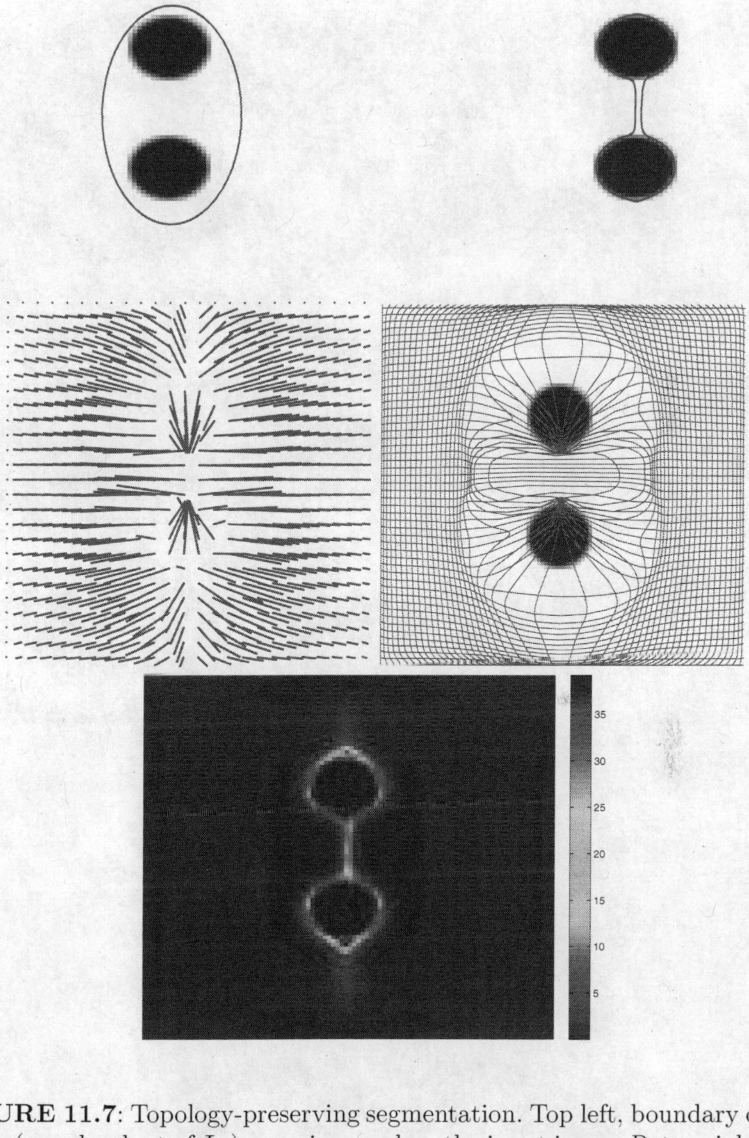

FIGURE 11.7: Topology-preserving segmentation. Top left, boundary of the ellipse (zero-level set of Φ_0) superimposed on the input image R; top right, the topology-preserving segmentation of the two disks; middle left, deformation field using nonlinear elasticity regularization; middle right, deformation grid from reference to template; bottom, $\det(\nabla\varphi)$.

FIGURE 11.8: Topology-preserving segmentation of a noisy synthetic image. Top left, the boundary of the disk (zero-level set of Φ_0) superimposed on the input data R; top right, the segmentation of the geometric figure (only one contour is detected due to the topology preservation); middle left, deformation field using nonlinear elasticity regularization; middle right, deformed grid from reference to template; bottom, $\det(\nabla\varphi)$.

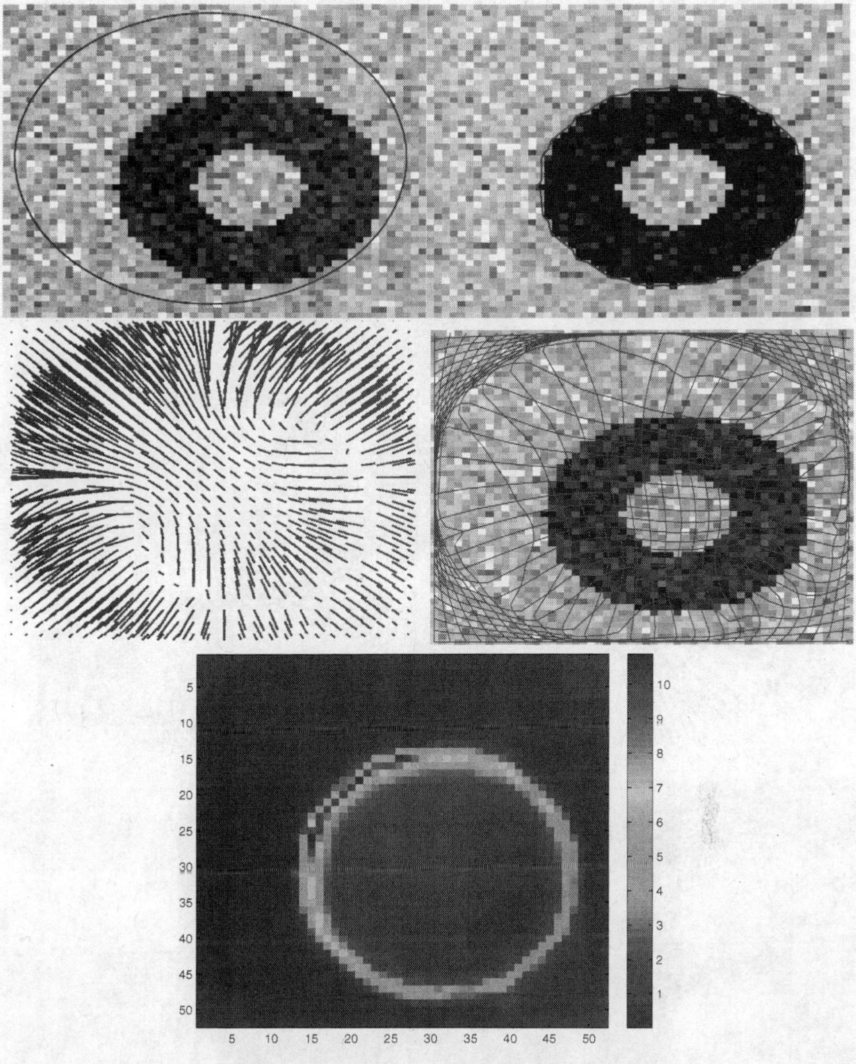

FIGURE 11.9: Topology-preserving segmentation of a noisy synthetic image (same data as in Figure 11.8, but with different initialization to illustrate larger deformation). Top left, the boundary of the disk (zero-level set of Φ_0) superimposed on the input data R; top right, the segmentation of the geometric figure (only one contour is detected due to the topology preservation); middle left, deformation field using nonlinear elasticity regularization; middle right, deformed grid from reference to template; bottom, det $(\nabla\varphi)$.

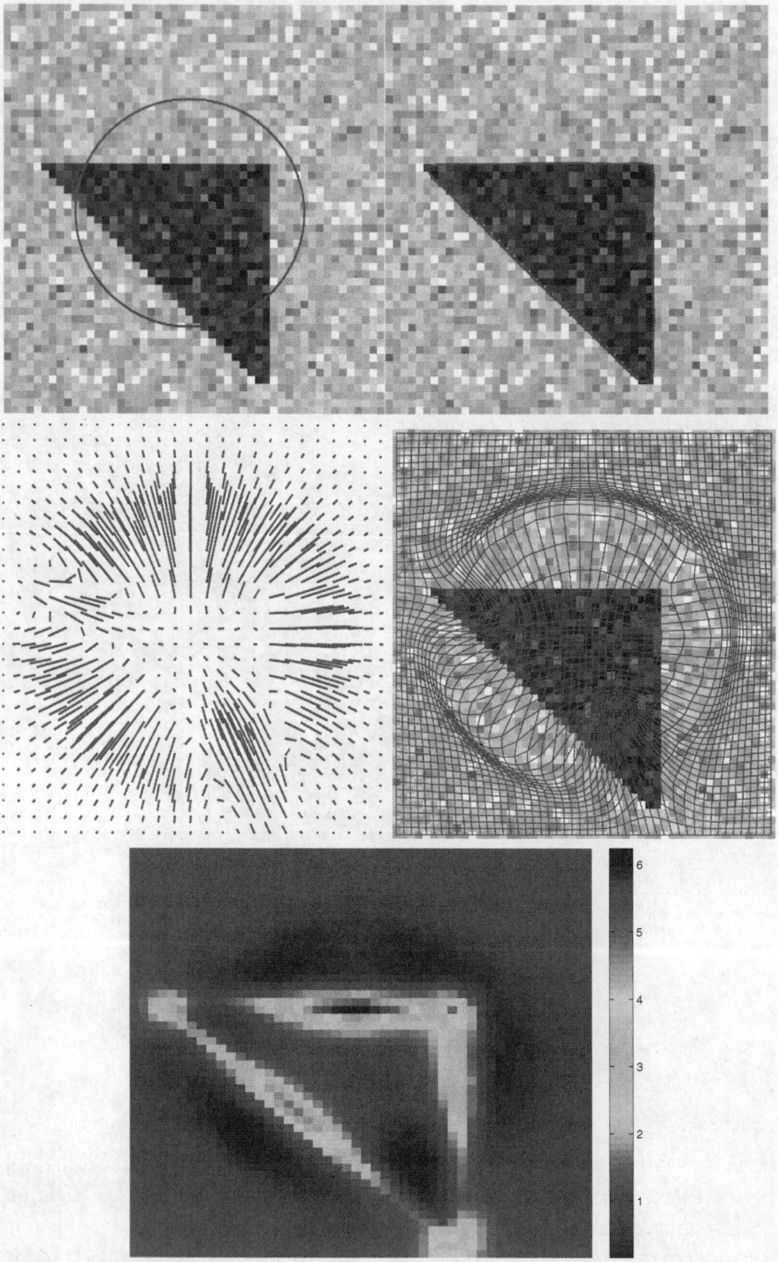

FIGURE 11.10: Topology-preserving segmentation of another noisy synthetic image. Top left, the boundary of the disk (zero-level set of Φ_0) superimposed on the input data R; top right, the segmentation of the geometric figure; middle left, deformation field using nonlinear elasticity regularization; middle right, deformed grid from reference to template; bottom, $\det(\nabla\varphi)$.

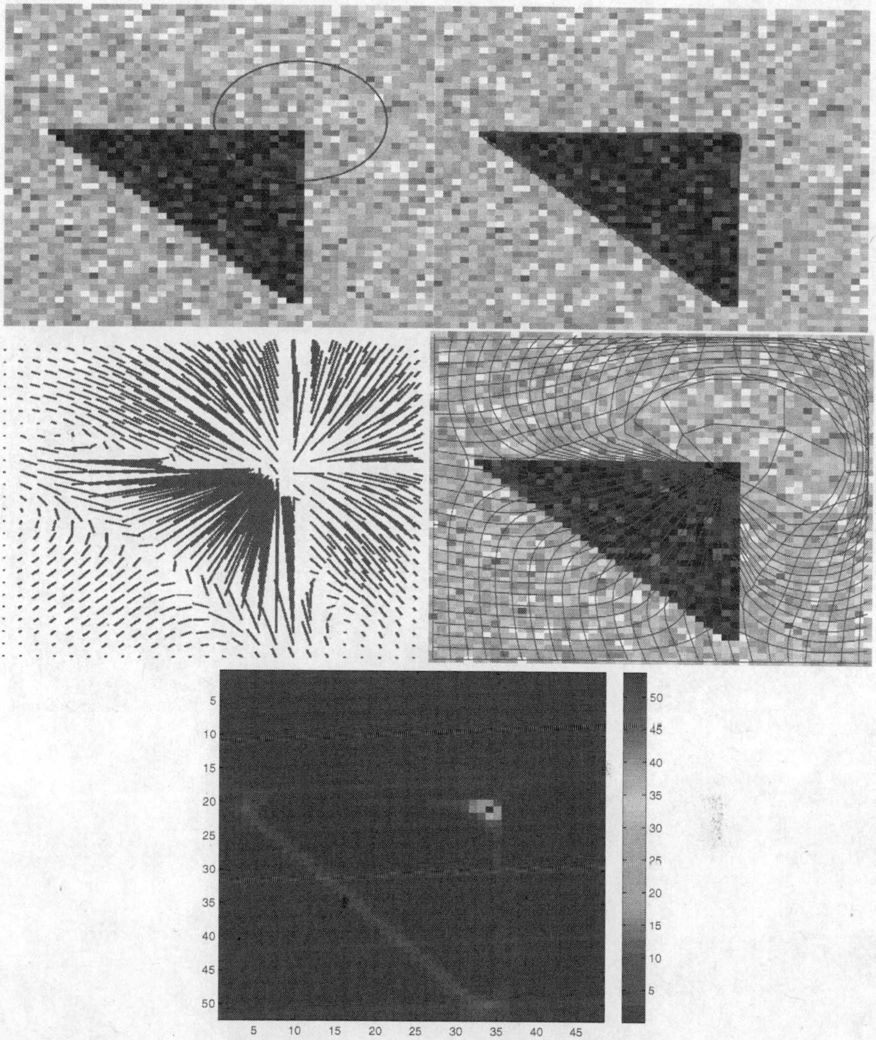

FIGURE 11.11: Topology-preserving segmentation of another noisy synthetic image (same data as in Figure 11.10, but with different initialization to illustrate larger deformation). Top left, the boundary of the disk (zero-level set of Φ_0) superimposed on the input data R; top right, the segmentation of the geometric figure; middle left, deformation field using nonlinear elasticity regularization; middle right, deformed grid from reference to template; bottom, $\det(\nabla\varphi)$.

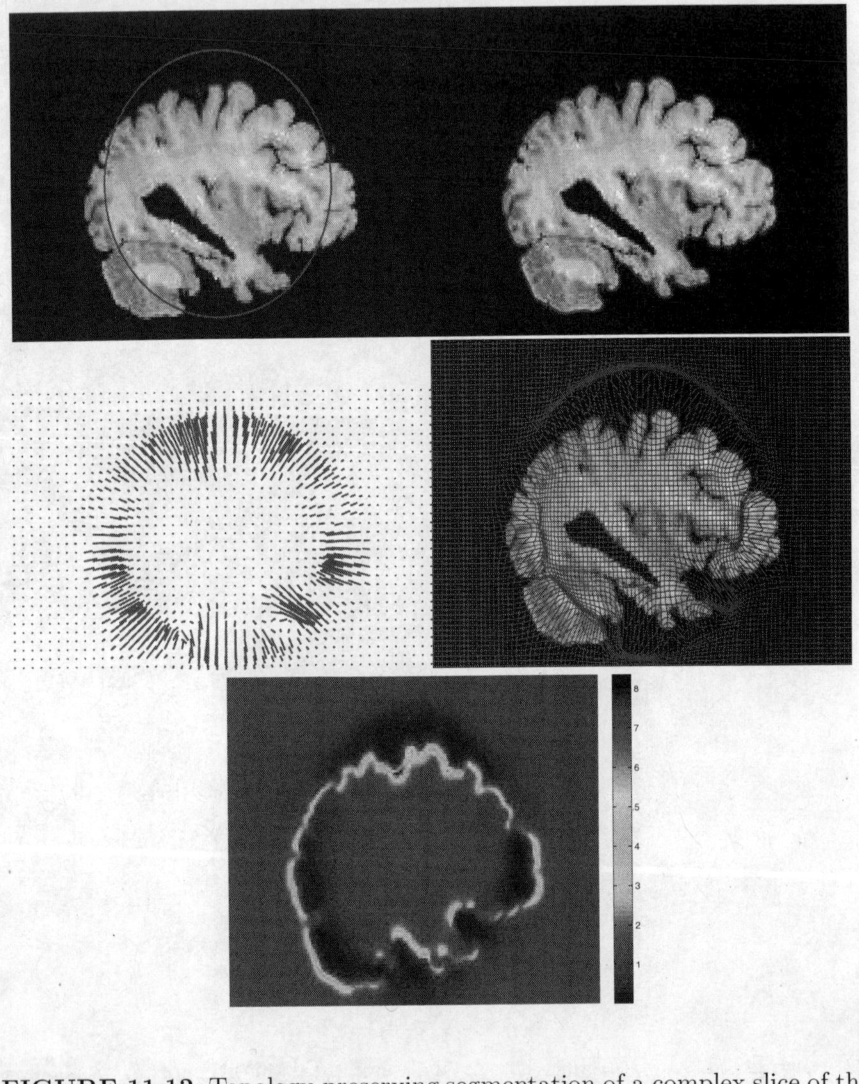

FIGURE 11.12: Topology-preserving segmentation of a complex slice of the brain. Top left, the boundary of the disk (zero-level set of Φ_0) superimposed on the reference image; top right, the segmentation of the slice of the brain; middle left, deformation field using nonlinear elasticity regularization; middle right, deformed grid from reference to template; bottom, det $(\nabla\varphi)$.

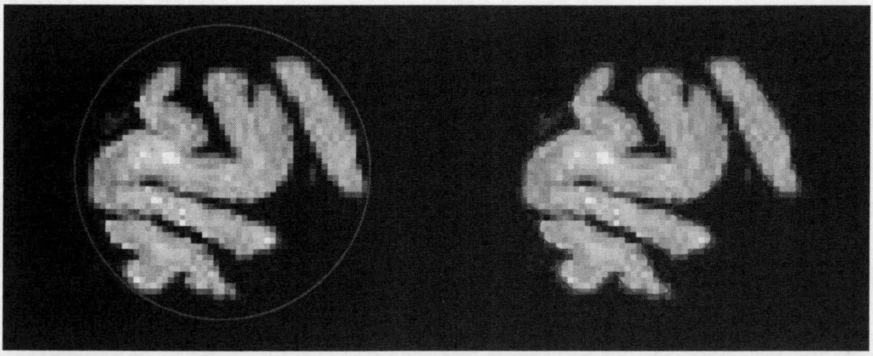

FIGURE 11.13: Topology-preserving segmentation of a complex slice of the brain. Left, the boundary of the disk (zero-level set of Φ_0) superimposed on the reference image. Right, the segmentation of the slice of the brain.

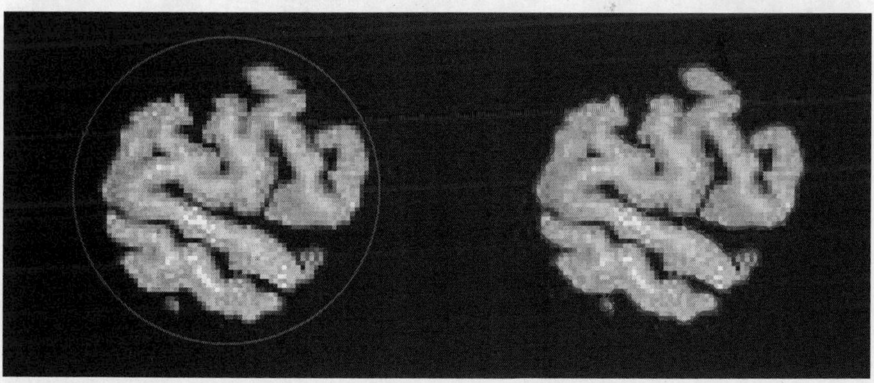

FIGURE 11.14: Topology-preserving segmentation of a complex slice of the brain. Left, the boundary of the disk (zero-level set of Φ_0) superimposed on the reference image. Right, the segmentation of the slice of the brain.

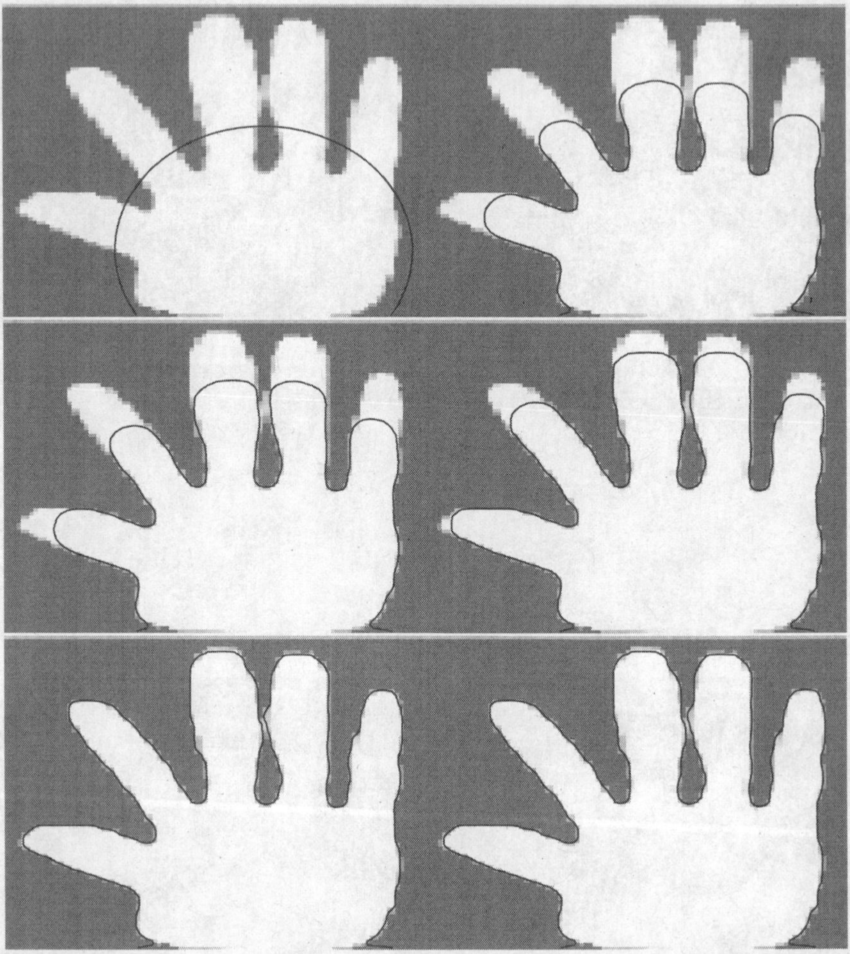

FIGURE 11.15: Topology-preserving segmentation of the hand image producing similar results with those found in [168] and [208]. From left to right, top to bottom: initial contour superimposed over the input image data R, and results at iterations 500, 1000, 1500, 2500, 3000.

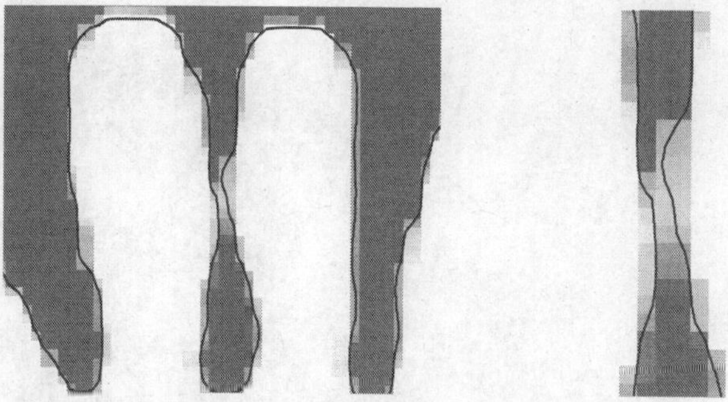

FIGURE 11.16: Two zooms on the final segmentation result from Figure 11.15, showing that the curve did not undergo merging or breaking between the fingers.

Chapter 12

Variational Image Registration Models

12.1 Introduction

Image registration is important in many applications such as computer vision and medical imaging. For medical applications, it is used in clinical studies of disease and for atlas-based identification and segmentation of anatomical structures. The registration process consists in deforming a template image T to match an unbiased reference image R through a smooth invertible transformation. The transformation may further provide information such as the changes of location and degree of deformation to help with diagnosis. For images of the same modality, a well-registered template has geometric features and intensity distribution matched with those of the reference. For images produced by different mechanisms and possessing distinct modalities, the goal of registration is to correlate the images while maintaining the modality of the template.

In this chapter we review several variational image registration methods and present an application relevant to neuroscience, the registration of gene expression data to a mouse atlas, where we want to match anatomically or geometrically significant points for the template with the corresponding ones for the reference. We consider several regularizations of the displacement field together with their finite-difference approximations: linear diffusion, linear elasticity, biharmonic, but we focus on the nonlinear elasticity one described in the previous chapter which has the advantage of allowing larger deformations.

We follow here existing texts on image registration by Modersitzki [238, 239] and also the work by Lin et al. [220, 219] on a nonlinear elasticity registration model that incorporates landmark matching.

An extensive overview of image registration models and numerical algorithms is given in two texts already mentioned by Modersitzki [238, 239], discussing parametric models such as landmark-based spline registration, and nonparametric models employing linear diffusion, linear elasticity, biharmonic and fluid regularization. Additionally, variational methods for regularization of the deformation by linear elasticity or by diffusion tensor using mutual information and other information theoretic approaches are presented in [171] in a theoretical framework. For models that deal with larger deformations, we refer to [101] for a well-known large deformation fluid registration method (not in variational form), and to a variational registration model for large deformations (large deformation diffeomorphic metric mapping or LDDMM) [44], [237]. The log-unbiased fluid registration method [349], [348] also handles large deformations. Besides fluid models, nonlinear elasticity regularization is implemented using the finite element method in [266] and [258]. Nonlinear elasticity principles have also been used with the regularized gradient flow in [131]. As for landmark-based registration methods, we refer to [178], in which a consistent landmark and intensity-based registration method is presented using thin-plate spline regularization (or biharmonic regularization). Another related reference is [298] in which data fidelity, spline regularization and soft landmark constraints are combined, as in this chapter. Finally, we refer to Chapter 11 for additional references and relevant work on variational image registration models.

In general, it is assumed that one is given two images T (template) and R (reference), both defined on the open and bounded domain Ω in the plane (a rectangle in general) and having real values. There are forward and backward registrations. The former is done in the Lagrangian framework in which a forward transformation Ψ is sought and grid points $\mathbf{x}$ with intensity values $T(\mathbf{x})$ are moved and arrive at non-grid points $\mathbf{y}$ with intensity values $T(\Psi^{-1}(\mathbf{y})) = T(\mathbf{x})$, $\forall \mathbf{x} \in \Omega$ or $\forall \mathbf{y} \in \Psi(\Omega)$. In the Eulerian framework (considered here) we find a backward transformation $\Phi = \Psi^{-1}$ such that grid points $\mathbf{y}$ in the deformed image arrive from non-grid points $\mathbf{x} = \Phi(\mathbf{y})$ and are assigned intensity values $T(\mathbf{x}) = T(\Phi(\mathbf{y}))$. For more detailed descriptions of the two frameworks, readers may refer to [238]. It is common to search for a smooth displacement vector field $\mathbf{u} = (u_1, u_2)$, instead of directly searching for the transformation $\Phi(\mathbf{x}) = \mathbf{x} + \mathbf{u}(\mathbf{x})$, $\mathbf{x} = (x_1, x_2) \in \Omega$.

In the variational framework, the registration problem can be solved by finding a smooth displacement vector field $\mathbf{u}(\mathbf{x})$, $\mathbf{x} = (x_1, x_2)$ that minimizes an energy functional consisting of data fidelity, landmark constraints, and regularization. The general form of the minimization problem is

$$\inf_{\mathbf{u}} \Big\{ J(\mathbf{u}) = Fid_{T,R}(\mathbf{u}) + \gamma D^{LM}(\mathbf{u}) + \alpha Reg(\mathbf{u}) \Big\}, \qquad (12.1)$$

where γ and α are weighting parameters. In the gradient descent framework, the time-dependent Euler–Lagrange equations in the displacement vector field $\mathbf{u} = (u_1, u_2)$ are

$$\frac{\partial u_l}{\partial t} = -\frac{\partial Fid(\mathbf{u})}{\partial u_l} - \gamma \frac{\partial D^{LM}(\mathbf{u})}{\partial u_l} - \alpha \frac{\partial Reg(\mathbf{u})}{\partial u_l}, \ l = 1, 2,$$

where it is assumed that all terms of energy J are Gâteaux differentiable.

We first review the most standard choices for each of the terms in the general energy functional J from (12.1). Then, following Lin et al. [220, 219], we will describe in more details a nonlinear elasticity regularization for mapping gene expression data to mouse atlas through larger deformations.

In what follows, h denotes the discretization space step in the numerical schemes. Also, we will use the notation $T1(\mathbf{x})$ for the transformed template $T(\mathbf{x}+\mathbf{u}(\mathbf{x}))$ and $T_{x_l}(\mathbf{x})$ will denote $\frac{\partial T}{\partial x_l}(\mathbf{x})$. The expression $T_{x_l}(\mathbf{x}+\mathbf{u}(\mathbf{x}))$ that appears in the computations is obtained by applying MATLAB® interpolation function to the discrete partial derivative of T.

12.1.1 Data fidelity in Eulerian framework

The simplest data matching term, most commonly used for images T and R of the same modality, is the standard L^2 distance as a dissimilarity measure between $T \circ \Phi$ and R, defined by

$$Fid(\mathbf{u}) = \frac{1}{2}\int_\Omega |T(\mathbf{x}+\mathbf{u}(\mathbf{x})) - R(\mathbf{x})|^2 d\mathbf{x},$$

using $\Phi(\mathbf{x}) = \mathbf{x} + \mathbf{u}(\mathbf{x})$, $\mathbf{x} = (x_1, x_2) \in \Omega$. Its Gâteaux derivative is given by

$$\frac{\partial Fid(\mathbf{u})}{\partial u_l} = (T(\mathbf{x}+\mathbf{u}(\mathbf{x})) - R(\mathbf{x}))T_{x_l}(\mathbf{x}+\mathbf{u}(\mathbf{x})),\ l = 1, 2.$$

12.1.2 Landmark constraints

Let $\mathbf{x}^{R,k}$ be manually selected landmark points for the reference R and $\mathbf{x}^{T,k}$ for the template T, with $k = 1, ..., m$ landmarks. We want to map $\mathbf{x}^{R,k}$ to $\mathbf{x}^{T,k}$ via a smooth deformation Φ such that $\Phi(\mathbf{x}^{R,k}) \sim \mathbf{x}^{T,k}$ by minimizing the following landmark distance function:

$$D^{LM}(\mathbf{u}) = \frac{1}{2}\sum_{k=1}^{m} \|\mathbf{x}^{T,k} - \Phi(\mathbf{x}^{R,k})\|^2 = \frac{1}{2}\sum_{k=1}^{m} \|\mathbf{x}^{T,k} - \mathbf{x}^{R,k} - \mathbf{u}(\mathbf{x}^{R,k})\|^2,$$

using $\Phi(\mathbf{x}^{R,k}) = \mathbf{x}^{R,k} + \mathbf{u}(\mathbf{x}^{R,k})$. Here $\|\cdot\|$ denotes the Euclidean norm in $\mathbb{R}^2$. Its Gâteaux derivative is given by:

$$\frac{\partial D^{LM}(\mathbf{u})}{\partial \mathbf{u}} = -\sum_{k=1}^{m} (\mathbf{x}^{T,k} - \mathbf{x}^{R,k} - \mathbf{u}(\mathbf{x}^{R,k}))\, \delta_{\mathbf{x}^{R,k}}(\mathbf{x}).$$

12.1.3 Regularization

For comparison purposes, we review here the most common and simplest regularization terms that lead to linear terms with respect to derivatives in the Euler–Lagrange equations. Such regularizations are most useful for small deformations. We refer the reader again to the instructive textbooks [238] and [239]. Standard regularizations are combined with the above L^2 similarity measure and landmark distance function to form the energy J from (12.1) to be minimized. We also give the discretization of the corresponding Euler–Lagrange equations using finite differences.

12.1.3.1 Image registration using diffusion regularization

The diffusion $\dot{H}^1$ regularization, defined as

$$Reg(\mathbf{u}) = \frac{1}{2}\sum_{l=1}^{2}\int_{\Omega}\|\nabla u_l\|^2 d\mathbf{x},$$

is motivated by its smoothing properties and also by its small number of operations required [143, 238]. Combined with the above L^2 similarity measure and landmark distance function, minimizing the energy J from (12.1) in this case leads to the time-dependent Euler–Lagrange equations:

$$\frac{\partial u_l}{\partial t} = -(T(\mathbf{x}+\mathbf{u}(\mathbf{x})) - R(\mathbf{x}))T_{x_l}(\mathbf{x}+\mathbf{u}(\mathbf{x})) - \gamma\frac{\partial D^{LM}(\mathbf{u})}{\partial u_l} + \alpha\triangle u_l,\ l = 1, 2.$$

These time-dependent Euler–Lagrange equations can be discretized by the following semi-implicit finite differences scheme:

$$\begin{aligned}\frac{{u_l}_{i,j}^{n+1} - {u_l}_{i,j}^{n}}{\Delta t} = &- \Big(T_{x_l}(\mathbf{x}+\mathbf{u}(\mathbf{x}))\Big)_{i,j}(T1_{i,j} - R_{i,j}) - \gamma\frac{\partial D^{LM}(\mathbf{u})}{\partial u_l}\\ &+ \alpha\left(\frac{{u_l}_{i+1,j}^{n} - 2{u_l}_{i,j}^{n+1} + {u_l}_{i-1,j}^{n}}{h^2}\right.\\ &\left.+ \frac{{u_l}_{i,j+1}^{n} - 2{u_l}_{i,j}^{n+1} + {u_l}_{i,j-1}^{n}}{h^2}\right),\end{aligned}$$

where $l = 1, 2$.

12.1.3.2 Image registration using biharmonic regularization

The biharmonic $\dot{H}^2$ regularization is defined as

$$Reg(\mathbf{u}) = \frac{1}{2}\sum_{l=1}^{2}\int_{\Omega}(\triangle u_l)^2 d\mathbf{x}.$$

The integrand $(\triangle u_l)^2$ approximates the curvature and thus this regularizer minimizes the curvature of the displacement vectors [144, 238]. According to [238], the biharmonic registration is less dependent on the initial alignment of the two images to be registered and thus it is more suitable when an affine linear pre-registration is not available. Combined with the above L^2 similarity measure and landmark distance function, minimizing the energy J from (12.1) in this case leads to the time-dependent Euler–Lagrange equations

$$\frac{\partial u_l}{\partial t} = -(T(\mathbf{x}+\mathbf{u}(\mathbf{x})) - R(\mathbf{x}))T_{x_l}(\mathbf{x}+\mathbf{u}(\mathbf{x})) - \gamma\frac{\partial D^{LM}(\mathbf{u})}{\partial u_l} - \alpha\Delta^2 u_l,\ l = 1, 2.$$

These time-dependent Euler–Lagrange equations can be discretized by the following semi-implicit finite differences scheme:

$$\begin{aligned}\frac{u_{l\,i,j}^{n+1} - u_{l\,i,j}^{n}}{\Delta t} = &- \Big(T_{x_l}(\mathbf{x} + \mathbf{u}(\mathbf{x}))\Big)_{i,j}(T1_{i,j} - R_{i,j}) - \gamma \frac{\partial D^{LM}(\mathbf{u})}{\partial u_l} \\ &- \alpha \frac{1}{h^4}\big(20u_{l\,i,j}^{n+1} + u_{l\,i+2,j}^{n} - 8u_{l\,i+1,j}^{n} - 8u_{l\,i-1,j}^{n} + u_{l\,i-2,j}^{n} \\ &\quad + u_{l\,i,j+2}^{n} - 8u_{l\,i,j+1}^{n} - 8u_{l\,i,j-1}^{n} + u_{l\,i,j-2}^{n} \\ &\quad + 2u_{l\,i+1,j+1}^{n} + 2u_{l\,i+1,j-1}^{n} + 2u_{l\,i-1,j+1}^{n} + 2u_{l\,i-1,j-1}^{n}\big),\end{aligned}$$

with $l = 1, 2$.

12.1.3.3 Image registration using linear elasticity regularization

It is appropriate to view the objects to be matched as the observations of a single elastic body before and after being subjected to a deformation. Viewing the shape change of the image after transformation as the deformation of an elastic material under external force was first adopted in [67] in developing a linear elastic registration method allowing for small smooth deformations. We recall its formulation here.

Using the strain-displacement relations

$$\epsilon(\mathbf{u}) = \frac{1}{2}(\nabla \mathbf{u}^t + \nabla \mathbf{u}),$$

and the stored energy

$$W(\epsilon) = \frac{\lambda}{2}(\text{trace}(\epsilon))^2 + \mu \text{trace}(\epsilon^2)$$

(where λ and μ are the Lamé coefficients of the material), the strain energy given by

$$Reg(\mathbf{u}) = \int_\Omega W(\epsilon) d\mathbf{x}$$

defines the regularization term in the registration problem. Combined with the above L^2 similarity measure and landmark distance function, minimizing the energy J from (12.1) in this case leads to the time-dependent Euler–Lagrange equations

$$\begin{aligned}\frac{\partial \mathbf{u}}{\partial t} = & - \quad (T(\mathbf{x} + \mathbf{u}(\mathbf{x})) - R(\mathbf{x}))\nabla T(\mathbf{x} + \mathbf{u}(\mathbf{x})) - \gamma \frac{\partial D^{LM}(\mathbf{u})}{\partial \mathbf{u}} \\ & + \quad \alpha[\mu \Delta \mathbf{u} + (\lambda + \mu)\nabla(\text{div}\mathbf{u})].\end{aligned}$$

These time-dependent Euler–Lagrange equations can be discretized by the following semi-implicit finite differences scheme:

$$\frac{u_{p_{i,j}}^{n+1} - u_{p_{i,j}}^{n}}{\triangle t} = -\left(T_{x_p}(\mathbf{x} + \mathbf{u}(\mathbf{x}))\right)_{i,j}(T1_{i,j} - R_{i,j}) - \gamma\frac{\partial D^{LM}(\mathbf{u})}{\partial u_p}$$
$$+\alpha\left[(\lambda+\mu)A_p + \mu\left(\frac{u_{p_{i+1,j}}^{n} - 2u_{p_{i,j}}^{n+1} + u_{p_{i-1,j}}^{n}}{h^2}\right.\right.$$
$$\left.+\frac{u_{p_{i,j+1}}^{n} - 2u_{p_{i,j}}^{n+1} + u_{p_{i,j-1}}^{n}}{h^2}\right)$$
$$\left.+(\lambda+\mu)\left(\frac{u_{q_{i+1,j+1}}^{n} - u_{q_{i+1,j-1}}^{n} - u_{q_{i-1,j+1}}^{n} + u_{q_{i-1,j-1}}^{n}}{4h^2}\right)\right]$$

where $p, q = 1, 2$, $p \neq q$, and

$$A_1 = \frac{u_{1_{i+1,j}}^{n} - 2u_{1_{i,j}}^{n+1} + u_{1_{i-1,j}}^{n}}{h^2}, \quad A_2 = \frac{u_{2_{i,j+1}}^{n} - 2u_{2_{i,j}}^{n+1} + u_{2_{i,j-1}}^{n}}{h^2}.$$

12.2 Variational image registration algorithm using nonlinear elasticity regularization

The focus of this chapter is on a variational image registration model that uses a nonlinear elasticity regularization. We have chosen to use the standard L^2 distance as a dissimilarity measure between $T \circ \Phi$ and R for simplicity of calculation even if the real application that we consider deals with template and reference images of different modalities. The L^2 measure is complemented by the use of additional landmarks as geometric constraints and it is the simplest by comparison to information theoretic matching measures such as the mutual information (appropriate for different modalities). We will see that even if we use such a simple similarity measure, the mutual information between the transformed template and the reference also increases within iterations.

The mapping of landmark points is done simply by minimizing the sum of the squared distances between the points without incorporating a spline model.

Since the linear elastic model works better for small deformations, we use a nonlinear elastic model to allow larger deformations. Among the various nonlinear elastic models, we have chosen the Saint-Venant–Kirchhoff material for its simplicity [105],[106]. As we mentioned, the finite element method has been used in prior work based on nonlinear elasticity principles. To achieve a simpler numerical algorithm, we introduce an auxiliary variable for the Jacobian matrix of the displacement to remove the nonlinearity in the derivatives. This idea has been inspired by a more theoretical work [249].

Compared with the experimental results on characteristic images using models with linear diffusion, linear elasticity and biharmonic regularization, the described nonlinear elasticity algorithm allows larger and smoother deformations without numerical correction such as regridding [101] most of the times.

The strain energy corresponding to Saint-Venant–Kirchhoff hyperelastic materials [105],[106], [317] is given by

$$Reg(\mathbf{u}) = \int_\Omega W(\epsilon) d\mathbf{x},$$

with tensor

$$\epsilon(\mathbf{u}) = \frac{1}{2}(\nabla \mathbf{u}^t + \nabla \mathbf{u} + \nabla \mathbf{u}^t \nabla \mathbf{u})$$

and the stored energy

$$W(\epsilon) = \frac{\lambda}{2}(\text{trace}(\epsilon))^2 + \mu \text{trace}(\epsilon^2),$$

where λ and μ are the Lamé coefficients of the material (to allow larger smooth deformations, we keep the nonlinear term in $\epsilon(\mathbf{u})$). The regularization can be expressed as

$$\begin{aligned} Reg(\mathbf{u}) = \int_\Omega \Big\{ \frac{\lambda}{8}[2(\text{div}\mathbf{u}) + \sum_{k=1}^{2} |\nabla u_k|^2]^2 + \frac{\mu}{4}\Big\{ \sum_{i=1}^{2} \Big[2\frac{\partial u_i}{\partial x_i} + \sum_{k=1}^{2} (\frac{\partial u_k}{\partial x_i})^2 \Big]^2 \\ + \sum_{i,j=1, i \neq j}^{2} \Big[\frac{\partial u_j}{\partial x_i} + \frac{\partial u_i}{\partial x_j} + \sum_{k=1}^{2} \frac{\partial u_k}{\partial x_i}\frac{\partial u_k}{\partial x_j} \Big]^2 \Big\} \Big\} d\mathbf{x}. \end{aligned} \tag{12.2}$$

It is cumbersome to directly compute and discretize the associated Euler–Lagrange equation in $\mathbf{u}$. To avoid this difficulty, we introduce a matrix variable $\mathbf{v} = \begin{pmatrix} v_{11} & v_{12} \\ v_{21} & v_{22} \end{pmatrix}$ for $\nabla \mathbf{u}$, and we substitute $\nabla \mathbf{u}$ by $\mathbf{v}$ in the term $Reg(\mathbf{u})$. To deal with the hard constraint $\nabla \mathbf{u} = \mathbf{v}$, we employ the quadratic penalty method [252] and we redefine the regularization for β large enough as $Reg_\beta(\mathbf{u}, \mathbf{v})$:

$$\begin{aligned} Reg_\beta(\mathbf{u}, \mathbf{v}) &= \int_\Omega \Big[W\Big(\frac{1}{2}(\mathbf{v}^t + \mathbf{v} + \mathbf{v}^t \mathbf{v})\Big) + \beta |\mathbf{v} - \nabla \mathbf{u}|^2 \Big] d\mathbf{x}, \\ &= \int_\Omega \Big\{ \frac{\lambda}{8}[2(v_{11} + v_{22}) + (v_{11}^2 + v_{12}^2 + v_{21}^2 + v_{22}^2)]^2 \\ &+ \frac{\mu}{4}[(2v_{11} + v_{11}^2 + v_{21}^2)^2 + (2v_{22} + v_{12}^2 + v_{22}^2)^2 \\ &+ 2(v_{12} + v_{21} + v_{11}v_{12} + v_{21}v_{22})^2] \Big\} d\mathbf{x} \\ &+ \beta \int_\Omega \Big(\Big| v_{11} - \frac{\partial u_1}{\partial x_1} \Big|^2 + \Big| v_{12} - \frac{\partial u_1}{\partial x_2} \Big|^2 + \Big| v_{21} - \frac{\partial u_2}{\partial x_1} \Big|^2 \\ &+ \Big| v_{22} - \frac{\partial u_2}{\partial x_2} \Big|^2 \Big) d\mathbf{x}. \end{aligned}$$

Thus the described registration method in the presence of landmarks can be expressed as the minimization

$$\inf_{\mathbf{u},\mathbf{v}}\Big\{J(\mathbf{u},\mathbf{v}) = \frac{1}{2}\int_\Omega |T(\mathbf{x}+\mathbf{u}(\mathbf{x})) - R(\mathbf{x})|^2 d\mathbf{x} + \frac{\gamma}{2}\sum_{k=1}^m \|\mathbf{x}^{T,k} - \mathbf{x}^{R,k} - \mathbf{u}(\mathbf{x}^{R,k})\|^2 + \alpha Reg_\beta(\mathbf{u},\mathbf{v})\Big\}.$$

Now we solve by gradient descent the Euler–Lagrange equations in u_l, $l = 1, 2$,

$$\begin{aligned}\frac{\partial u_l}{\partial t} =& -(T(\mathbf{x}+\mathbf{u}(\mathbf{x})) - R(\mathbf{x}))T_{x_l}(\mathbf{x}+\mathbf{u}(\mathbf{x})) \\ &- \gamma\frac{\partial D^{LM}(\mathbf{u})}{\partial u_l} + 2\alpha\beta(\triangle u_l - \frac{\partial v_{l1}}{\partial x_1} - \frac{\partial v_{l2}}{\partial x_2}),\ l = 1, 2,\end{aligned}$$

and then update the approximation matrix $\mathbf{v}$ by solving the four (nonlinear) Euler–Lagrange equations in $\mathbf{v}$:

$$\begin{aligned}\frac{\partial v_{11}}{\partial t} =& \alpha\{2\beta(\frac{\partial u_1}{\partial x_1} - v_{11}) - \lambda\mathcal{I}(1+v_{11}) - \mu[(2v_{11}+v_{11}^2+v_{21}^2)(1+v_{11}) \\ &+ \mathcal{J}v_{12}]\},\\ \frac{\partial v_{12}}{\partial t} =& \alpha\{2\beta(\frac{\partial u_1}{\partial x_2} - v_{12}) - \lambda\mathcal{I}v_{12} - \mu[(2v_{22}+v_{12}^2+v_{22}^2)v_{12} + \mathcal{J}(1+v_{11})]\},\\ \frac{\partial v_{21}}{\partial t} =& \alpha\{2\beta(\frac{\partial u_2}{\partial x_1} - v_{21}) - \lambda\mathcal{I}v_{21} - \mu[(2v_{11}+v_{11}^2+v_{21}^2)v_{21} + \mathcal{J}(1+v_{22})]\},\\ \frac{\partial v_{22}}{\partial t} =& \alpha\{2\beta(\frac{\partial u_2}{\partial x_2} - v_{22}) - \lambda\mathcal{I}(1+v_{22}) - \mu[(2v_{22}+v_{12}^2+v_{22}^2)(1+v_{22}) \\ &+ \mathcal{J}v_{21}]\},\end{aligned}$$

where $\mathcal{I} = v_{11} + v_{22} + \frac{1}{2}v_{11}^2 + \frac{1}{2}v_{21}^2 + \frac{1}{2}v_{12}^2 + \frac{1}{2}v_{22}^2$ and $\mathcal{J} = v_{12} + v_{21} + v_{11}v_{12} + v_{21}v_{22}$. The following equations are the semi-implicit finite difference schemes for the time-dependent Euler–Lagrange equations for the regularization term in $\mathbf{u}$ and in $\mathbf{v}$:

$$\begin{aligned}\frac{{u_l}_{i,j}^{n+1} - {u_l}_{i,j}^n}{\triangle t} =& -\Big(T_{x_l}(\mathbf{x}+\mathbf{u}(\mathbf{x}))\Big)_{i,j}(T1_{i,j} - R_{i,j}) - \gamma\frac{\partial D^{LM}(\mathbf{u})}{\partial u_l} \\ &+ 2\alpha\beta\Big(\frac{{u_l}_{i+1,j}^n - 2{u_l}_{i,j}^{n+1} + {u_l}_{i-1,j}^n}{h^2} + \frac{{u_l}_{i,j+1}^n - 2{u_l}_{i,j}^{n+1} + {u_l}_{i,j-1}^n}{h^2} \\ &- \frac{{v_{l1}}_{i+1,j}^n - {v_{l1}}_{i-1,j}^n + {v_{l2}}_{i,j+1}^n - {v_{l2}}_{i,j-1}^n}{2h}\Big),\end{aligned}$$

and

$$\frac{v_{11}^{n+1}-v_{11}^{n}}{\triangle t} =\alpha[2\beta(\frac{\partial u_1^n}{\partial x_1}-v_{11}^{n+1})-\lambda E_1E_5-\mu(E_2E_5+E_4v_{12})],$$
$$\frac{v_{12}^{n+1}-v_{12}^{n}}{\triangle t} =\alpha[2\beta(\frac{\partial u_1^n}{\partial x_2}-v_{12}^{n+1})-\lambda E_1v_{12}-\mu(E_3v_{12}+E_4E_5)],$$
$$\frac{v_{21}^{n+1}-v_{21}^{n}}{\triangle t} =\alpha[2\beta(\frac{\partial u_2^n}{\partial x_1}-v_{21}^{n+1})-\lambda E_1v_{21}-\mu(E_2v_{21}+E_4E_6)],$$
$$\frac{v_{22}^{n+1}-v_{22}^{n}}{\triangle t} =\alpha[2\beta(\frac{\partial u_2^n}{\partial x_2}-v_{22}^{n+1})-\lambda E_1E_6-\mu(E_3E_6+E_4v_{21})],$$

where

$$E_1 = v_{11}+\frac{1}{2}v_{11}^2+\frac{1}{2}v_{21}^2+v_{22}+\frac{1}{2}v_{12}^2+\frac{1}{2}v_{22}^2,$$
$$E_2 = 2v_{11}+v_{11}^2+v_{21}^2, \quad E_3 = 2v_{22}+v_{12}^2+v_{22}^2,$$
$$E_4 = v_{12}+v_{21}+v_{11}v_{12}+v_{21}v_{22}, \quad E_5 = 1+v_{11}, \quad E_6 = 1+v_{22}.$$

12.3 Experimental results

We present here numerical results for image registration using the described nonlinear elasticity algorithm and comparisons with the diffusion, biharmonic and linear elasticity models on synthetic and real data. As a real application, we consider the problem of mapping gene expression data to the mouse atlas motivated by research in neuroscience.

12.3.1 Numerical correction: regridding

An admissible deformation field $\Phi : \overline{\Omega} \to \overline{\Omega}$, $\Phi(\mathbf{x}) = \mathbf{x} + \mathbf{u}(\mathbf{x})$, should satisfy $det(\nabla\Phi) > 0$ in Ω, $\Phi(\mathbf{x}) = \mathbf{x}$ on $\partial\Omega$, and Φ is one-to-one and onto. To enforce such a constraint, some numerical corrections such as regridding are introduced [101]. In the present work, if $det(\nabla\Phi^{n+1}) < 0.025$ $(= tol)$, we set the displacement field $\mathbf{u}^{n+1} = 0$, the template $T(\mathbf{x}) = T(\mathbf{x} + \mathbf{u}^n)$, and the landmarks, if any, $\mathbf{x}^{R,k} = \mathbf{x}^{T,k} - \mathbf{u}^n(\mathbf{x}^{R,k})$. After the iteration is done, we calculate the composite displacement field by interpolating each of the intermediate displacement fields saved during the regridding process.

The main algorithm is as follows:

(1) Formulate identity matrices S_1 and S_2 so that $S_1(x, y) = x$, $S_2(x, y) = y$, $\forall x, y \in \Omega$; initialize $\mathbf{u}^0 = (u_1^0, u_2^0)$; initialize $regrid.count = 0$.

(2) Iteration starts: compute $\mathbf{u}^{n+1}$, $n = 0, 1, 2, ...$; update displacement matrices S_{11} and S_{22} so that $S_{11}^{n+1} = S_1 + u_1^{n+1}, S_{22}^{n+1} = S_2 + u_2^{n+1}$.

(3) If $det(Jacobian(\Phi)) < tol$, then $regrid.count = regrid.count + 1$; if

regrid.count $<$ *regrid.number.limit*, then $T = T1$; save $\mathbf{u}^n$ as data files u_k(regrid.count), $k = 1, 2$; $\mathbf{u}^{n+1} = \mathbf{0}$; else iteration ends;

(4) If landmark point distance converges, iteration ends; go to (5); otherwise, go to (2).

(5) If *regrid.count* > 0, then composite.S_{kk}(*regrid.count* + 1) = S_{kk}(*final.iteration*);
composite.u_k(*regrid.count* + 1) = u_k(*final.iteration*), $k = 1, 2$;
for i = *regrid.count* : -1 : 1, read and load data files u_k(*regrid.count*);
$U0_k$=u_k(*regrid.count*), $k = 1, 2$;
composite.u_k(*regrid.count*)=composite.u_k(*regrid.count* + 1)
+interpolation($U0_k$,composite.S_{22}(*regrid.count* + 1),
composite.S_{11}(*regrid.count* + 1));
composite.S_{kk}(*regrid.count*) =S_k+composite.u_k(*regrid.count*);
regrid.count = *regrid.count* − 1;
end
u_k=composite.u_k(1); S_{kk}=composite.S_{kk}(1).

One of the criteria for evaluating non-rigid registration models could be the number of regridding steps (fewer regridding steps may be preferable). Most of the following results using nonlinear elasticity regularization were obtained without regridding.

12.3.2 Synthetic images

We compare the four models presented on synthetic data without using the landmark distance.

12.3.2.1 Disk to letter C

We first compare the linear elasticity, linear diffusion, biharmonic, and nonlinear elasticity models for registration from disk to letter C (Figure 12.1), which is the "academic" example used in [238].

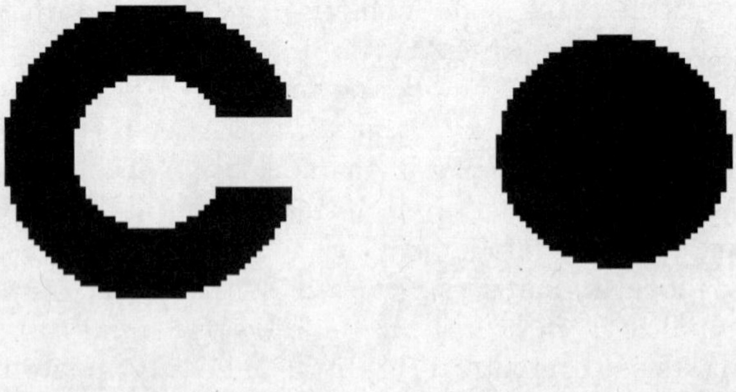

FIGURE 12.1: Reference (left) and template (right).

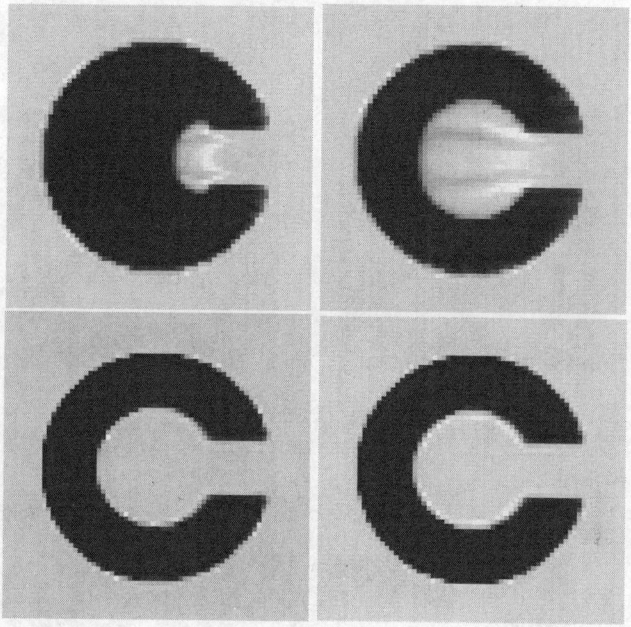

FIGURE 12.2: Transformed images by (from left to right and top to bottom) linear elasticity, linear diffusion, biharmonic, and nonlinear elasticity regularizations. The dissimilarity values between the transformed template $T1$ and the reference image R are: 2201.8619 (linear elasticity), 196.7257 (linear diffusion), 43.1154 (biharmonic regularization), and 29.0903 (nonlinear elasticity).

From the registration results shown in Figure 12.2, we can see that linear elasticity regularization is only suitable for small deformations. The linear diffusion regularization works better but introduces a lot of artifacts. The biharmonic model produces a more satisfactory result (even though a darker background is observed) together with the nonlinear elasticity model. In terms of the regridding numbers, the biharmonic model and the nonlinear elasticity model require three regridding steps while the linear diffusion model requires four.

12.3.2.2 Other binary images

We further compare the four models for the registration of other binary images (Figure 12.3) artificially obtained from the real data used later. We found from the registration results (Figure 12.4) that the nonlinear elasticity model does not introduce artifacts (e.g., under the right ear where we expect the largest deformation) as other models do. In terms of regridding numbers, the nonlinear elasticity model is the only one which does not require any regridding step, and the biharmonic model requires fewer regridding steps than the other two linear models.

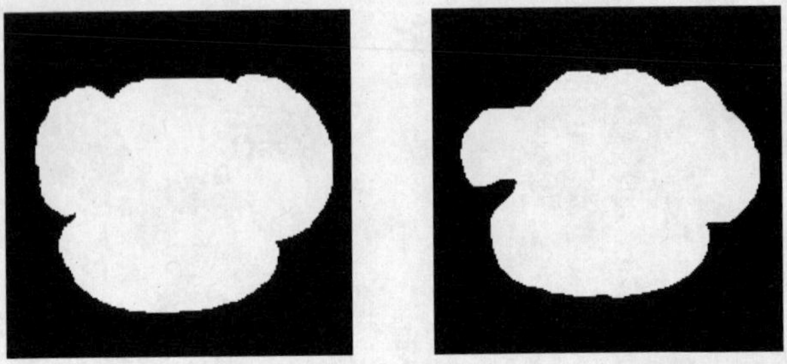

FIGURE 12.3: Reference (left) and template (right).

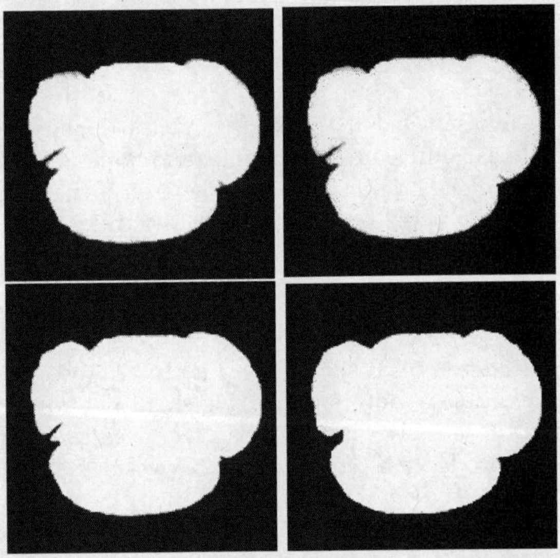

FIGURE 12.4: From left to right and top to bottom: transformed images by linear elasticity, linear diffusion, biharmonic, and nonlinear elasticity regularizations.

Since the biharmonic model is more comparable to the nonlinear elasticity model, we want to further examine the two models using real data. We note that the nonlinear elastic model requires longer computational time than the biharmonic model.

12.3.3 Real data

12.3.3.1 Ground truth test

In this experiment (without landmark distance), we start with two images T and R, such that T has been obtained from R via a known artificial deformation. We have a magneto-resonance image R (Figure 12.5, left) and its artificially deformed version T (Figure 12.5, right) using a known distortion map from R to T shown in Figure 12.6, left (data test kindly provided by H. Tagare [310]). We want to see whether the distortion maps rendered by the nonlinear elasticity and biharmonic models are similar to the true map. The true distortion map plots the vector fields from R to T, and the error between the true map and the maps after registration by the two models is measured by the Euclidean norm. The difference from the ground truth is smaller for the nonlinear elasticity model (Figure 12.6, right) than for the biharmonic model (Figure 12.6, middle) after the same number of iterations. Therefore, the nonlinear elasticity model results in a closer match to the true map than the biharmonic model. Thus we expect that the nonlinear elasticity model will produce better results for the gene expression data to atlas registration.

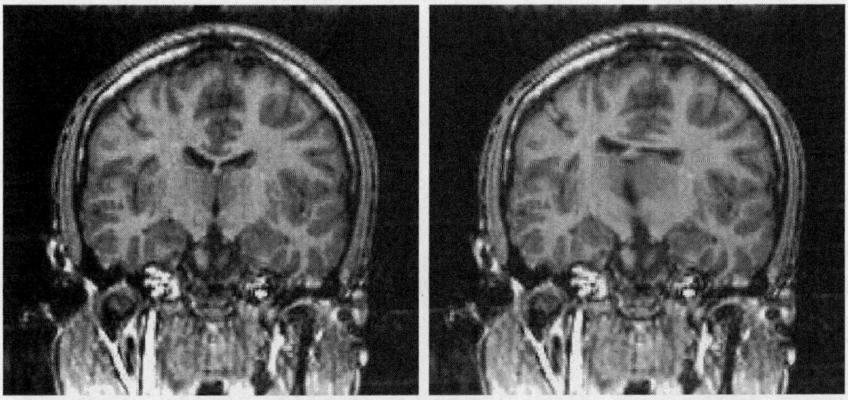

FIGURE 12.5: Reference (left) and template (right).

The biharmonic and nonlinear elasticity regularizations are then applied to the registration of mouse brain atlas to gene expression data in the presence of manually selected landmarks, for comparison. Thus data fidelity, landmark matching term and regularization are combined. Figure 12.7 shows a mouse atlas image as reference R (left), mouse gene expression data as template T (right) and gene expression after histogram equalization with specified landmarks (bottom). Figures 12.8 and 12.9 show the numerical results obtained using the biharmonic regularization and the nonlinear elasticity one, respectively. In each case we view the deformed template (left), the deformation field

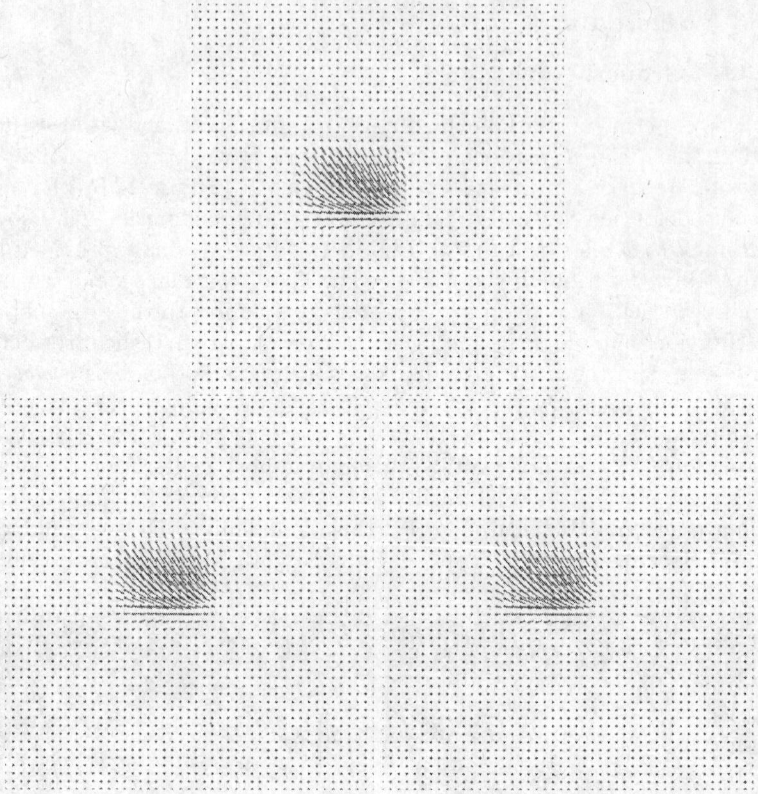

FIGURE 12.6: True distortion map; distortion map by the biharmonic model (error = 0.12692); and distortion map by the nonlinear elasticity model (error =0.12413).

plus the landmarks transformation (right), and $1/det(\nabla\Phi)$ plus the deformed grid (bottom).

Table 12.1 summarizes the comparison between the two sets of results from where we can see that using the nonlinear elasticity regularization gives better registration results.

Moreover, we can see that the nonlinear elasticity model renders better landmark matching, smaller dissimilarity measure, and larger deformation. The range of the determinant of the Jacobian matrix is also larger corresponding to larger deformation. Since the lower bound of the range of determinant of Jacobian remains positive and no regridding step is triggered, the transformation is well defined and desirable. The distortion maps draw the vectors from the grid points of the reference image to the non-grid points after registration; the original reference/template landmarks are marked in red/green, the

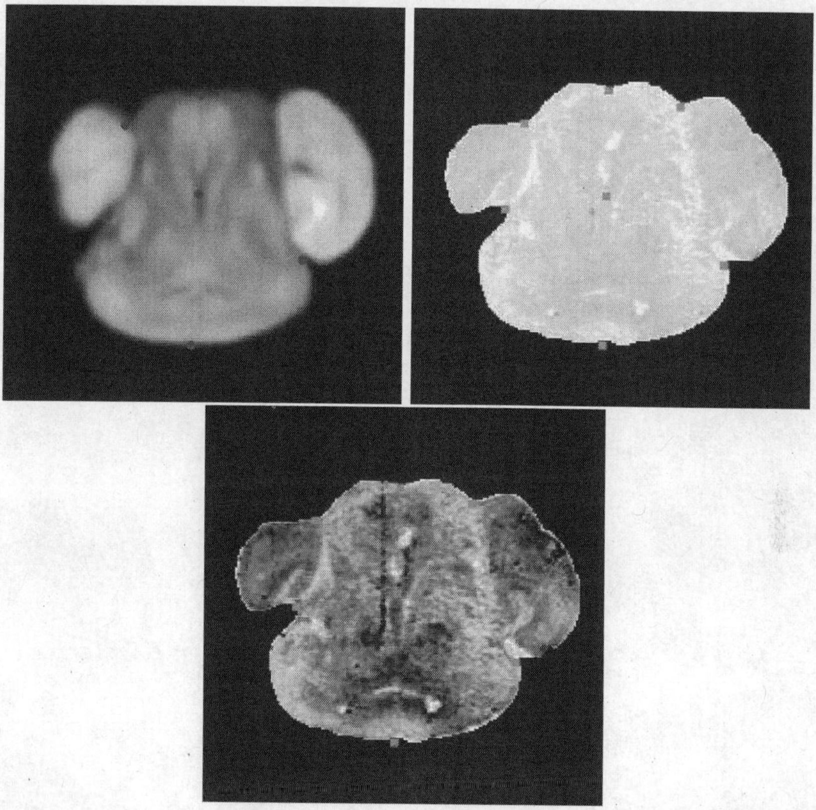

FIGURE 12.7: Mouse atlas as reference R (left), mouse gene expression data as template T (right) and gene expression after histogram equalization with specified landmarks (bottom).

reference landmarks after registration are marked in blue. We can see that the landmarks converge (moving from red spots to blue spots to approach the green spots) in accordance with the distortion field. As for the deformed grids, where the grid area expands/shrinks, we observe lighter/darker gray levels corresponding to larger/smaller value of the inverse of determinant of Jacobian which indicates a real expansion/shrinkage.

In order to evaluate the registration results from Figures 12.8 and 12.9 in a different way, we compute another cost functional measuring image alignment called the mutual information between two random variables and defined as follows:

$$MI_{\mathbf{X},\mathbf{Y}} = \int_{\mathbf{Y}} \int_{\mathbf{X}} p_{\mathbf{X},\mathbf{Y}}(x,y) log \frac{p_{\mathbf{X},\mathbf{Y}}(x,y)}{p_{\mathbf{X}}(x) p_{\mathbf{Y}}(y)} \, dx \, dy,$$

where $\mathbf{X}$ and $\mathbf{Y}$ are random variables, $p_{\mathbf{X},\mathbf{Y}}(x,y)$ is the joint probability den-

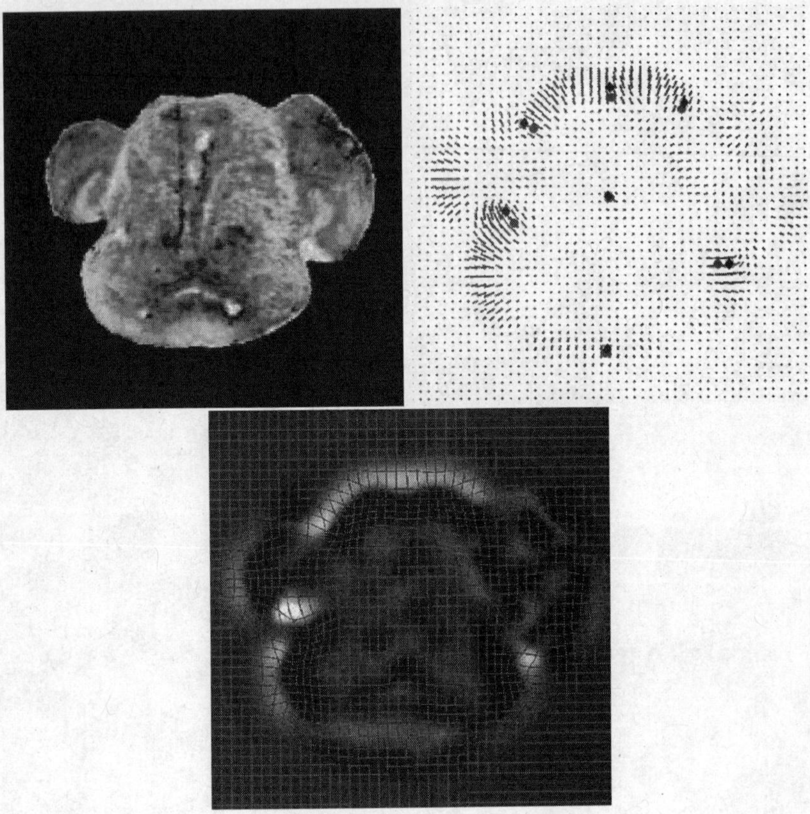

FIGURE 12.8: Deformed template (left), deformation field plus the landmarks transformation (right) and $1/det(\nabla\Phi)$ plus the deformed grid (bottom) using biharmonic regularization.

sity function of $\mathbf{X}$ and $\mathbf{Y}$, and $p_{\mathbf{X}}(x)$ and $p_{\mathbf{Y}}(y)$ are the marginal probability density functions of $\mathbf{X}$ and $\mathbf{Y}$, respectively.

Mutual information quantifies the dependence between $\mathbf{X}$ and $\mathbf{Y}$, which in our case are the intensity maps of R and T. Considering that larger mutual information indicates better registration, we see in Figure 12.10 that the nonlinear elasticity model indicated by the black line that is above is more desirable in this respect.

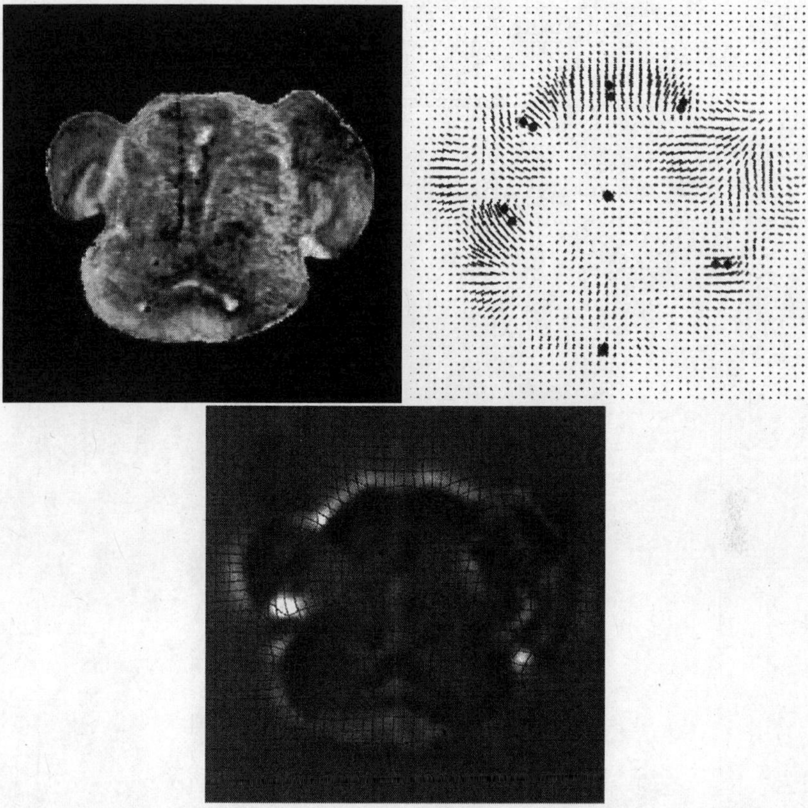

FIGURE 12.9: Deformed template (left), deformation field plus the landmarks transformation (right) and $1/det(\nabla\Phi)$ plus the deformed grid (bottom) using nonlinear elasticity regularization.

12.4 Exercises

Exercise 12.1 *Derive in two dimensions the Euler–Lagrange equations associated with the minimization of*

$$E(\mathbf{u}) = Fid(\mathbf{u}) + \gamma D^{LM}(\mathbf{u}) + \frac{\alpha}{2}\sum_{l=1}^{2}\int_{\Omega}\|\nabla u_l\|^2 d\mathbf{x}.$$

Exercise 12.2 *Implemement the numerical algorithm given in Subsection 12.1.3.1 for* $\gamma = 0$.

Exercise 12.3 *Derive in two dimensions the Euler–Lagrange equations asso-*

Regularization	LD	DM	RJ
BH	0.11232	1,255	(0.23, 2.30)
NE	0.10712	1,177	(0.15, 2.40)

TABLE 12.1: Landmark points averaged distances after registration (LD), L^2 dissimilarity measures (DM), and ranges of determinant of Jacobian matrix $\nabla \mathbf{u}$ (RJ) for the biharmonic model (BH) and the nonlinear elasticity model (NE), respectively.

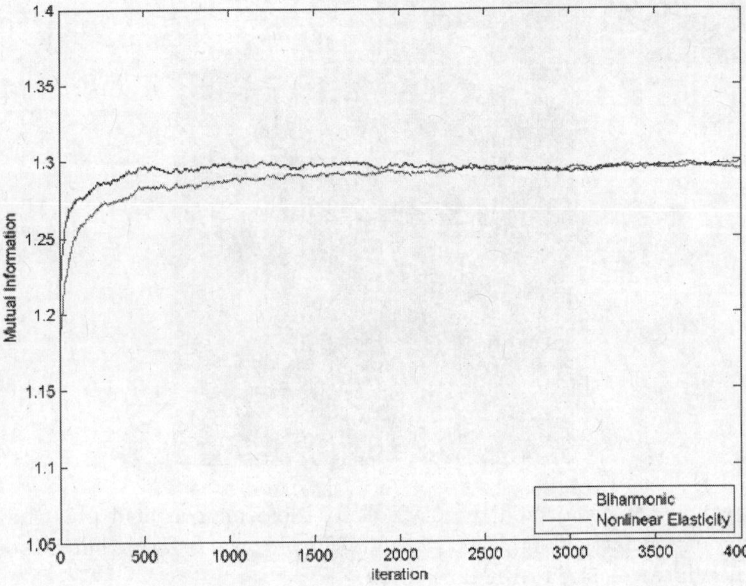

FIGURE 12.10: Mutual information increasing with time using biharmonic regularization and nonlinear elasticity regularization for the mouse registration.

ciated with the minimization of

$$E(\mathbf{u}) = Fid(\mathbf{u}) + \gamma D^{LM}(\mathbf{u}) + \frac{\alpha}{2}\sum_{l=1}^{2}\int_{\Omega}(\triangle u_l)^2 d\mathbf{x}.$$

Exercise 12.4 *Implemement the numerical algorithm given in Subsection 12.1.3.2 for $\gamma = 0$.*

Exercise 12.5 *Derive in two dimensions the Euler–Lagrange equations asso-*

ciated with the minimization of

$$E(\mathbf{u}) = Fid(\mathbf{u}) + \gamma D^{LM}(\mathbf{u}) + \alpha \int_{\Omega} \left[\frac{\lambda}{2}(trace(\epsilon(\mathbf{u})))^2 + \mu trace(\epsilon^2(\mathbf{u}))\right] d\mathbf{x}.$$

Exercise 12.6 *Implemement the numerical algorithm given in Subsection 12.1.3.3 for* $\gamma = 0$.

Exercise 12.7 *Verify formula (12.2).*

Exercise 12.8 *Derive in two dimensions the Euler–Lagrange equations associated with the minimization of* $J(\mathbf{u}, \mathbf{v})$ *given in Section 12.2.*

Exercise 12.9 *Implement the numerical algorithm given in Section 12.2 for* $\gamma = 0$.

Chapter 13

Piecewise-Constant Binary Model for Electrical Impedance Tomography

Electrical impedance tomography (EIT) is a non-invasive inverse method which attempts to determine the electrical conductivity σ of a medium in a domain Ω, by making voltage and current measurements at the boundary $\partial\Omega$ of the medium. In mathematical terms, the EIT problem is the recovery of the coefficient σ of an elliptic partial differential equation, defined for $x \in \Omega$, given knowledge of the Cauchy data, i.e., the Neumann-to-Dirichlet map or the Dirichlet-to-Neumann map. The EIT problem has important applications in fields such as medical imaging, nondestructive testing of materials, environmental cleaning and geophysics. In the last two decades, it has been the topic of many theoretical and numerical studies. However, there are still important questions, such as improving the stability of reconstruction algorithms, improving the resolution and reliability of reconstructions of σ, and increasing the speed of inversion algorithms so σ can be imaged in real time [55].

This chapter presents an application of the binary piecewise-constant segmentation model presented in Chapter 8 to electrical impedance tomography. It follows the work of Tanushev and Vese from [316].

13.1 Introduction

Let Ω be an open, bounded and simply connected domain of $\mathbb{R}^2$.

13.1.1 Forward problem

For a known isotropic electrical conductivity function $\sigma \in L^\infty(\Omega)$ (scalar valued, strictly positive and bounded in Ω), we can define the Neumann-to-Dirichlet map Λ_σ for a domain Ω in the following way. Let u, called the potential, be the solution to the partial differential equation,

$$\begin{aligned} \nabla \cdot \sigma \nabla u = \operatorname{div}(\sigma \nabla u) &= 0 && \text{in } \Omega \\ \sigma \frac{\partial u}{\partial \nu} &= I && \text{on } \partial\Omega \\ \int_{\partial\Omega} u \, dS &= 0, \end{aligned} \tag{13.1}$$

where ν is the unit outward normal to $\partial\Omega$ and $\frac{\partial u}{\partial \nu} = \nabla u \cdot \nu$. The function I is restricted such that $\int_{\partial\Omega} I \, dS = 0$. Based on the Lax-Milgram theorem from Chapter 2, this Neumann boundary value problem, referred to as the forward problem, has a unique solution $u \in H^1(\Omega)$ (at least in the weak sense), given that $I \in H^{-1/2}(\partial\Omega)$ (the potential is unique up to an additive constant that we fix by imposing the condition $\int_{\partial\Omega} u dS = 0$).

The Neumann-to-Dirichlet operator $\Lambda_\sigma : \{I \in H^{-1/2}(\partial\Omega), \int_{\partial\Omega} I dS = 0\} \to H^{1/2}(\partial\Omega)$, maps the Neumann boundary data I to the restriction (trace) of u to the boundary of Ω:

$$\Lambda_\sigma I = u|_{\partial\Omega}.$$

This map depends nonlinearly on the conductivity σ. The Dirichlet-to-Neumann map can also be considered, $\Lambda_\sigma^{-1} : H^{1/2}(\partial\Omega) \to H^{-1/2}(\partial\Omega)$ with $\Lambda_\sigma^{-1} V = \sigma \frac{\partial u}{\partial \nu}$, where $u = V$ on $\partial\Omega$.

13.1.2 Inverse problem

The inverse conductivity problem, as formulated by Calderón [77], is to find a bounded, strictly positive function $\sigma(x)$, given the map Λ_σ^{-1}. Theoretically, this problem can be solved uniquely for a large class of functions σ, as established in [26], [307], [248], [196], [197], [68], and [257] (we highlight the more recent work [26] in two dimensions, where only the above assumptions on Ω and σ are imposed). We will assume knowledge of Λ_σ rather than Λ_σ^{-1}, since in practice Λ_σ is less sensitive to noise. Thus we want to recover a function $\sigma \in L^\infty(\Omega)$, satisfying $\sigma(x) \geq \sigma_0 > 0$ in Ω, given I and $V = u|_{\partial\Omega}$.

We introduce the adjoint potential, τ, as the unique solution to the following problem (called the adjoint problem),

$$\begin{aligned} \nabla \cdot \sigma \nabla \tau &= 0 && \text{in } \Omega \\ \sigma \frac{\partial \tau}{\partial \nu} &= u|_{\partial\Omega} - V && \text{on } \partial\Omega \\ \int_{\partial\Omega} \tau \, dS &= 0. \end{aligned} \tag{13.2}$$

Here u is the solution to the forward problem (13.1) and $V \in H^{1/2}(\partial\Omega)$ is such that $\int_{\partial\Omega} V \, dS = 0$. The adjoint potential will be useful in later sections.

13.1.3 Related work

L. Rondi and F. Santosa have previously applied the Ambrosio–Tortorelli approximations of the general Mumford–Shah functional from Chapters 6–7 to the inverse conductivity problem in an elegant work [273] in which σ is recovered from an energy minimization formulation with data fidelity and regularizing terms. The authors give theoretical results of existence of piecewise-smooth minimizers σ in $SBV(\Omega)$, and show convergence of the elliptic approximations by Γ-convergence.

Segmentation techniques for recovering the conductivity σ and other elliptic equation coefficients have been proposed by T. Chan, E. Chung, and X.C. Tai [91], [102] using approaches similar to the approach in this chapter. The work [91] considers a different but related inverse problem; it applies a slightly modified version of the piecewise-constant segmentation method from Chapter 8 to recover the coefficient $q(x)$ from an elliptic PDE, by using the total variation, $\int_\Omega |Dq|$, as a regularization instead of the length regularization. The work [102] (that much inspired this work) addresses the problem of inverse conductivity; it uses a binary piecewise-constant segmentation method, as in Chapter 8, but the regularization is again the total variation. In this chapter we show that the length term of the discontinuity set of σ is sufficient to recover σ with high accuracy and smaller jumps, and is simpler in the piecewise-constant case; also, the computational results need about 200 to 250 iterations for convergence to steady state instead of 200 to 50000 iterations in [102]; finally, we also show that interior contours (or holes) of the conductivity σ can also be detected by the described approach.

The work of L. Borcea et al. [57] also uses a variational approach with regularization to recover the conductivity σ, but in a different way.

Other related works can be found in [297], [198], [199], [250], [75], [57], [55], and [56].

To sum up, the binary piecewise constant segmentation model from Chapter 8 will be used to recover the conductivity σ in the next section. This method was motivated by the above-mentioned approaches [273], [91], [102], [57]. Thus the level set approach is used to minimize the Mumford and Shah functional in the simpler (although more restrictive) case of binary piecewise-constant conductivity.

13.2 Formulation of the minimization

In practice, we do not completely know the Neumann-to-Dirichlet or Dirichlet-to-Neumann map. Instead, we are given a set of N evaluations of the map. Therefore, we assume that Λ_{σ^*} is known for N functions for the true conductivity σ^*. That is, there are N pairs of functions (I_n, V_n) such that

$\Lambda_{\sigma^*} I_n = V_n$. Experimentally, this is accomplished by setting a current excitation pattern I_n and measuring the resulting voltage V_n at discrete locations of the electrodes along the boundary $\partial\Omega$. In practice, the EIT problem is to find σ^* from partial and usually noisy knowledge of Λ_{σ^*}. A significant difficulty is the severe ill-posedness of EIT. This problem is ill-posed in the sense that small perturbations of the boundary data are exponentially amplified in the image of σ inside Ω [55]. Therefore, in the reconstruction process, we have to restrict σ to a subset of $L^\infty(\Omega)$ of smoother functions as in [57], [273], [91], [102], among others.

In order to obtain a numerical solution to the inverse conductivity problem, we formulate it as a minimization: the functional to be optimized will consist of a data fidelity term (fitting term), and of a regularization term. Using the ideas of the binary piecewise-constant reconstruction from Chapter 8 in a level set framework in the simplified case of piecewise constant σ taking two values, we let

$$\sigma(\phi, c_1, c_2) = c_1 + (c_2 - c_1)H(\phi),$$

where ϕ is the level set function, c_1 and c_2 are the constant values of the conductivity, and H is the Heaviside function. Therefore, motivated by the classical Mumford and Shah functional from Chapter 6 and its level set formulations from Chapter 8, we minimize the energy functional,

$$E(\phi, c_1, c_2) = F(\sigma(\phi, c_1, c_2)) + \alpha(F)L(\phi),$$

where the fitting term and the regularizing length term are given respectively by,

$$\begin{aligned} F(\sigma) &= \sum_{n=1}^{N} \frac{\|\Lambda_\sigma I_n - V_n\|^2_{L^2(\partial\Omega)}}{N\|V_n\|^2_{L^2(\partial\Omega)}}, \\ L(\phi) &= \int_\Omega |DH(\phi)| = \int_\Omega \delta(\phi)|\nabla\phi| dx, \end{aligned}$$

with $\alpha(s)$ a non-decreasing function such that $\alpha(0) = 0$. This function allows us to control the size of the regularization. Note that under these restrictions on α, the true conductivity $\sigma^* = \sigma(\phi^*, c_1^*, c_2^*)$ is an absolute minimum of the functional, as $E(\phi^*, c_1^*, c_2^*) = 0$ (at least in the noiseless case). By this minimization, $\sigma \in SBV(\Omega)$ (σ will be a piecewise-constant function, with the discontinuity set of finite length).

We now introduce an artificial time parameter t and let $\phi(\cdot) = \phi(\cdot, t)$, $c_1 = c_1(t)$ and $c_2 = c_2(t)$. Differentiating the energy, we obtain:

$$\frac{dE}{dt} = \frac{\partial E}{\partial c_1}\frac{dc_1}{dt} + \frac{\partial E}{\partial c_2}\frac{dc_2}{dt} + \int_\Omega \frac{\partial E}{\partial \phi}\frac{\partial \phi}{\partial t}\, dx.$$

Furthermore,

$$\begin{aligned}
\frac{\partial E}{\partial \phi} &= (1+\alpha'(F)L)\frac{\partial F}{\partial \phi} + \alpha(F)\frac{\partial L}{\partial \phi}, \\
&= (1+\alpha'(F)L)\frac{\partial F}{\partial \sigma}\frac{\partial \sigma}{\partial \phi} - \alpha(F)\delta(\phi)\nabla\cdot\frac{\nabla\phi}{|\nabla\phi|}, \\
&= \delta(\phi)\Big((1+\alpha'(F)L)(c_2-c_1)\frac{\partial F}{\partial \sigma} - \alpha(F)\nabla\cdot\frac{\nabla\phi}{|\nabla\phi|}\Big),
\end{aligned}$$

and

$$\begin{aligned}
\frac{\partial E}{\partial c_1} &= (1+\alpha'(F)L)\int_\Omega \frac{\partial F}{\partial \sigma}\frac{\partial \sigma}{\partial c_1}dx, \\
&= (1+\alpha'(F)L)\int_\Omega \frac{\partial F}{\partial \sigma}(1-H(\phi))dx.
\end{aligned}$$

Similarly,

$$\frac{\partial E}{\partial c_2} = (1+\alpha'(F)L)\int_\Omega \frac{\partial F}{\partial \sigma}H(\phi)dx.$$

Let u_n and τ_n be the nth potential and adjoint potential, so that they respectively solve (13.1) and (13.2) with $I = I_n$ and $V = V_n$. It can be shown that the Fréchet derivative of the fitting term is given by (see for example [57] or [102]),

$$\frac{\partial F}{\partial \sigma} = -2\sum_{n=1}^{N}\frac{\nabla u_n\cdot\nabla\tau_n}{N\|V_n\|^2_{L^2(\partial\Omega)}}.$$

Thus if we set,

$$\begin{aligned}
\frac{\partial \phi}{\partial t} &= -\delta(\phi)\Big((1+\alpha'(F)L)(c_2-c_1)\frac{\partial F}{\partial \upsilon} - \alpha(F)\nabla\cdot\frac{\nabla\phi}{|\nabla\phi|}\Big), \\
\frac{dc_1}{dt} &= -(1+\alpha'(F)L)\int_\Omega \frac{\partial F}{\partial \sigma}(1-H(\phi))\,dx, \\
\frac{dc_2}{dt} &= -(1+\alpha'(F)L)\int_\Omega \frac{\partial F}{\partial \sigma}H(\phi)\,dx,
\end{aligned}$$

we have that $\frac{dE}{dt} \leq 0$. Hence, these equations give the minimization formulation of the inverse conductivity problem.

13.3 Numerical details

We take the domain to be the unit square $\Omega = [0,1]^2$. To derive the linear system of equations that represents $\nabla\cdot\sigma\nabla$ at the grid points away from the

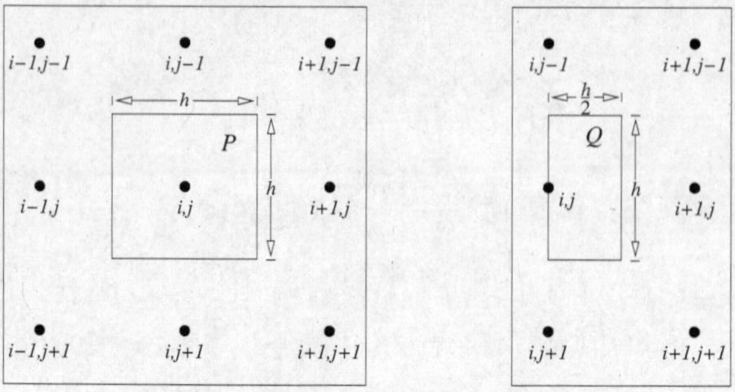

FIGURE 13.1: Integration domains.

boundary, we look at the integral of equation (13.1) (or (13.2)) over a square, P, centered at a grid point (see Figure 13.1):

$$0 = \int_P \nabla \cdot \sigma \nabla u dx = \int_{\partial P} \sigma \frac{\partial u}{\partial n} dS,$$

n denoting the unit outward normal to ∂P.

Using finite differences to approximate the normal derivative of u in this formula, we obtain:

$$\begin{aligned} 0 &= \sigma_{i+\frac{1}{2},j}\left(\frac{u_{i+1,j} - u_{i,j}}{h}\right)h + \sigma_{i,j-\frac{1}{2}}\left(\frac{u_{i,j-1} - u_{i,j}}{h}\right)h \\ &\quad + \sigma_{i-\frac{1}{2},j}\left(\frac{u_{i-1,j} - u_{i,j}}{h}\right)h + \sigma_{i,j+\frac{1}{2}}\left(\frac{u_{i,j+1} - u_{i,j}}{h}\right)h, \\ 0 &= \frac{-(\sigma_{i+\frac{1}{2},j} + \sigma_{i,j-\frac{1}{2}} + \sigma_{i-\frac{1}{2},j} + \sigma_{i,j+\frac{1}{2}})}{h^2}u_{i,j} \\ &\quad + \frac{\sigma_{i+\frac{1}{2},j}}{h^2}u_{i+1,j} + \frac{\sigma_{i,j-\frac{1}{2}}}{h^2}u_{i,j-1} + \frac{\sigma_{i-\frac{1}{2},j}}{h^2}u_{i-1,j} + \frac{\sigma_{i,j+\frac{1}{2}}}{h^2}u_{i,j+1}. \end{aligned}$$

At the boundary nodes, but not the corner nodes, we use the boundary condition,

$$\sigma \frac{\partial u}{\partial n} = f,$$

where f is either I or $u|_{\partial\Omega} - V$, and we integrate over a domain Q (see

Figure 13.1):

$$\begin{aligned}
0 &= \int_Q \nabla\cdot\sigma\nabla u dx = \int_{\partial Q} \sigma\frac{\partial u}{\partial n} dS, \\
&= \sigma_{i+\frac{1}{2},j}\left(\frac{u_{i+1,j}-u_{i,j}}{h}\right)h + \sigma_{i,j-\frac{1}{2}}\left(\frac{u_{i,j-1}-u_{i,j}}{h}\right)\frac{h}{2} \\
&\quad + f_{i,j}h + \sigma_{i,j+\frac{1}{2}}\left(\frac{u_{i,j+1}-u_{i,j}}{h}\right)\frac{h}{2}, \\
-\frac{f_{i,j}}{h} &= \frac{-(2\sigma_{i+\frac{1}{2},j}+\sigma_{i,j-\frac{1}{2}}+\sigma_{i,j+\frac{1}{2}})}{2h^2}u_{i,j} \\
&\quad + \frac{2\sigma_{i+\frac{1}{2},j}}{2h^2}u_{i+1,j} + \frac{\sigma_{i,j-\frac{1}{2}}}{2h^2}u_{i,j-1} + \frac{\sigma_{i,j+\frac{1}{2}}}{2h^2}u_{i,j+1}.
\end{aligned}$$

In a similar fashion, we obtain the equations at the four corner nodes. Note that in this discretization of the operator we need the values of the conductivity at points which lie in between the grid points. We take a value to be the minimum of the two nearest values to preserve the discontinuous nature of σ. This approximation ignores isolated points where σ is bigger than at its surrounding neighbors. This is taken into consideration in evaluating dc/dt and in re-normalizing ϕ.

We will assume that the conductivity constant $c_1^* = 1$ is fixed and that it is the value of σ^* on the boundary of the domain. In this case, the evolution equations for ϕ and the unknown conductivity constant $c_2 = c$ are given by:

$$\frac{\partial\phi}{\partial t} = -\delta(\phi)\Big((1+\alpha'(F))(c-1)\frac{\partial F}{\partial\sigma} - \alpha(F)\nabla\cdot\frac{\nabla\phi}{|\nabla\phi|}\Big), \tag{13.3}$$

$$\frac{dc}{dt} = -(1+\alpha'(F)L)\int_\Omega \frac{\partial F}{\partial\sigma}H(\phi)\,dx. \tag{13.4}$$

Following Chapter 8, we use a semi-implicit finite difference scheme to discretize the divergence term $\nabla\cdot\frac{\nabla\phi}{|\nabla\phi|}$. Since we are assuming that one of the values of the conductivity is known, we impose Dirichlet boundary conditions on ϕ. We use the approximations of the Heaviside and Dirac delta functions given in Chapter 2,

$$H_\epsilon(x) = \frac{1}{2}\Big[1+\frac{2}{\pi}\arctan\Big(\frac{x}{\epsilon}\Big)\Big], \quad \delta_\epsilon(x) = \frac{1}{\pi}\frac{\epsilon}{\epsilon^2+x^2}.$$

To compute the Fréchet derivative of the fitting term, we need the potential and the adjoint potential in Ω for each one of the configurations. To find these potentials, we use the conjugate gradient method [173] to solve the system of linear equations that results from the finite volume discretization of (13.1) and (13.2). In discretizing the $\nabla\cdot\sigma\nabla$ operator, we also use the above approximations H_ϵ and δ_ϵ. However, since the finite volume discretization requires the values of σ at the half grid points, we take its value to be the minimum of the two nearest values to preserve the discontinuous nature of σ, as explained earlier.

Since there is no analytic solution to the equation that governs the evolution of c, the differential equation (13.4) has to be solved numerically. However, instead of integrating this equation, we evaluate $\frac{dc}{dt}$ near the previous value of c and approximate $\frac{dc}{dt}$ by a quadratic polynomial. The next value of c is taken to be the value that minimizes the absolute value of this polynomial (if there are two such points, we take the one closest to the previous value). We alternate minimizing the functional with respect to ϕ and c.

To generate artificial data, we compute the Neumann-to-Dirichlet map Λ_{σ^*} by applying the conjugate gradient method to the finite volume discretization of the forward problem. The Neumann boundary data are chosen to be sines and cosines of higher and higher frequency on the boundary.

For the cases with regularizations, we use

$$\alpha(F) = \frac{10^{-7}}{\pi} \arctan(10^7 F),$$

and for no regularization, we use $\alpha(F) = 0$.

13.4 Numerical reconstruction results

13.4.1 Test 1 – Two inclusions

The true conductivity for this experiment has two inclusions (see Figure 13.2). Both inclusions have conductivity $c_2^* = 2$ and the background conductivity $c_1^* = 1$. The reconstructions in Figure 13.4 were carried out using six configurations with regularization and without regularization. Figure 13.5 shows the same reconstructions using twelve configurations. We also show the value of the fidelity term $F(t)$ versus iteration number. The overall behavior of the total energy $E(t)$ is qualitatively very similar to the behavior of the fidelity term. The final values from the minimization are listed in Table 13.1.

TABLE 13.1: Final values for test 1 reconstructions.

	N	c	F
without regularization	6	2.0219	$2.6691 * 10^{-8}$
with regularization	6	2.0256	$2.8221 * 10^{-8}$
without regularization	12	2.1400	$3.8258 * 10^{-8}$
with regularization	12	2.0270	$5.1410 * 10^{-8}$

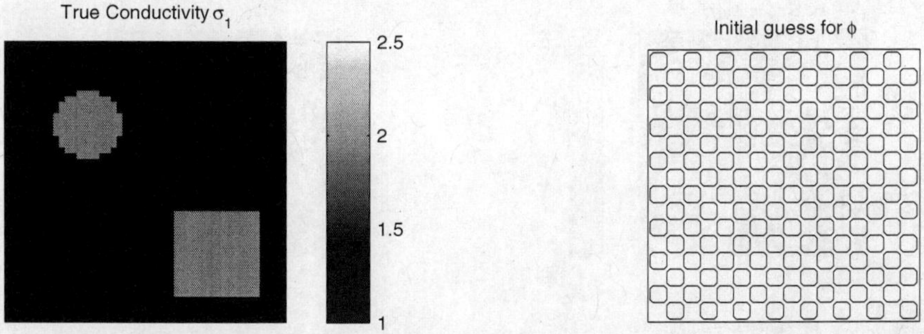

FIGURE 13.2: The true conductivity σ_1^* and the initial guess for ϕ. The conductivity has two inclusions, a square and a circle. The initial guess for the unknown conductivity constant c is 1.

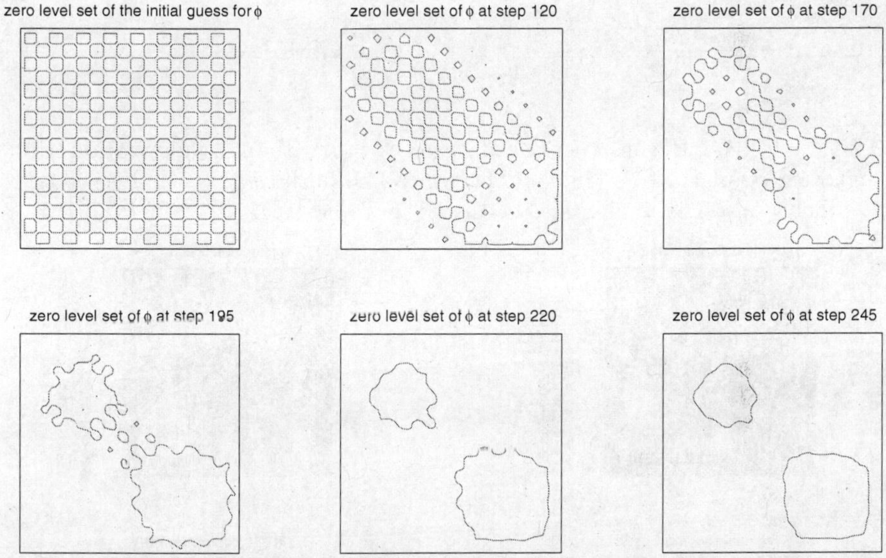

FIGURE 13.3: Evolution of the zero level line of ϕ over time for the $N = 6$ case with regularization.

13.4.2 Test 2 – Inclusion with empty interior

The true conductivity for this experiment has an inclusion that contains a hole (see Figure 13.6). The inclusion has a conductivity $c_2^* = 2$ and background conductivity $c_1^* = 1$. The reconstructions in Figure 13.7 were carried out using six configurations both with regularization and without regularization. Figure 13.8 shows the same reconstructions using twelve configurations. The final values from the minimization are listed in Table 13.2.

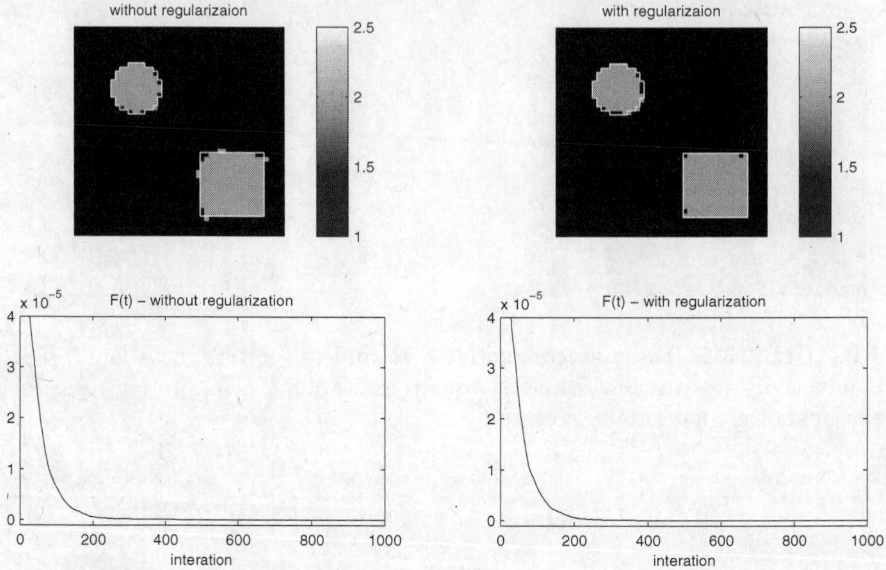

FIGURE 13.4: Reconstructions of σ_1^* for $N = 6$, without regularization (left) and with regularization (right). The white line outlines the true locations of the inclusions. The final values are listed in Table 13.1.

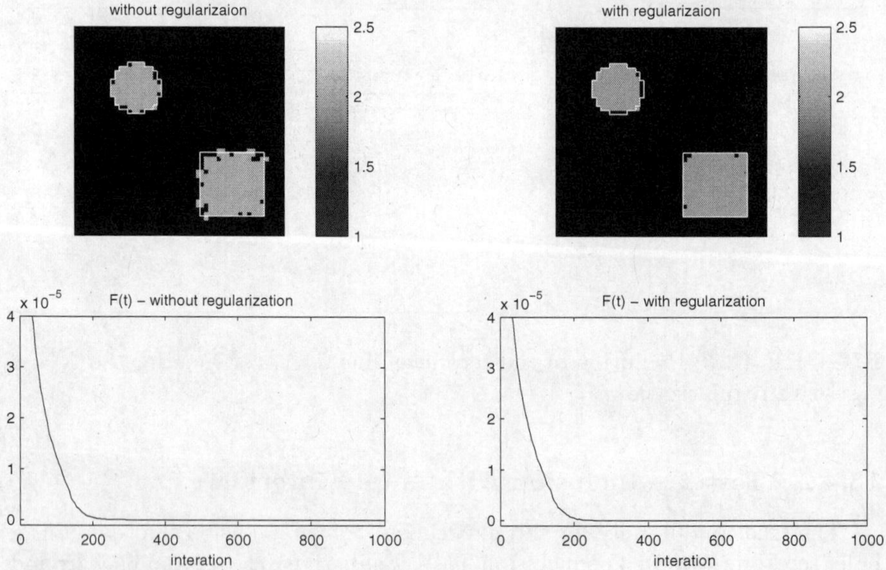

FIGURE 13.5: Reconstructions of σ_1^* for $N = 12$, without regularization (left) and with regularization (right). The white line outlines the true locations of the inclusions. The final values are listed in Table 13.1.

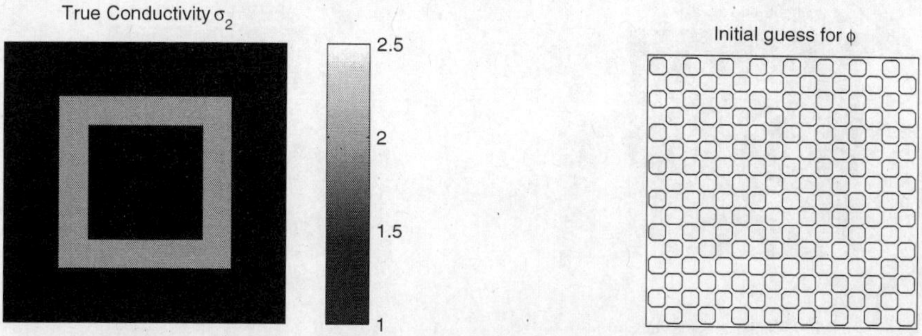

FIGURE 13.6: The true conductivity σ_2^* and the initial guess for ϕ. The conductivity has an inclusion with an empty interior. The initial guess for the unknown conductivity constant c is 1.

TABLE 13.2: Final values for test 2 reconstructions.

	N	c	F
without regularization	6	2.1517	$6.7434 * 10^{-8}$
with regularization	6	1.8598	$3.7849 * 10^{-8}$
without regularization	12	2.3290	$5.1093 * 10^{-8}$
with regularization	12	2.0335	$4.0141 * 10^{-8}$

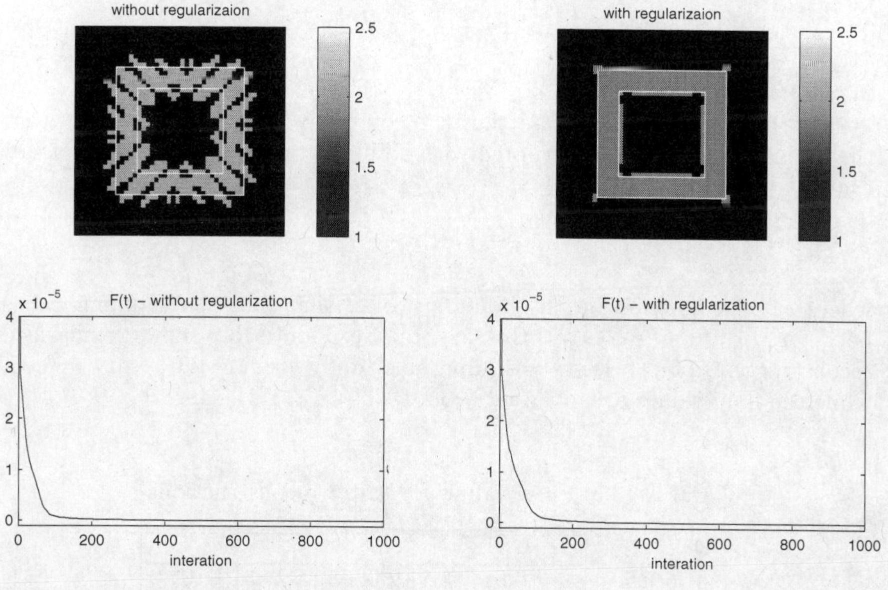

FIGURE 13.7: Reconstructions of σ_2^* for $N = 6$, without regularization (left) and with regularization (right). The white line outlines the true location of the inclusion. The final values are listed in Table 13.2.

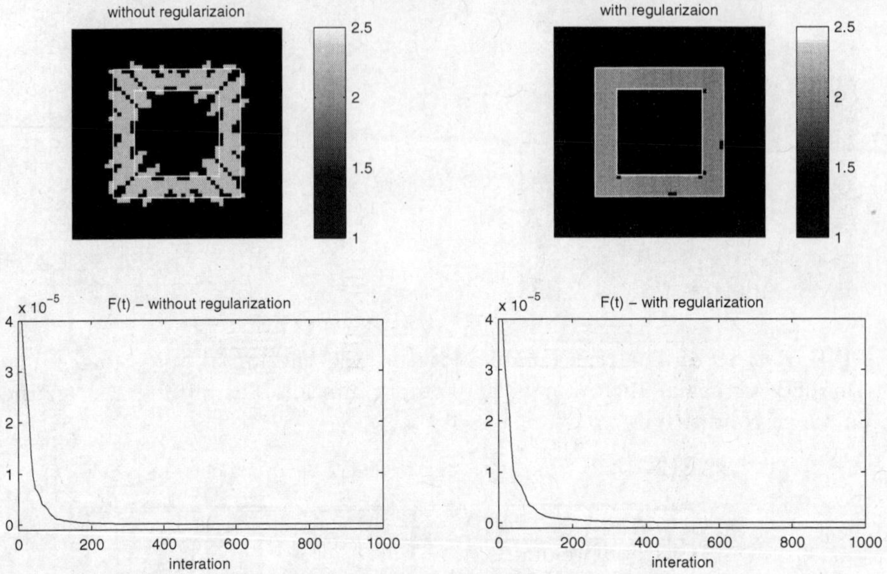

FIGURE 13.8: Reconstructions of σ_2^* for $N = 12$, without regularization (left) and with regularization (right). The white line outlines the true location of the inclusion. The final values are listed in Table 13.2.

13.4.3 Test 3 – Noise

For this experiment, we use the true conductivity σ_1^* of test 1. However, the measurements V_n are corrupted with additive uniformly distributed noise, that is,

$$V_n^c = (1 + \epsilon s_n)V_n,$$

where s_n is a function on the boundary of Ω that takes on random values between $[-1, 1]$ and ϵ controls the size of noise. Figure 13.9 shows regularized reconstructions of σ_1^* with $N = 12$ and noisy data with $\epsilon = 0.01$ and $\epsilon = 0.05$, while the final values are given in Table 13.3.

TABLE 13.3: Final values for test 3 reconstructions.

	N	c	F
$\epsilon = 0.01$	12	2.0363	$7.2851 * 10^{-7}$
$\epsilon = 0.05$	12	2.3865	$1.6377 * 10^{-5}$

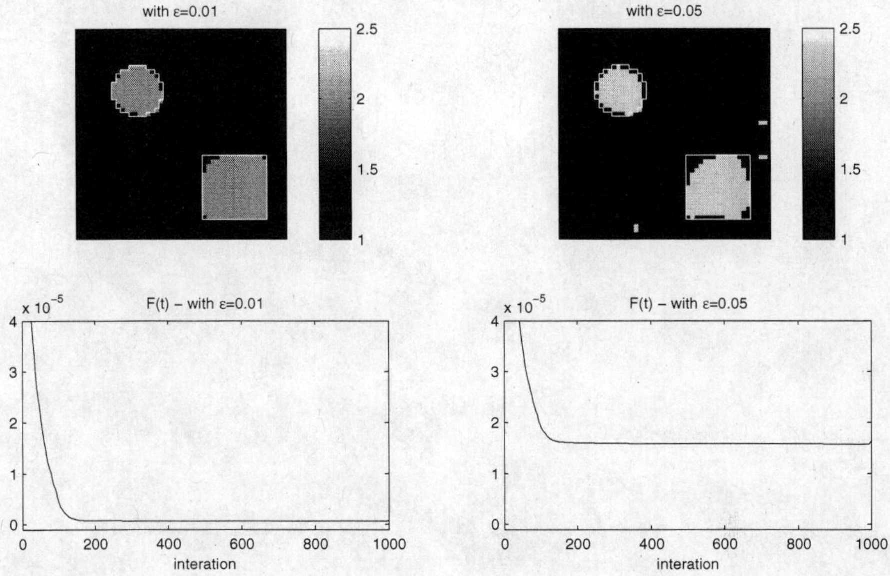

FIGURE 13.9: Reconstructions of σ_1^* for $N = 12$ with regularization for noisy data with $\epsilon = 0.01$ (left) and $\epsilon = 0.05$ (right).

13.5 Exercises

Exercise 13.1 *Prove that problem (13.1) admits a unique solution* $u \in H^1(\Omega)$.

Exercise 13.2 *Recover the expressions of* $\dfrac{\partial E}{\partial \Phi}$ *and* $\dfrac{\partial E}{\partial c_i}$, $i \in \{1, 2\}$ *of Section 13.2.*

Exercise 13.3 *Retrieve the finite difference schemes proposed in Section 13.3.*

Chapter 14

Additive and Multiplicative Piecewise-Smooth Segmentation Models in a Functional Minimization Approach

This chapter combines techniques presented in Chapters 2, 5 and 8 to illustrate applications to denoising, segmentation, image decomposition, and intensity inhomogeneity removal, in a single variational approach with theoretical results and numerical experiments. This chapter follows the work of Le and Vese from [204].

First, we decompose a given image u_0 into the sum $u_0 = v + w + noise$, where v is piecewise-constant, modeling sharp edges, while w is smooth, capturing global intensity inhomogeneities, smooth variations, variable lightning. The restored image is $u = v + w$. A related prior work in a variational approach, while proposed for image restoration and not for image segmentation, is by Chambolle and Lions in [86], where, given an image data $u_0 \in L^2(\Omega)$, two components v and w are extracted from u_0, such that $u_0 = v + w + noise$, by solving

$$\inf_{v,w,u=v+w} \Big\{ \int_\Omega |Dv| + \alpha \int_\Omega |D^2 w| + \lambda \int_\Omega |u_0 - v - w|^2 dx \Big\}.$$

Thus v will be a function of bounded variation, and w a smooth function of bounded Hessian. In this chapter, we will explicitly impose that v is piecewise-constant, thus edges will also be extracted from u_0 using segmentation and curve evolution. The degradation model will be $u_0 = v + w + noise$, with Gaussian additive noise of zero mean, and with v being piecewise-constant and w smooth.

Second, we consider the multiplicative case in a curve evolution segmentation approach. Here, we use a fidelity term inspired by the image restoration model of Rudin, Lions and Osher [276], [275]. In [276], [275], a model for image restoration in the presence of multiplicative noise was proposed, by solving

$$\inf_u \Big\{ \lambda \int_\Omega \Big| \frac{u_0}{u} - 1 \Big|^2 dx + \int_\Omega |Du| \Big\}.$$

In this chapter, the restored image u will be represented by $u = vw$, with v

piecewise-constant and w smooth. Thus we assume that the data u_0 follows a degradation model with multiplicative noise (of the form $noise = 1 +$ "noise", where "noise" is Gaussian noise of zero mean), given by $u_0 = v \cdot w \cdot noise$. In medical imaging, the factor w is called a bias field or intensity inhomogeneity. A related work has been proposed in [221], where the bias field is also smooth, multiplicative and represented by w, u is piecewise-constant as in [92], [94], but the noise is additive, and not multiplicative: $u_0 = v \cdot w + noise$. In this chapter, the described multiplicative model $u_0 = v \cdot w \cdot noise$ will have a different fidelity term inspired from [276], [275], and it applies to a different statistical model. Other related work that deal with intensity inhomogeneities and bias fields by different approaches are [343] and [5].

In Chapter 8, computational models for piecewise-constant (PC) segmentation have been presented, which can represent n phases in an image using $m = \log_2(n)$ level set functions, in the framework of the Mumford and Shah segmentation model from Chapter 6. Let $u_0 \in L^\infty(\Omega) \subset L^2(\Omega)$ be a given noisy image, $\Omega \subset \mathbb{R}^2$ open, bounded, and connected, and u the desired piecewise-constant image to recover and segment, given that $u_0 \approx u + \eta$, $\int_\Omega \eta dx = 0$. Denote $\vec{\Phi} = (\phi_1, ..., \phi_m)$ the vector of implicit functions $\phi_i : \Omega \to \mathbb{R}$, and $H(\vec{\Phi}) = (H(\phi_1), ..., H(\phi_m))$ the vector of Heaviside functions. Two pixels $(x_1, x_2), (y_1, y_2) \in \Omega$ belong to the same phase, if and only if $H(\vec{\Phi}(x_1, x_2)) = H(\vec{\Phi}(y_1, y_2))$, and there are up to $n = 2^m$ phases. Let $P = \{P_i : 1 \leq i \leq n\}$ be the set of disjoint phases or subsets of the partition, and $\vec{c} = (c_1, ..., c_n)$ be an unknown constant vector. Then the energy minimization problem for (PC) segmentation in the case of additive noise can be written as (Chapter 8):

$$\inf_{\vec{c}, \vec{\Phi}} F_n^{APC}(\vec{c}, \vec{\Phi}) = \sum_{1 \leq j \leq m} \mu_j \int_\Omega |DH(\phi_j)| + \sum_{1 \leq i \leq n = 2^m} \lambda_i \int_\Omega |u_0 - c_i|^2 \chi_{P_i} dx.$$

For $m = 1$ and $n = 2$ we have:

$$\chi_{P_1} = H(\phi_1), \quad \chi_{P_2} = H(-\phi_1).$$

For $m = 2$ and $n = 4$ we have:

$$\chi_{P_1} = H(\phi_1)H(\phi_2), \quad \chi_{P_2} = H(\phi_1)H(-\phi_2),$$

$$\chi_{P_3} = H(-\phi_1)H(\phi_2), \quad \chi_{P_4} = H(-\phi_1)H(-\phi_2).$$

In the general case, we have:

$$\chi_{P_i} = \Pi_{1 \leq j \leq m} H(+ \text{ or } - \phi_j), \quad 1 \leq i \leq n = 2^m,$$

and there are $n = 2^m$ different possibilities to define χ_{P_i}, satisfying $\sum_{1 \leq i \leq n} \chi_{P_i} \equiv 1$ everywhere in Ω.

We recall that H is the one-dimensional Heaviside function. The segmented image is given by $u(x) = \sum_{i=1}^n c_i \chi_{P_i}(x)$.

In this chapter we present extensions of this model to piecewise-smooth restoration in the presence of additive or multiplicative noise and smooth bias field. Theoretical results for existence of minimizers will be presented, together with numerical results.

14.1 Piecewise-smooth model with additive noise (APS)

For the first model, we will consider the unknown $u = v + w$, where v is piecewise-constant and w is a smooth function on Ω modeling global intensity inhomogeneities. We present the following energy minimization model, given the data u_0 and the degradation model $u_0 = v + w + noise$: minimize

$$F_n^{APS}(\vec{c}, \vec{\Phi}, w) = \sum_{1\leq j\leq m} \mu_j \int_\Omega |DH(\phi_j)| + \sum_{1\leq i\leq n=2^m} \lambda_i \int_\Omega |u_0 - c_i - w|^2 \chi_{P_i} dx + \nu_1 \int_\Omega |\nabla w|^p dx + \nu_2 \int_\Omega |D^2 w|^p dx, \tag{14.1}$$

where $p \in \{1, 2\}$, $D^2 w = \begin{pmatrix} w_{x_1x_1} & w_{x_1x_2} \\ w_{x_2x_1} & w_{x_2x_2} \end{pmatrix}$ is the Hessian of w, and $|D^2 w| = \sqrt{w_{x_1x_1}^2 + 2w_{x_1x_2}^2 + w_{x_2x_2}^2}$.

Remark 15 *We recall that in two dimensions, if Ω is an open, bounded subset of $\mathbb{R}^2$, with a C^1 boundary, and if $w \in H^2(\Omega) = W^{2,2}(\Omega)$, then based on the general Sobolev inequality [139], we have that $w \in C^{0,\gamma}(\overline{\Omega})$, with $0 < \gamma < 1$, and $\|w\|_{C^{0,\gamma}(\overline{\Omega})} \leq C\|w\|_{W^{2,2}(\Omega)}$ (the constant C depending only on Ω). Thus w is a continuous function in $\overline{\Omega}$, $w \in L^\infty(\overline{\Omega})$, and $\|w\|_{L^\infty(\overline{\Omega})} \leq C\|w\|_{W^{2,2}(\Omega)}$. This corresponds to the case $p = 2$ in (14.1).*

14.1.1 Existence of minimizers for APS model when $p = 2$

Keeping $\vec{\Phi}$, w fixed, and minimizing (14.1) with respect to c_i, we obtain when $P_i \neq \emptyset$:

$$c_i(\chi_{P_i}, w) = \frac{\int_\Omega (u_0 - w)\chi_{P_i} dx}{\int_\Omega \chi_{P_i} dx}, \text{ with } \chi_{P_i} = \chi_{P_i}(H(\phi_1), ..., H(\phi_m)). \tag{14.2}$$

Therefore, using $\chi_{E_j} := H(\phi_j)$, $1 \le j \le m$, (14.1) can be written as

$$\begin{aligned}\mathcal{F}_n^{APS}(\chi_{E_1}, ..., \chi_{E_m}, w) = &\sum_{1\le j\le m} \mu_j \int_\Omega |D\chi_{E_j}| \\ &+ \sum_{1\le i\le n} \lambda_i \int_\Omega |u_0 - c_i(\chi_{P_i}, w) - w|^2 \chi_{P_i} dx \\ &+ \nu_1 \int_\Omega |\nabla w|^2 dx + \nu_2 \int_\Omega |D^2 w|^2 dx,\end{aligned}$$

and we are led to consider the minimization

$$\inf_{(\chi_{E_1},...,\chi_{E_m},w)} \left\{ \mathcal{F}_n^{APS}(\chi_{E_1}, ..., \chi_{E_m}, w) : \chi_{E_i} \in \{0,1\},\ \chi_{E_i} \in BV(\Omega), \right. \\ \left. w \in H^2(\Omega) \right\}. \tag{14.3}$$

Theorem 40 *Assume that Ω is an open, bounded and connected subset of $\mathbb{R}^2$, with a C^1 boundary $\partial\Omega$. Let $u_0 \in L^2(\Omega)$, $\mu_j > 0$, $j = 1, ..., m$, $\lambda_i > 0$, $i = 1, ..., n = 2^m$ and $\nu_1, \nu_2 > 0$. Then there are functions $\chi_{E_j} \in BV(\Omega)$, with $\chi_{E_j}(x) \in \{0,1\}$ dx-a.e. in Ω, and $w \in H^2(\Omega)$ as solutions of problem (14.3).*

Proof. Let $(\chi_{E_{1,k}}, ..., \chi_{E_{m,k}}, w_k)$ be a minimizing sequence of (14.3). Without loss of generality, we can assume that $w_{k,\Omega} := \int_\Omega w_k dx = 0$. Denote $c_{i,k} = c_i(\chi_{P_{i,k}}, w_k)$, $i = 1, ..., n$. We have, C denoting a positive constant that may change from line to line,

$$\int_\Omega |D\chi_{E_{j,k}}| = |\chi_{E_{j,k}}|_{BV(\Omega)} \le C, \quad 1 \le j \le m, \tag{14.4}$$

$$\int_\Omega |u_0 - c_{i,k} - w_k|^2 \chi_{P_{i,k}}\, dx \le C, \quad 1 \le i \le n, \tag{14.5}$$

$$\int_\Omega |\nabla w_k|^2 dx \le C, \quad \int_\Omega |D^2 w_k|^2\, dx \le C. \tag{14.6}$$

By Poincaré's inequality, we have for all k,

$$\|w_k\|_{L^2(\Omega)} = \|w_k - w_{k,\Omega}\|_{L^2(\Omega)} \le C' \|\nabla w_k\|_{L^2(\Omega)} \le C. \tag{14.7}$$

Therefore, (14.6) and (14.7) imply that $\|w_k\|_{H^2(\Omega)} \le C$, for all k. Based on Remark 15, we deduce that w_k is also uniformly bounded in $L^\infty(\overline{\Omega})$.

Note that $\|\chi_{E_{j,k}}\|_{L^1(\Omega)} \le |\Omega|$, for all j, k. By passing to subsequences if necessary, there exist functions $\chi_{E_j} \in BV(\Omega)$, $j = 1, ..., m$, and $w \in H^2(\Omega)$ such that $\chi_{E_{j,k}}$ converges to χ_{E_j} weak* in $BV(\Omega)$ [140] (and strongly in $L^1(\Omega)$ and a.e.), and w_k converges to w weakly in $H^2(\Omega)$. We have

$$\int_\Omega |D\chi_{E_j}| \le \liminf_{k\to\infty} \int_\Omega |D\chi_{E_{j,k}}|,\ 1 \le j \le m,$$

$$\int_\Omega |\nabla w|^2 dx \le \liminf_{k\to\infty} \int_\Omega |\nabla w_k|^2 dx, \quad \int_\Omega |D^2 w|^2 dx \le \liminf_{k\to\infty} \int_\Omega |D^2 w_k|^2 dx.$$

Note that $|c_i(\chi_{P_{i,k}}, w_k)| \leq \frac{1}{\sqrt{|\Omega|}} ||u_0 - w_k||_{L^2(\Omega)} \leq C$. From Rellich–Kondrachov's theorem, we have, up to a subsequence, $w_k \to w$ strongly in $L^2(\Omega)$ and $dx-$ a.e. in Ω. It is easy to verify that $c_{i,k} \to c_i$ as $k \to \infty$. From the Lebesgue dominated convergence theorem, we obtain

$$\int_\Omega |u_0 - c_i - w|^2 \chi_{P_i} dx = \lim_{k\to\infty} \int_\Omega |u_0 - c_{i,k} - w_k|^2 \chi_{P_{i,k}} dx,$$

for $i = 1, ..., n$, where c_i is computed using (14.2) and w_k are uniformly bounded in $L^\infty(\Omega)$. Therefore,

$$\mathcal{F}_n^{APS}(\chi_{E_1}, ..., \chi_{E_m}, w) \leq \liminf_{k\to\infty} \mathcal{F}_n^{APS}(\chi_{E_{1,k}}, ..., \chi_{E_{m,k}}, w_k),$$

and $(\chi_{E_1}, ..., \chi_{E_m}, w)$ is a minimizer. Moreover, each $\chi_{E_j} \in BV(\Omega)$ is a characteristic function of some set $E_j \subset \Omega$ with finite perimeter, since $\chi_{E_{j,k}} \to \chi_{E_j}$ dx-a.e. in Ω and $\chi_{E_{j,k}} \in \{0, 1\}$. Also, we have $\sum_{i=1}^n \chi_{P_i} = 1$ dx-a.e. □

14.1.2 Existence of minimizers for APS model when $p = 1$

Definition 23 *[124] Let $u \in W^{1,1}(\Omega)$. We say that u is a function of bounded Hessian in Ω if $u \in W^{1,1}(\Omega)$ and $D^2u = (D_{x_1x_1}u, D_{x_1x_2}u, D_{x_2x_1}u, D_{x_2x_2}u)$ is representable by a finite Radon measure in Ω, i.e.,*

$$\int_\Omega u \frac{\partial^2 \varphi}{\partial x_i \partial x_j} dx = \int_\Omega \varphi dD_{x_i x_j} u, \forall \varphi \in C_c^\infty(\Omega), \text{ and } |D^2u|(\Omega) < \infty, \text{ where}$$

$$|D^2u|(\Omega)$$

$$= \sup\left\{ \int_\Omega u \left(\frac{\partial^2 \varphi_1}{\partial x_1^2} + 2\frac{\partial^2 \varphi_2}{\partial x_1 \partial x_2} + \frac{\partial^2 \varphi_3}{\partial x_2^2} \right) dx \,;\, \vec{\varphi} \in C_c^\infty(\Omega)^3, \||\vec{\varphi}|\|_\infty \leq 1 \right\}.$$

The vector space of all functions of bounded Hessian in Ω is denoted by $BH(\Omega)$. Equipped with the norm $\|u\|_{BH(\Omega)} = \|u\|_{W^{1,1}(\Omega)} + |D^2u|(\Omega)$, $BH(\Omega)$ becomes a Banach space.

Remark 16 *(lower semicontinuity) For any choice of $\vec{\varphi} \in C_c^\infty(\Omega)^3$, the functional*

$$L(u) = \int_\Omega u \left(\frac{\partial^2 \varphi_1}{\partial x_1^2} + 2\frac{\partial^2 \varphi_2}{\partial x_1 \partial x_2} + \frac{\partial^2 \varphi_3}{\partial x_2^2} \right) dx$$

is continuous in the $W^{1,1}(\Omega)$ topology. Therefore $|D^2u|(\Omega)$ is lower semicontinuous with respect to the $W^{1,1}(\Omega)$ topology.

Remark 17 *We recall the following result from [124]: if Ω is an open and bounded subset of $\mathbb{R}^2$, with a C^2 uniform boundary, there is a linear continuous injection from $BH(\Omega)$ into $C(\overline{\Omega})$.*

Let us recall the minimization problem when $p = 1$,

$$\inf_{(\chi_{E_1},...,\chi_{E_m},w)} \{\mathcal{G}_n^{APS}(\chi_{E_1},...,\chi_{E_m},w) : \chi_{E_j} \in \{0,1\},\ \chi_{E_j} \in BV(\Omega),$$
$$w \in BH(\Omega)\}, \tag{14.8}$$

where

$$\begin{aligned}\mathcal{G}_n^{APS}(\chi_{E_1},...,\chi_{E_m},w) = &\sum_{1\le j\le m} \mu_j \int_\Omega |D\chi_{E_j}| \\ &+ \sum_{1\le i\le n} \lambda_i \int_\Omega |u_0 - c_i(\chi_{P_i},w) - w|^2 \chi_{P_i} dx \\ &+ \nu_1 \int_\Omega |\nabla w| dx + \nu_2 \int_\Omega |D^2 w|,\end{aligned}$$

and c_i are given as before by (14.2). Recall that $\int_\Omega |D^2 w| = |D^2 w|(\Omega)$.

Theorem 41 *Assume that $\Omega \subset \mathbb{R}^2$ is open and bounded, with a C^2 uniform boundary. Let $u_0 \in L^2(\Omega)$, $\mu_j > 0$, $j = 1,...,m$, $\lambda_i > 0$, $i = 1,...,n = 2^m$ and $\nu_1, \nu_2 > 0$. Then there are functions $\chi_{E_j} \in BV(\Omega)$, with $\chi_{E_j}(x) \in \{0,1\}$ dx-a.e. in Ω, and $w \in BH(\Omega)$ as solutions of problem (14.8).*

Proof. The proof is almost similar to the case $p = 2$. By Poincaré–Wirtinger's inequality, we will deduce that w_k will be uniformly bounded in $L^2(\Omega)$ and thus in $L^1(\Omega)$. Thus w_k will be now uniformly bounded in $BH(\Omega)$, and a subsequence w_k will converge to $w \in BH(\Omega)$ weak* in $BH(\Omega)$ and strongly in $W^{1,1}(\Omega)$. This sequence will also be uniformly bounded in $L^\infty(\overline{\Omega})$ and in $C(\overline{\Omega})$, based on Remark 17. The other steps are similar and we omit them. □

The difference between the cases $p = 1$ and $p = 2$ is that for $p = 1$, w can have, roughly speaking, discontinuities along curves in the first order derivatives, because in this case $\nabla w \in BV(\Omega)^2$ only.

14.1.3 Two-phase APS model

The minimization for two-phases with $p = 2$ is

$$\begin{aligned}\inf_{c_1,c_2,\phi,w} F_2^{APS}(c_1,c_2,\phi,w) =&\mu \int_\Omega |DH(\phi)| + \lambda_1 \int_\Omega |u_0 - c_1 - w|^2 H(\phi) dx \\ &+ \lambda_2 \int_\Omega |u_0 - c_2 - w|^2 H(-\phi) dx + \nu_1 \int_\Omega |\nabla w|^2 dx \\ &+ \nu_2 \int_\Omega |D^2 w|^2 dx.\end{aligned} \tag{14.9}$$

Keeping ϕ and w fixed, and minimizing F_2^{APS} with respect to the constants c_1 and c_2, we obtain $c_1(\phi,w) = \frac{\int_\Omega (u_0-w)H(\phi)dx}{\int_\Omega H(\phi)dx}$, $c_2(\phi,w) = \frac{\int_\Omega (u_0-w)H(-\phi)dx}{\int_\Omega H(-\phi)dx}$.

Formally minimizing F_2^{APS} with respect to w and keeping c_1, c_2, and ϕ fixed, we obtain the Euler–Lagrange equation for w, parameterized in the gradient descent direction by an artificial time $t \geq 0$, with $w(0,x) = w_0(x)$,

$$\frac{\partial w}{\partial t} = \lambda_1(u_0 - c_1 - w)H(\phi) + \lambda_2(u_0 - c_2 - w)H(-\phi) + \nu_1 \triangle w - \nu_2 \triangle^2 w,$$

and with zero boundary conditions for first, second and third order partial derivatives of w on $\partial\Omega$. To obtain the Euler–Lagrange equation for ϕ, we will replace H in (14.9) with a more regular H_ϵ (see Chapter 2) such that $H_\epsilon \to H$ as $\epsilon \to 0$. Therefore, using $\delta_\epsilon = H_\epsilon'$, keeping c_1, c_2, and w fixed, formally minimizing with respect to ϕ the regularized energy, we obtain the Euler–Lagrange equation for $\phi(t,x)$, also parameterized in the gradient descent direction by an artificial time $t \geq 0$, with $\phi(0,x) = \phi_0(x)$,

$$\frac{\partial \phi}{\partial t} = \delta_\epsilon(\phi)\big[\mu \text{div}\big(\frac{\nabla \phi}{|\nabla \phi|}\big) - \lambda_1|u_0 - c_1 - w|^2 + \lambda_2|u_0 - c_2 - w|^2\big],$$

with boundary condition $\frac{\partial \phi}{\partial \vec{n}}|_{\partial\Omega} = 0$, $\vec{n}$ denoting the outward unit normal to $\partial\Omega$.

In Figures 14.1 through 14.6 we show experimental results on synthetic and real images, some corrupted by noise, with the model (14.9). We denote by $v = c_1H(\phi) + c_2H(-\phi)$ the piecewise-constant component extracted from u_0. We present the final detected contours superposed over the initial data u_0. In the experimental results, we have $\lambda_1 = \lambda_2$. Semi-implicit finite difference schemes were used to discretize the equations in ϕ and w. In the case of noisy images, the noise is white, Gaussian, additive, of zero mean and variance 20.

14.1.4 Four-phase APS model

With two level set functions [335], consider

$$\begin{aligned}
\inf F_4^{APS}(c_1, c_2, c_3, c_4, \phi_1, \phi_2, w) = & \mu_1 \int_\Omega |DH(\phi_1)| + \mu_2 \int_\Omega |DH(\phi_2)| \\
& + \lambda_1 \int_\Omega |u_0 - c_1 - w|^2 H(\phi_1)H(\phi_2)dx \\
& + \lambda_2 \int_\Omega |u_0 - c_2 - w|^2 H(\phi_1)H(-\phi_2)dx \\
& + \lambda_3 \int_\Omega |u_0 - c_3 - w|^2 H(-\phi_1)H(\phi_2)dx \\
& + \lambda_4 \int_\Omega |u_0 - c_4 - w|^2 H(-\phi_1)H(-\phi_2)dx \\
& + \nu_1 \int_\Omega |\nabla w|^2 dx + \nu_2 \int_\Omega |D^2 w|^2 dx.
\end{aligned}$$

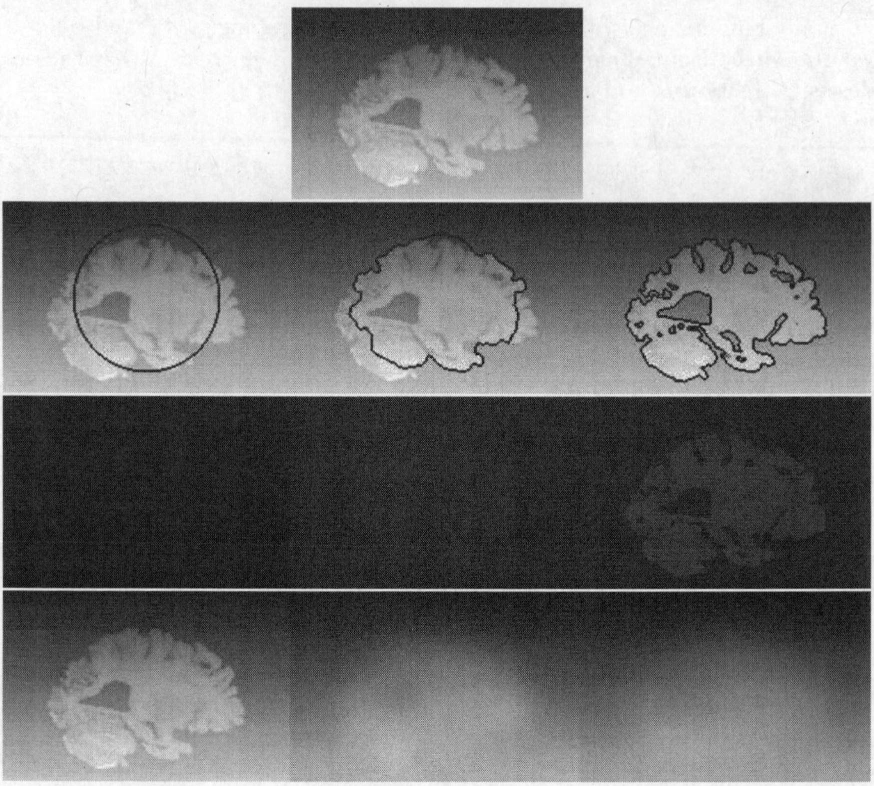

FIGURE 14.1: Two-phase APS model. Top: initial MRI image u_0 with global intensity inhomogeneity. Second row: initial, intermediate and final detected contours. Third row: initial, intermediate and final v. Bottom row: initial, intermediate and final w. Parameters: $\mu = 400$, $\lambda_1 = \lambda_2 = 1$, $\nu_1 = 1000$, $\nu_2 = 0.1$. Iterations: 200.

Minimizing F_4^{APS} with respect to its variables, and replacing H by H_ϵ, we obtain:

$$
\begin{aligned}
c_1(\phi_1, \phi_2, w) &= \frac{\int_\Omega (u_0 - w)H(\phi_1)H(\phi_2)dx}{\int_\Omega H(\phi_1)H(\phi_2)dx}, \\
c_2(\phi_1, \phi_2, w) &= \frac{\int_\Omega (u_0 - w)H(\phi_1)H(-\phi_2)dx}{\int_\Omega H(\phi_1)H(-\phi_2)dx}, \\
c_3(\phi_1, \phi_2, w) &= \frac{\int_\Omega (u_0 - w)H(-\phi_1)H(\phi_2)dx}{\int_\Omega H(-\phi_1)H(\phi_2)dx}, \\
c_4(\phi_1, \phi_2, w) &= \frac{\int_\Omega (u_0 - w)H(-\phi_1)H(-\phi_2)dx}{\int_\Omega H(-\phi_1)(H(-\phi_2)dx},
\end{aligned}
$$

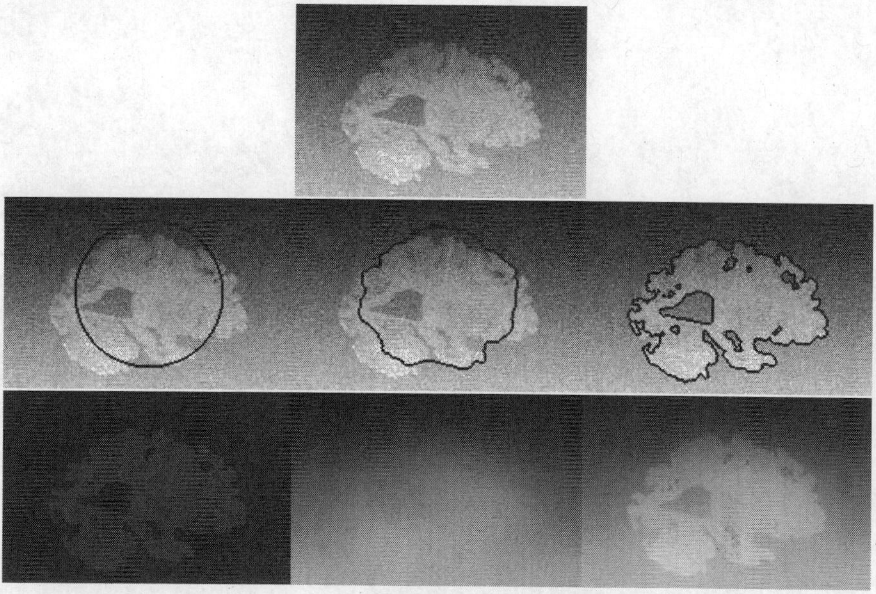

FIGURE 14.2: Two-phase APS model. Top: initial MRI image u_0 with additive noise and bias field. Middle row: initial, intermediate and final detected contours. Bottom row: final v, w, and $v + w$. Parameters: $w_0 = u_0$, $\mu = 509$, $\lambda_1 = \lambda_2 = 1$, $\nu_1 = 1500$, $\nu_2 = 1$.

$$
\begin{aligned}
\frac{\partial w}{\partial t} =& \lambda_1(u_0 - c_1 - w)H(\phi_1)H(\phi_2) + \lambda_2(u_0 - c_2 - w)H(\phi_1)H(-\phi_2) \\
&+ \lambda_3(u_0 - c_3 - w)H(-\phi_1)H(\phi_2) + \lambda_4(u_0 - c_4 - w)H(-\phi_1)H(-\phi_2) \\
&+ \nu_1 \triangle w - \nu_2 \triangle^2 w, \\
\frac{\partial \phi_1}{\partial t} =& \delta_\epsilon(\phi_1)\big[\mu_1 \mathrm{div}\big(\frac{\nabla \phi_1}{|\nabla \phi_1|}\big) - \lambda_1 |u_0 - c_1 - w|^2 H(\phi_2) \\
&- \lambda_2 |u_0 - c_2 - w|^2 H(-\phi_2) + \lambda_3 |u_0 - c_3 - w|^2 H(\phi_2) \\
&+ \lambda_4 |u_0 - c_4 - w|^2 H(-\phi_2)\big], \\
\frac{\partial \phi_2}{\partial t} =& \delta_\epsilon(\phi_2)\big[\mu_2 \mathrm{div}\big(\frac{\nabla \phi_2}{|\nabla \phi_2|}\big) - \lambda_1 |u_0 - c_1 - w|^2 H(\phi_1) \\
&+ \lambda_2 |u_0 - c_2 - w|^2 H(\phi_1) - \lambda_3 |u_0 - c_3 - w|^2 H(-\phi_1) \\
&+ \lambda_4 |u_0 - c_4 - w|^2 H(-\phi_1)\big].
\end{aligned}
$$

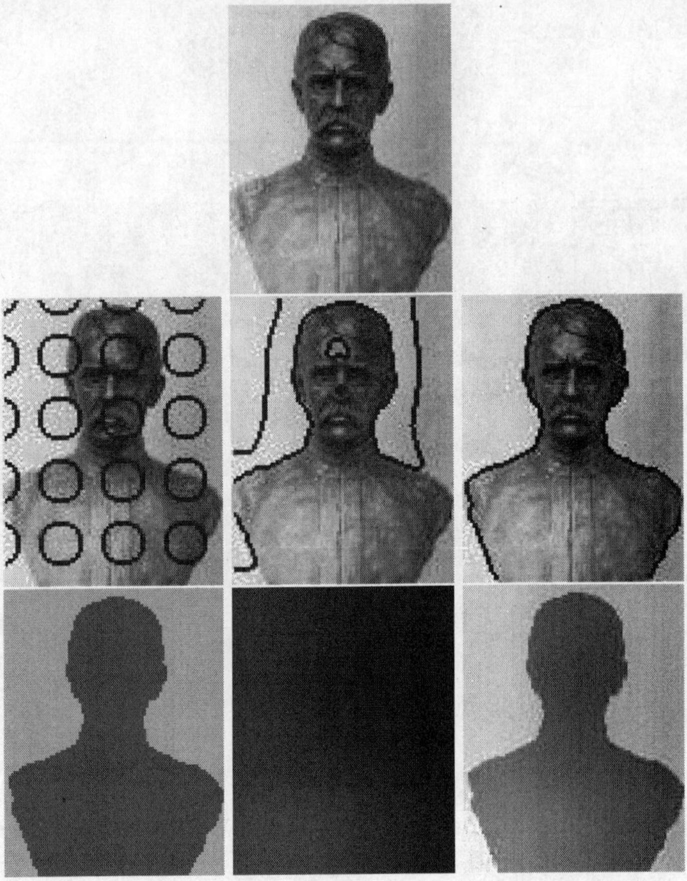

FIGURE 14.3: Two-phase APS model. Top: initial u_0 with slight intensity inhomogeneity as a lightning variation. Middle row: initial, intermediate and final contours. Bottom row: final v, w, $v + w$. Parameters: $\mu = 2000$, $\lambda_1 = \lambda_2 = 0.7$, $\nu_1 = 300$, $\nu_2 = 0.1^{-5}$. Iterations: 2000.

14.2 Piecewise-smooth model with multiplicative noise (MPS)

In this last case, the unknown u is piecewise-smooth, i.e., $u = v{\cdot}w$, where v is piecewise-constant and w is smooth. Therefore, $u_0 = u{\cdot}noise = (v{\cdot}w){\cdot}noise$, with $|\Omega| = 1$, and $\int_\Omega noise = 1$. Similarly, we can express the energy for this model to represent $n = 2^m$ phases in an image written using characteristic

FIGURE 14.4: Two-phase APS model. Top: initial u_0 with additive noise and slight intensity inhomogeneity. Middle row: initial, intermediate, detected final contours. Bottom row: v, w, $v + w$. Parameters: $\mu = 6000$, $\lambda_1 = \lambda_2 = 1$, $\nu_1 = 120$, $\nu_2 = 0.005$. Iterations: 1500.

functions as

$$\begin{aligned}\mathcal{L}_n^{MPS}(\chi_{E_1}, ..., \chi_{E_m}, w) = &\sum_{1\le j\le m} \mu_j \int_\Omega |D\chi_{E_j}| \\ &+ \sum_{1\le i\le n} \lambda_i \int_\Omega \left|\frac{u_0}{c_i w} - 1\right|^2 \chi_{P_i} dx \\ &+ \nu_1 \int_\Omega |w|^2 dx + \nu_2 \int_\Omega |\nabla w|^2 dx \\ &+ \nu_3 \int_\Omega |D^2 w|^2 dx, \end{aligned} \tag{14.10}$$

FIGURE 14.5: Two-phase APS model. Left to right, top to bottom: initial u_0 with additive global intensity inhomogeneity and detected contours after processing; extracted piecewise-constant component v; extracted smooth component w; $v+w$. Parameters: $w_0 = mean(u_0)$, $\mu = 2.1{\cdot}255^2{\cdot}10^{-2}$, $\lambda_1 = \lambda_2 = 1$, $\nu_1 = 1$, $\nu_2 = 150$.

where again D^2w is the Hessian matrix of w, constrained to be a smooth function, belonging to $H^2(\Omega)$, and χ_{P_i} is made of products of χ_{E_j} or $1 - \chi_{E_j}$ of m factors, $1 \leq i \leq n = 2^m$. For images, it is natural to assume that $u_0 \in L^\infty(\Omega) \subset L^2(\Omega)$. We assume in addition that $u_0 \geq \alpha > 0$, where α is a (small) positive constant.

Minimizing $\mathcal{L}_n^{MPS}$ with respect to c_i, for fixed χ_{E_j} and fixed w, we obtain

$$c_i = \frac{\displaystyle\int_\Omega \left(\frac{u_0}{w}\right)^2 \chi_{P_i} dx}{\displaystyle\int_\Omega \left(\frac{u_0}{w}\right) \chi_{P_i} dx}. \tag{14.11}$$

FIGURE 14.6: Two-phase APS model. Left to right, top to bottom: u_0 with additive noise and smooth inhomogeneity and extracted contours after processing; extracted piecewise-constant component v, extracted smooth component w; denoised image $v+w$. Parameters: $w_0 = mean(u_0)$, $\mu = 1.5{\cdot}255^2{\cdot}10^{-2}$, $\lambda_1 = \lambda_2 = 1$, $\nu_1 = 60$, $\nu_2 = 1$.

14.2.1 Existence of minimizers for the MPS model

Consider the minimization

$$\inf_{(\chi_{E_1},...,\chi_{E_m},w)} \left\{ \mathcal{L}_n^{MPS}(\chi_{E_1},...,\chi_{E_m},w) : \chi_{E_j} \in \{0,1\}, \right.$$
$$\left. w \in H^2(\Omega), w \geq \alpha > 0 \right\}. \tag{14.12}$$

Theorem 42 *Let Ω be an open and bounded subset of $\mathbb{R}^2$, with a C^1 boundary $\partial\Omega$. Let also $u_0 \in L^\infty(\Omega)$, such that $u_0(x) \geq \alpha > 0$, for all $x \in \Omega$, and $\mu_j > 0$, $j = 1,...,m$, $\lambda_i > 0$, $i = 1,...,n = 2^m$ and $\nu_1, \nu_2, \nu_3 > 0$. Then there are* *functions* $\chi_{E_j} \in BV(\Omega)$, $\chi_{E_j}(x) \in \{0,1\}$ *dx-a.e. in* Ω, *and* $w \in H^2(\Omega)$ *with* $w(x) \geq \alpha > 0$ *for all* $x \in \Omega$, *as solutions of (14.12).*

Proof. Consider a minimizing sequence $(\chi_{E_{1,k}}, ..., \chi_{E_{m,k}}, w_k)$ of (14.12). We have for $j = 1, ..., m$, $i = 1, ..., n$, and all k, and with $C > 0$ denoting a positive constant that may change line to line

$$\int_\Omega |D\chi_{E_{j,k}}| = |\chi_{E_{j,k}}|_{BV(\Omega)} \leq C, \quad \int_\Omega (\frac{u_0}{c_{i,k}w_k} - 1)^2 \chi_{P_{i,k}} dx \leq C,$$

$$\int_\Omega |w_k|^2 dx \leq C, \quad \int_\Omega |\nabla w_k|^2 dx \leq C, \quad \int_\Omega |D^2 w_k|^2 \, dx \leq C,$$

where $c_{i,k} = \frac{\int_\Omega (\frac{u_0}{w_k})^2 \chi_{P_{i,k}} dx}{\int_\Omega (\frac{u_0}{w_k}) \chi_{P_{i,k}} dx}$. Notice that $c_{i,k} = 0$ iff $\chi_{P_{i,k}} = 0$ or $\frac{u_0}{w_k}\chi_{P_{i,k}} = 0$.

We have $\|\chi_{E_{j,k}}\|_{L^1(\Omega)} \leq |\Omega|$, for all $j = 1, ..., m$ and all k. By passing to subsequences if necessary, there exist functions $\chi_{E_j} \in BV(\Omega)$, $j = 1, ..., m$, and $w \in H^2(\Omega)$ such that $\chi_{E_{j,k}}$ converges to χ_{E_j} weak* in $BV(\Omega)$ for all j (and strongly in $L^1(\Omega)$ and $dx-$a.e. in Ω), and $w_k \rightharpoonup w$ weakly in $H^2(\Omega)$. We obtain

$$\int_\Omega |D\chi_{E_j}| \leq \liminf_{k\to\infty} \int_\Omega |D\chi_{E_{j,k}}|, \quad \int_\Omega |w|^2 dx \leq \liminf_{k\to\infty} \int_\Omega |w_k|^2 dx$$

$$\int_\Omega |\nabla w|^2 dx \leq \liminf_{k\to\infty} \int_\Omega |\nabla w_k|^2 dx, \quad \int_\Omega |D^2 w|^2 dx \leq \liminf_{k\to\infty} \int_\Omega |D^2 w_k|^2 dx.$$

Based on Remark 15, w_k and w are uniformly bounded in $L^\infty(\overline{\Omega})$ and in $C^0(\overline{\Omega})$. Since, up to a subsequence, $w_k \to w$ a.e. and $w_k \geq \alpha$, we obtain $w \geq \alpha$ and

$$0 < \frac{\alpha^3}{\|u_0\|_{L^\infty(\Omega)} \|w\|^2_{L^\infty(\Omega)}} \leq |c_i| \leq \frac{\|u_0\|^2_{L^\infty(\Omega)} \|w\|_{L^\infty(\Omega)}}{\alpha^3}. \tag{14.13}$$

Since $u_0 \geq \alpha > 0$, and $w_k \geq \alpha > 0$ we have,

$$0 < \frac{\alpha^3}{\|u_0\|_{L^\infty(\Omega)} \|w_k\|^2_{L^\infty(\Omega)}} \leq |c_{i,k}| \leq \frac{\|u_0\|^2_{L^\infty(\Omega)} \|w_k\|_{L^\infty(\Omega)}}{\alpha^3}. \tag{14.14}$$

Again, since w_k are uniformly bounded in $L^\infty(\Omega)$, the sequence $c_{i,k}$ is bounded, and $\frac{u_0}{c_{i,k}w_k}, \frac{u_0}{c_i w} \in L^\infty(\Omega)$. It is easy to show that $c_{i,k} \to c_i$ as $k \to \infty$. Finally, from the Lebesgue dominated convergence theorem, we obtain

$$\int_\Omega (\frac{u_0}{c_i w} - 1)^2 \chi_{P_i} dx = \lim_{k\to\infty} \int_\Omega (\frac{u_0}{c_{i,k}w_k} - 1)^2 \chi_{P_{i,k}} dx,$$

for $i = 1, ..., n$. Moreover, χ_{E_j} are characteristic functions of sets $E_j \subset \Omega$, and

$$\mathcal{L}_n^{MPS}(\chi_{E_1}, ..., \chi_{E_m}, w) \leq \liminf_{k\to\infty} \mathcal{L}_n^{MPS}(\chi_{E_{1,k}}, ..., \chi_{E_{m,k}}, w_k).$$

□

14.2.2 Two-phase MPS model

The energy corresponding to the two-phase model can be written as:

$$\begin{aligned} L_2^{MPS}(c_1, c_2, \phi, w) =&\mu \int_\Omega |DH(\phi)| + \lambda_1 \int_\Omega |\frac{u_0}{c_1 w} - 1|^2 H(\phi)dx \\ &+ \lambda_2 \int_\Omega |\frac{u_0}{c_2 w} - 1|^2 H(-\phi)dx + \nu_1 \int_\Omega |w|^2 dx \\ &+ \nu_2 \int_\Omega |\nabla w|^2 dx + \nu_3 \int_\Omega |D^2 w|^2 dx. \end{aligned} \tag{14.15}$$

Keeping ϕ and w fixed, minimizing L_2^{MPC} with respect to c_1 and c_2, we obtain again explicit expressions for these constants:

$$c_1(\phi, w) = \frac{\int_\Omega (\frac{u_0}{w})^2 H(\phi)dx}{\int_\Omega (\frac{u_0}{w}) H(\phi)dx}, \ c_2(\phi, w) = \frac{\int_\Omega (\frac{u_0}{w})^2 H(-\phi)dx}{\int_\Omega (\frac{u_0}{w}) H(-\phi)dx}.$$

The Euler–Lagrange equation for w, parameterized in the gradient descent direction, is

$$\frac{\partial w}{\partial t} = \frac{\lambda_1 u_0}{c_1 w^2}(\frac{u_0}{c_1 w} - 1)H(\phi) + \frac{\lambda_2 u_0}{c_2 w^2}(\frac{u_0}{c_2 w} - 1)H(-\phi) - \nu_1 w + \nu_2 \triangle w - \nu_3 \triangle^2 w.$$

Replacing H by a more regular H_ϵ in (14.15), we obtain the Euler–Lagrange equation for ϕ, parameterized in the gradient descent direction,

$$\frac{\partial \phi}{\partial t} = \delta_\epsilon(\phi)\big[\mu \text{div}(\frac{\nabla \phi}{|\nabla \phi|}) - \lambda_1(\frac{u_0}{c_1 w} - 1)^2 + \lambda_2(\frac{u_0}{c_2 w} - 1)^2\big].$$

In Figure 14.7, we apply the piecewise-smooth multiplicative model from (14.15) to denoise and segment an image corrupted by multiplicative noise of mean 1 (random noise defined by $1 + \eta$, with η a white Gaussian noise of zero mean) and by a smooth bias field.

FIGURE 14.7: Two-phase MPS model. Left to right, top to bottom: initial u_0 with multiplicative noise of mean 1 and global intensity inhomogeneity; final contour over u_0; rescaled version of v, with $10c_i + 100$; smooth $w + 100$; denoised image $v \cdot w$.

14.2.3 Four-phase MPS model

The energy corresponding to the four-phase multiplicative piecewise-smooth model can be written as:

$$
\begin{aligned}
L_4^{MPS}(c_1, c_2, c_3, c_4, \phi_1, \phi_2, w) =& \mu_1 \int_\Omega |DH(\phi_1)| + \mu_2 \int_\Omega |DH(\phi_2)| \\
&+ \lambda_1 \int_\Omega |\frac{u_0}{c_1 w} - 1|^2 H(\phi_1)H(\phi_2)dx \\
&+ \lambda_2 \int_\Omega |\frac{u_0}{c_2 w} - 1|^2 H(\phi_1)H(-\phi_2)dx \\
&+ \lambda_3 \int_\Omega |\frac{u_0}{c_3 w} - 1|^2 H(-\phi_1)H(\phi_2)dx \\
&+ \lambda_4 \int_\Omega |\frac{u_0}{c_4 w} - 1|^2 H(-\phi_1)H(-\phi_2)dx \\
&+ \int_\Omega \Big[\nu_1 |w|^2 + \nu_2 |\nabla w|^2 + \nu_3 |D^2 w|^2\Big] dx.
\end{aligned}
$$

14.3 Exercises

Exercise 14.1 *Derive the Euler–Lagrange equations associated with the minimization of the functional F_2^{APS} given in equation (14.9). Verify that the functional decreases under the obtained time-dependent system of equations.*

Exercise 14.2 *Using discretizations presented in Chapter 8 for piecewise-constant segmentation, write a numerical algorithm for implementing the Euler Lagrange equations derived in Exercise 14.1.*

Exercise 14.3 *Derive the Euler–Lagrange equations associated with the minimization of the functional F_4^{APS} given in Subsection 14.1.4. Verify that the functional decreases under the obtained time-dependent system of equations.*

Exercise 14.4 *Using discretizations presented in Chapter 8 for piecewise-constant segmentation, write a numerical algorithm for implementing the Euler–Lagrange equations derived in Exercise 14.3.*

Exercise 14.5 *Derive the Euler–Lagrange equations associated with the minimization of the functional L_2^{MPS} given in equation (14.15). Verify that the functional decreases under the obtained time-dependent system of equations.*

Exercise 14.6 *Using discretizations presented in Chapter 8 for piecewise-constant segmentation, write a numerical algorithm for implementing the Euler–Lagrange equations derived in Exercise 14.5.*

Exercise 14.7 . *Derive the Euler–Lagrange equations associated with the minimization of the functional* L_4^{MPS} *given in Subsection 14.2.3. Verify that the functional decreases under the obtained time-dependent system of equations.*

Exercise 14.8 *Using discretizations presented in Chapter 8 for piecewise-constant segmentation, write a numerical algorithm for implementing the Euler–Lagrange equations derived in Exercise 14.7.*

Chapter 15

Numerical Methods for p–Harmonic Flows and Applications to Image Processing

15.1 Introduction

This chapter is concerned with the minimization of constrained functionals, in particular with the computation of p–harmonic maps. This problem has applications to liquid crystals, as well as to directional diffusion and chromaticity denoising. We follow here the work of Osher and the first author of the present text from [337].

Let $\Omega \subset \mathbb{R}^M$ be an open and bounded domain, and let S^{N-1} be the unit sphere in $\mathbb{R}^N$, for $M \geq 1$ and $N \geq 2$.

We first recall the following notations and terminology. The Euclidean norm of a vector y will be denoted by $|\ |$. We say that a vector-valued function $U = (U_1, ..., U_N) : \Omega \to \mathbb{R}^N$ belongs to S^{N-1} if and only if $|U(x)| = 1$, a.e. (for almost every) $x \in \Omega$.

For $1 \leq i \leq N$, the component gradient ∇U_i and its Euclidean norm are, respectively, defined by:

$$\nabla U_i = \Big(\frac{\partial U_i}{\partial x_1}, \frac{\partial U_i}{\partial x_2}, ..., \frac{\partial U_i}{\partial x_M}\Big),\ |\nabla U_i| = \sqrt{\Big(\frac{\partial U_i}{\partial x_1}\Big)^2 + \Big(\frac{\partial U_i}{\partial x_2}\Big)^2 + ... + \Big(\frac{\partial U_i}{\partial x_M}\Big)^2}.$$

The Jacobian matrix of the vector-valued function U and its Frobenius norm are respectively defined by:

$$\nabla U = \begin{pmatrix} \nabla U_1 \\ \vdots \\ \nabla U_N \end{pmatrix} = \begin{pmatrix} \frac{\partial U_1}{\partial x_1} & \dots & \frac{\partial U_1}{\partial x_M} \\ \vdots & & \vdots \\ \frac{\partial U_N}{\partial x_1} & \dots & \frac{\partial U_N}{\partial x_M} \end{pmatrix},\ |\nabla U| = \sqrt{\sum_{i=1}^{N}\sum_{j=1}^{M}\Big(\frac{\partial U_i}{\partial x_j}\Big)^2}.$$

For $U : \Omega \to S^{N-1}$ and $p \geq 1$, we consider the p–energy

$$E_p(U) = \int_\Omega |\nabla U|^p dx, \tag{15.1}$$

which is finite if U belongs to the Sobolev class

$$W^{1,p}(\Omega, S^{N-1}) = \{U \in W^{1,p}(\Omega, \mathbb{R}^N),\ |U| = 1 \text{ a.e.}\}.$$

Minimizing E_p over $W^{1,p}(\Omega, S^{N-1})$ with associated boundary conditions on $\partial\Omega$ is a constrained minimization problem. Mappings which are stationary for E_p are called p–harmonic maps.

The associated boundary conditions can be, for example: $U|_{\partial\Omega}$ equals a given map in $S^{N-1}(\partial\Omega)$, or the Neumann boundary conditions $\frac{\partial U}{\partial \vec{n}}|_{\partial\Omega} = 0$, where $\vec{n}$ denotes the exterior unit normal to $\partial\Omega$.

Many authors have studied harmonic maps between manifolds (existence, uniqueness or non-uniqueness, regularity; essentially most of them worked on the case $p = 2$): F. Bethuel, H. Brezis and J.M. Coron [49]; F. Bethuel, H. Brezis and F. Helein [51], [50]; R. Schoen and K. Uhlenbeck [282], [283], [284]; M. Struwe [301], [304], [302]; P. Courilleau and F. Demengel [112]; J.-M. Coron and R. Gulliver [111]; H. Brezis, J.-M. Coron and E.H. Lieb [66], and others. There are fewer results for the case $p = 1$ (for example, Giaquinta et al. [153]).

There are difficulties of finding numerically the minimizers or the p–harmonic maps, due to non-convexity (the constraint $|U(x)| = 1$ a.e. is not convex), non-regularity and non-uniqueness of minimizers.

There are several classical approaches used to solve the minimization problem (15.1).

A first approach is to solve the Euler–Lagrange equations associated with the minimization problem. These consist of a set of coupled partial differential equations:

$$-\text{div}(|\nabla U|^{p-2}\nabla U) = U|\nabla U|^p. \tag{15.2}$$

The above system of equations holds if and only if $U \in S^{N-1}$. However in practice, the numerical solution does not necessarily satisfy the constraint $|U| = 1$ everywhere. To correct the numerical error, several authors ([342], [314], [315]) replace the solution U_*^n obtained at each iteration n by $U^n = \frac{U_*^n}{|U_*^n|}$, but then the question is whether one still decreases the energy. In this framework, we also refer the reader to [110]. It is known [14] that the energy decrease is guaranteed after this re-normalization if $|U_*^n| \geq 1$, but the behavior of the energy is not known if $|U_*^n| < 1$. Also, if we would like to extend this numerical procedure involving the projection at each step to other manifolds, then the energy decrease is guaranteed only when the manifold is the boundary of a convex set, and again, if in addition, U_*^n does not belong to the interior of that convex domain.

This problem has been solved in [14] for the S^2 case and in three dimensions; an interesting convergent algorithm is proposed, but it still involves a re-normalization step at each iteration (ensuring now that the energy decreases after the re-normalization step). Numerical methods for p–harmonic flows are also proposed in [342] and [110], again based on the re-normalization procedure at each step.

The second classical approach is given by the Ginzburg–Landau functionals [51], [50]. Here, the problem is solved by approximation to eliminate the constraint. The minimization of the energy E_p from (15.1) under the constraint $|U(x)| = 1$ a.e., is approximated by the unconstrained minimization of the following energies, as $\varepsilon \to 0$:

$$E_\varepsilon(U) = \int_\Omega |\nabla U|^p dx + \frac{1}{\varepsilon}\int_\Omega (1 - |U|^2)^2 dx. \tag{15.3}$$

This chapter deals with alternative formulations and numerical methods for computing harmonic maps. A different approach for solving minimization problems on S^{N-1} is presented. This method has been proposed by Osher and Vese in [337]. An unconstrained minimization problem is solved instead, on the entire space of functions, and not only on S^{N-1}. The method uses the projection of an arbitrary function V to the sphere S^{N-1}. We present this alternative approach in the S^{N-1} case. Then we discuss how this approach can be extended to more general manifolds, and in particular to manifolds defined implicitly, via a level set function. The numerical schemes are presented for the S^1 and S^2 cases. We will also see how these can be used in practice, and in particular for applications to directional diffusion and color image denoising.

In the framework of image processing and directional diffusion, related works are [259], [295], [289], [314], [193], [315], [330], and [296], [194]. We also refer to [115] for manifold constrained variational problems. In the framework of energy minimization with values in S^2, we refer the reader to [134], where the algorithm from [14] is applied in the presence of a data term.

The main idea is as follows. For $U : \Omega \to S^{N-1}$, with $\Omega \subset \mathbb{R}^M$, consider $V : \Omega \to \mathbb{R}^N \setminus \vec{0}$, such that

$$U = \frac{V}{|V|}.$$

We minimize without constraint the corresponding energy with respect to V:

$$\inf_V \left\{ F(V) = \int_\Omega \left|\nabla\left(\frac{V}{|V|}\right)\right|^p dx \right\}, \tag{15.4}$$

and then recover U, a minimizer of (15.1), projecting back on S^{N-1}, by $U = \frac{V}{|V|}$, where V is a minimizer of (15.4).

The idea of solving constrained minimization problems for harmonic maps by associating unconstrained minimization problems has been used as a theoretical tool by Y. Chen, F.-H. Lin and M. Struwe [97], [98], [303]. They find a smooth energy-minimizing harmonic map U as a weak limit of minimizers U_L to an unconstrained variational problem, as $L \to \infty$. They construct in a different way the unconstrained variational problems.

15.2 S^1 case

To present the main idea introduced in [337], let us first consider the case $N = 2$ of S^1. Then for $U : \Omega \to S^1$, consider $V = (u, v) : \Omega \to \mathbb{R}^2$, such that $U = \frac{V}{|V|}$.

In order to obtain in an elegant way the Euler–Lagrange equations associated with the minimization problem (15.4), we consider the orientation formulation (but which is not always equivalent with the directional formulation). Let $U = (\cos\theta, \sin\theta)$, and $V = (r\cos\theta, r\sin\theta)$ with $r = r(x_1, x_2, \cdots, x_M)$ and $\theta = \theta(x_1, x_2, \cdots, x_M)$. Then $u^2 + v^2 = r^2$ and we have

$$\left|\nabla\Big(\frac{V}{|V|}\Big)\right|^2 = |\nabla\theta|^2.$$

For $p = 2$ (the heat flow for harmonic maps), solving

$$\inf_\theta \int_\Omega |\nabla\theta|^2 dx,$$

and parameterizing the descent direction by an artificial time t, we obtain (denoting $u_t = \frac{\partial u}{\partial t}$ and $v_t = \frac{\partial v}{\partial t}$):

$$\theta_t = \triangle\theta,\ r_t = 0.$$

Using

$$\theta = \tan^{-1}\Big(\frac{v}{u}\Big),\quad \nabla\theta = \frac{u\nabla v - v\nabla u}{u^2 + v^2},$$

we first deduce that:

$$\frac{uv_t - vu_t}{u^2 + v^2} = \mathrm{div}\Big(\frac{u\nabla v - v\nabla u}{u^2 + v^2}\Big).$$

Now, using $uu_t + vv_t = 0$ (from $r_t = 0$), we obtain the associated Euler–Lagrange equations for $p = 2$:

$$u_t = -v\mathrm{div}\Big(\frac{u\nabla v - v\nabla u}{u^2 + v^2}\Big),\quad v_t = +u\mathrm{div}\Big(\frac{u\nabla v - v\nabla u}{u^2 + v^2}\Big). \tag{15.5}$$

For $p = 1$ (the total variation minimization), on solving

$$\inf_\theta \int_\Omega |\nabla\theta| dx,$$

we obtain:

$$\theta_t = \mathrm{div}\Big(\frac{\nabla\theta}{|\nabla\theta|}\Big),\ r_t = 0.$$

Then, in a similar way, the associated Euler–Lagrange equations for $p = 1$ are:

$$u_t = -v\mathrm{div}\Big(\frac{u\nabla v - v\nabla u}{|u\nabla v - v\nabla u|}\Big),\quad v_t = +u\mathrm{div}\Big(\frac{u\nabla v - v\nabla u}{|u\nabla v - v\nabla u|}\Big). \tag{15.6}$$

In the general case, i.e., for any $p \geq 1$, the corresponding linear system in u_t and v_t is:

$$\begin{aligned} uu_t + vv_t &= 0, \\ \frac{uv_t - vu_t}{u^2 + v^2} &= \mathrm{div}\Big[\Big(\frac{|u\nabla v - v\nabla u|}{u^2 + v^2}\Big)^{p-2}\Big(\frac{u\nabla v - v\nabla u}{u^2 + v^2}\Big)\Big]. \end{aligned}$$

Solving this linear system in the unknowns u_t and v_t yields similar equations in u and v, like those for the cases $p = 2$ and $p = 1$ from (15.5) and (15.6), respectively.

We will associate with the problems (15.5) and (15.6) initial conditions in the form: $u(0,x) = u_0(x)$ and $v(0,x) = v_0(x)$ in Ω. At the boundary, we can prescribe either Dirichlet boundary conditions $V(t,x)/|V(t,x)| = F(x)$ with $F : \partial\Omega \to S^1$ given for $t \geq 0$ and $x \in \partial\Omega$, or Neumann boundary conditions $\frac{\partial u}{\partial \vec{n}}(t,\cdot) = 0$ and $\frac{\partial v}{\partial \vec{n}}(t,\cdot) = 0$ on $\partial\Omega$, $t \geq 0$ (where $\vec{n}$ denotes the exterior unit normal to $\partial\Omega$).

We could add data terms in the energy, as in [289] or [134].

Remark 18 *With these formulations, with both $p = 1$ and $p = 2$ (and in fact for any $p \geq 1$), we always have, for any fixed $x \in \Omega$: $u(t,x)u_t(t,x) + v(t,x)v_t(t,x) = 0$, or $u^2(t,x) + v^2(t,x) =$ constant in time, for fixed x.*

Remark 19 *Note that an artificial time t has been used for the computation of a stationary solution of the problem. This is a common technique, and this artificial time represents a parametrization of the descent direction. It can be shown (as in Chapter 2)) that the energy is decreasing in time under such a time-dependent flow for both Dirichlet and Neumann boundary conditions.*

15.2.1 Numerical algorithm for S^1 case

To discretize the systems (15.5) and (15.6), we use finite differences. Assume for simplicity that $U : [0,1]^M \to S^1$, let h be the space step, and $\triangle t$ the time step. We denote by u^n and v^n the approximations of $u(n\triangle t, x)$ and of $v(n\triangle t, x)$, respectively, where x is a grid point. To simplify the notation, we will not explicitly indicate the discrete point $x_{i,j}$ where the approximation is considered; for instance, if $M = 2$, u^n means $u^n_{i,j}$, etc; similarly, any expression of the form $(E)^n$ denotes an approximation of the quantity E at $(n\triangle t, x)$, at the same discrete point x; this notational convention will allow us to consider any dimension $M \geq 1$.

We use the following semi-implicit scheme for (15.5) with $p = 2$:

$$\frac{u^{n+1} - u^n}{\triangle t} = -\frac{v^{n+1} + v^n}{2}\Big[\mathrm{div}\Big(\frac{u\nabla v - v\nabla u}{u^2 + v^2}\Big)\Big]^n,$$
$$\frac{v^{n+1} - v^n}{\triangle t} = +\frac{u^{n+1} + u^n}{2}\Big[\mathrm{div}\Big(\frac{u\nabla v - v\nabla u}{u^2 + v^2}\Big)\Big]^n.$$

Similarly, for (15.6) with $p = 1$, we use:

$$\frac{u^{n+1} - u^n}{\triangle t} = -\frac{v^{n+1} + v^n}{2}\Big[\mathrm{div}\Big(\frac{u\nabla v - v\nabla u}{|u\nabla v - v\nabla u|}\Big)\Big]^n,$$
$$\frac{v^{n+1} - v^n}{\triangle t} = +\frac{u^{n+1} + u^n}{2}\Big[\mathrm{div}\Big(\frac{u\nabla v - v\nabla u}{|u\nabla v - v\nabla u|}\Big)\Big]^n.$$

Denoting by $(Div)^n$ an approximation of one of the expressions $\operatorname{div}\left(\frac{u\nabla v - v\nabla u}{u^2+v^2}\right)$ (for $p = 2$) and $\operatorname{div}\left(\frac{u\nabla v - v\nabla u}{|u\nabla v - v\nabla u|}\right)$ (for $p = 1$), evaluated at $(n\triangle t, ih, jh)$, and solving the previous algebraic systems in u^{n+1} and v^{n+1}, we obtain, for both $p = 1$ and $p = 2$:

$$u^{n+1} = \frac{u^n - \left(2v^n + u^n \frac{\triangle t(Div)^n}{2}\right)\frac{\triangle t(Div)^n}{2}}{1 + \left(\frac{\triangle t(Div)^n}{2}\right)^2},$$

$$v^{n+1} = \frac{v^n + \left(2u^n - v^n \frac{\triangle t(Div)^n}{2}\right)\frac{\triangle t(Div)^n}{2}}{1 + \left(\frac{\triangle t(Div)^n}{2}\right)^2}.$$

To discretize the expressions $\operatorname{div}\left(\frac{u\nabla v - v\nabla u}{u^2+v^2}\right)$ (for $p = 2$) and $\operatorname{div}\left(\frac{u\nabla v - v\nabla u}{|u\nabla v - v\nabla u|}\right)$ (for $p = 1$), we use the finite difference scheme proposed in [277] for $\operatorname{div}\left(\frac{\nabla u}{|\nabla u|}\right)$ which has also been used in [31] for a more general case.

Remark 20 *As in the continuous case, it is easy to verify that the numerical solution exactly satisfies:*

$$(u^{n+1})^2 + (v^{n+1})^2 = (u^n)^2 + (v^n)^2,$$

at any grid point x. This proves that the scheme produces bounded solutions independent of the relation between $\triangle t$ and h.

Remark 21 *Note that there is no need to apply a renormalization step at every iteration. Only in the end of the algorithm we let $U = \frac{V}{|V|}$, with $V = (u, v)$. Note also that if the initial data $V_0 = (u_0, v_0)$ already satisfies $|V_0| = 1$ everywhere, due to the previous remark, this equality will be preserved in time, and therefore the numerical solution U will be directly given by V (in other words, in this case, there is no need to renormalize V at the steady state; we will simply have $U = V$).*

Remark 22 *Although the solutions remain bounded regardless of the magnitude of $\triangle t$, the numerical domain of dependence of u^{n+1}, v^{n+1} is such that convergence for $p = 2$ is only possible if $\triangle t \le Ch^2$. This follows from the fact that θ satisfies the heat equation. Convergence for $p = 1$ requires a more restrictive constraint on $\triangle t$, typical of that for total variation minimization [277] in θ.*

Note that additional penalty terms obtained by imposing constraints on V or on $\frac{V}{|V|}$ could be added to the energy or to the Euler–Lagrange equations without any difficulty.

15.3 S^2 case

We will follow the same concept as in the previous case to derive the Euler–Lagrange equations associated with the unconstrained minimization problem (15.4), for any $M \geq 1$ and $N = 3$.

Using spherical coordinates, we let

$$U = (\cos\theta_1 \cos\theta_2, \cos\theta_1 \sin\theta_2, \sin\theta_1) \in S^2,$$

and

$$V = (r\cos\theta_1 \cos\theta_2, r\cos\theta_1 \sin\theta_2, r\sin\theta_1) = (u, v, w).$$

We then have $r^2 = u^2 + v^2 + w^2$,

$$\theta_1 = \tan^{-1}\left(\frac{w}{\sqrt{u^2+v^2}}\right),\ \theta_2 = \tan^{-1}\left(\frac{v}{u}\right),$$

and it can be shown that

$$|\nabla U|^2 = |\nabla\theta_1|^2 + \cos^2\theta_1 |\nabla\theta_2|^2.$$

Let us consider first the case $p = 2$. From

$$\inf_{\theta_1,\theta_2} \int_\Omega |\nabla\theta_1|^2 + \cos^2\theta_1 |\nabla\theta_2|^2 dx,$$

we obtain (parameterizing the descent directions by an artificial time t):

$$\theta_{1,t} = \triangle\theta_1 + \sin\theta_1\cos\theta_1 |\nabla\theta_2|^2, \tag{15.7}$$
$$\theta_{2,t} = \operatorname{div}(\cos^2\theta_1 \nabla\theta_2). \tag{15.8}$$

Let us denote by E_1 and E_2, respectively, the expressions on the right-hand sides of (15.7) and (15.8), i.e.,

$$\theta_{1,t} = E_1,\ \theta_{2,t} = E_2. \tag{15.9}$$

Again, from $r_t = 0$, we deduce that

$$uu_t + vv_t + ww_t = 0. \tag{15.10}$$

Computing and using

$$\nabla\theta_1 = \frac{(u^2+v^2)(\nabla w) - uw(\nabla u) - vw(\nabla v)}{(u^2+v^2+w^2)\sqrt{u^2+v^2}}, \tag{15.11}$$

$$\nabla\theta_2 = \frac{u(\nabla v) - v(\nabla u)}{u^2+v^2}, \tag{15.12}$$

we can then express E_1 and E_2 as functions of (u, v, w) by

$$E_1 = \triangle\theta_1 + \frac{w\sqrt{u^2+v^2}}{u^2+v^2+w^2}|\nabla\theta_2|^2, \quad E_2 = \operatorname{div}\Big(\frac{u(\nabla v) - v(\nabla u)}{u^2+v^2+w^2}\Big).$$

On the other hand, we have:

$$\theta_{1,t} = \frac{(u^2+v^2)w_t - uwu_t - vwv_t}{(u^2+v^2+w^2)\sqrt{u^2+v^2}}, \quad \theta_{2,t} = \frac{uv_t - vu_t}{u^2+v^2}.$$

We consider now the system formed by the equations (15.9), (15.10), in the unknowns u_t, v_t and w_t:

$$uu_t + vv_t + ww_t = 0, \quad \frac{(u^2+v^2)w_t - uwu_t - vwv_t}{(u^2+v^2+w^2)\sqrt{u^2+v^2}} = E_1, \quad \frac{uv_t - vu_t}{u^2+v^2} = E_2.$$

Solving this linear system in the unknowns u_t, v_t and w_t, we deduce the associated Euler–Lagrange equations:

$$u_t = -\frac{uw}{\sqrt{u^2+v^2}}E_1 - vE_2, \tag{15.13}$$

$$v_t = -\frac{vw}{\sqrt{u^2+v^2}}E_1 + uE_2, \tag{15.14}$$

$$w_t = \sqrt{u^2+v^2}E_1. \tag{15.15}$$

For the case $p = 1$ of the total variation minimization, we consider first the problem in $\theta = (\theta_1, \theta_2) \in [-\frac{\pi}{2}, \frac{\pi}{2}]^2$:

$$\inf_{\theta_1,\theta_2} \int_\Omega \sqrt{|\nabla\theta_1|^2 + \cos^2\theta_1|\nabla\theta_2|^2}dx,$$

which yields the equations:

$$\theta_{1,t} = \operatorname{div}\Big(\frac{\nabla\theta_1}{\sqrt{|\nabla\theta_1|^2 + \cos^2\theta_1|\nabla\theta_2|^2}}\Big) + \frac{\sin\theta_1\cos\theta_1|\nabla\theta_2|^2}{\sqrt{|\nabla\theta_1|^2 + \cos^2\theta_1|\nabla\theta_2|^2}}, \tag{15.16}$$

$$\theta_{2,t} = \operatorname{div}\Big(\cos^2\theta_1\frac{\nabla\theta_2}{\sqrt{|\nabla\theta_1|^2 + \cos^2\theta_1|\nabla\theta_2|^2}}\Big). \tag{15.17}$$

Denoting again by E_1 and E_2 the expressions on the right-hand sides of the above equations (15.16) and (15.16) respectively (corresponding now to the case $p = 1$), these can be expressed as functions of (u, v, w) using (15.11) and (15.12). The Euler–Lagrange equations for the case $p = 1$, in (u, v, w), are therefore as in (15.13) through (15.15), but with the corresponding differential operators E_1 and E_2 for $p = 1$.

15.3.1 Numerical algorithm for S^2 case

The expressions E_1 and E_2 are discretized following [277] and [31] for both $p = 1$ and $p = 2$ (we will still denote their discretizations at a given point by E_1 and E_2).

Let us denote by u^n, v^n, w^n the discrete solutions at a discrete point in two or three spatial dimensions (but without writing $u^n_{i,j}$ or $u^n_{i,j,k}$, for simplicity). The system (15.13) through (15.15) is discretized using the following implicit scheme:

$$\begin{aligned}
u^{n+1} &= u^n - \frac{\Delta t}{\sqrt{(u^n)^2 + (v^n)^2}} u^n \Big(\frac{w^{n+1} + w^n}{2}\Big) E_1 - \Big(\frac{v^{n+1} + v^n}{2}\Big) E_2 \Delta t, \\
v^{n+1} &= v^n - \frac{\Delta t}{\sqrt{(u^n)^2 + (v^n)^2}} v^n \Big(\frac{w^{n+1} + w^n}{2}\Big) E_1 + \Big(\frac{u^{n+1} + u^n}{2}\Big) E_2 \Delta t, \\
w^{n+1} &= w^n + \Delta t \sqrt{(u^n)^2 + (v^n)^2} E_1.
\end{aligned}$$

We will use the notations:

$$A = \frac{E_1 \Delta t}{2\sqrt{(u^n)^2 + (v^n)^2}}, \quad B = \frac{E_2 \Delta t}{2}, \quad C = \Delta t \sqrt{(u^n)^2 + (v^n)^2} E_1.$$

The linear system in $u^{n+1}, v^{n+1}, w^{n+1}$ is non-singular and has the unique solution:

$$u^{n+1} = \frac{R_1 - BR_2}{1 + B^2}, \quad v^{n+1} = \frac{R_2 + BR_1}{1 + B^2}, \quad w^{n+1} = w^n + C,$$

where $R_1 = u^n - Au^n(2w^n + C) - v^n B$ and $R_2 = v^n - Av^n(2w^n + C) + u^n B$.

Remark 23 *The numerical scheme will exactly satisfy the relation:*

$$(u^{n+1})^2 + (v^{n+1})^2 + (w^{n+1})^2 = (u^n)^2 + (v^n)^2 + (w^n)^2$$

at each grid point, if in the above discretizations the expression $\sqrt{(u^n)^2 + (v^n)^2}$ *is replaced by* $\sqrt{u^n\left(\frac{u^n + u^{n+1}}{2}\right) + v^n\left(\frac{v^n + v^{n+1}}{2}\right)}$, *but this yields a nonlinear system in the unknowns* u^{n+1}, v^{n+1} *and* w^{n+1} *which could be solved by a fixed-point iteration.*

15.4 Numerical experiments

In this section we present numerical experiments from [337] in the cases ($M = 2$, $N = 2$), and ($M = 2$ or 3 and $N = 3$). We will consider the cases with Dirichlet boundary conditions (Subsection 15.4.1) and Neumann boundary conditions (Subsection 15.4.2).

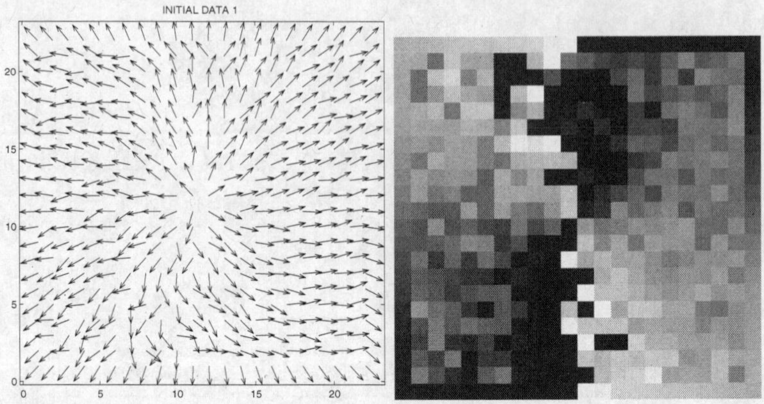

FIGURE 15.1: Left: initial condition for the Dirichlet problem, as a perturbation of $\frac{x-x_0}{|x-x_0|}$ in $(0,1)^2$, given by (15.18)-(15.19), and agreeing with $\frac{x-x_0}{|x-x_0|}$ at the boundary, where $x_0 = (0.5, 0.5)$. Right: corresponding initial angle $\theta = \tan^{-1}\left(\frac{v^0}{u^0}\right)$.

15.4.1 Numerical results for prescribed boundary conditions

In the S^1 case, we first consider the Dirichlet problem with the boundary condition $U(x) = \frac{x-x_0}{|x-x_0|}$ on $\partial\Omega$, with $x_0 = (0.5, 0.5)$ where $\Omega = (0,1)^2$. In this case, it is known that the map $x \mapsto \frac{x-x_0}{|x-x_0|}$ is an exact solution and minimizer in $\overline{\Omega}$. We will show that the numerical solution has the correct behavior, approximating very well the exact solution.

Following [110], an initial condition $V^0 = (u^0, v^0)$ inside Ω can be a perturbation of $\frac{x-x_0}{|x-x_0|}$ (shown in Figure 15.1, after normalization):

$$u^0(x_1, x_2) = \frac{x_1 - .5}{|x - x_0|} + .6(1 + x_1^2 - x_2^2) - .8\eta, \tag{15.18}$$

$$v^0(x_1, x_2) = \frac{x_2 - .5}{|x - x_0|} + .6(x_1 - 2x_2) + .8\eta, \tag{15.19}$$

for all $(x_1, x_2) \in \Omega$, where η is random noise (called "initial data 1").

We will also consider another initial condition in this case, defined using the distance function to the boundary: for $(x, y) \in \Omega$, find $(x_b, y_b) \in \partial\Omega$ as the closest point to the boundary $\partial\Omega$ from (x, y). Then let $(u^0(x, y), v^0(x, y)) = U(x_b, y_b)$, where U defines the boundary conditions on $\partial\Omega$ (this second initial condition is shown in Figure 15.2 and it is called "initial data 2").

We consider now the case $p = 2$, for these two initial conditions. For the initial data 1 from Figure 15.1, we also compare the results (the error and the energy decrease) with the classical harmonic map formulation with numerical renormalization at each time step by solving the semi-discrete problem (using

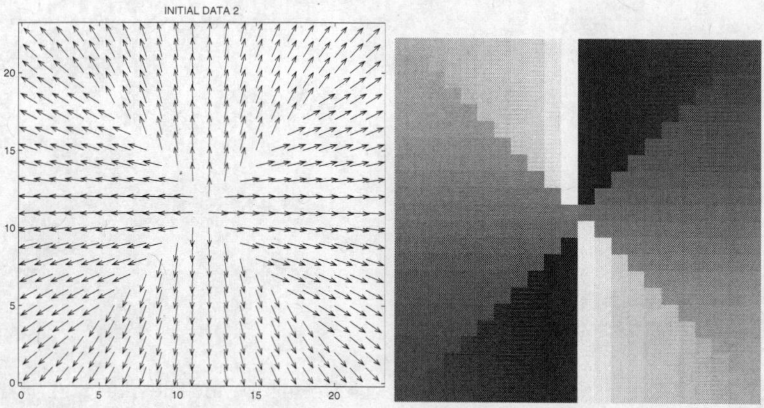

FIGURE 15.2: Left: another initial condition for the Dirichlet problem constructed using the closest point to the boundary, and agreeing with $\frac{x-x_0}{|x-x_0|}$ at the boundary, where $x_0 = (0.5, 0.5)$. Right: corresponding initial angle $\theta = \tan^{-1}\left(\frac{v^0}{u^0}\right)$.

central difference approximations for the space derivatives, and with the same prescribed boundary conditions and the same time and space steps):

$$\begin{aligned}
\frac{u_*^{n+1} - u^n}{\triangle t} &= \triangle u^n + u^n\Big[(u_x^n)^2 + (u_y^n)^2 + (v_x^n)^2 + (v_y^n)^2\Big], \\
\frac{v_*^{n+1} - v^n}{\triangle t} &= \triangle v^n + v^n\Big[(u_x^n)^2 + (u_y^n)^2 + (v_x^n)^2 + (v_y^n)^2\Big], \\
(u^{n+1}, v^{n+1}) &= \frac{(u_*^{n+1}, v_*^{n+1})}{|(u_*^{n+1}, v_*^{n+1})|}.
\end{aligned}$$

We show the energy decrease and the error versus iterations for the results obtained with the classical harmonic maps applied to the initial data 1, and with the presented model applied to both initial data 1 and 2 (see Figure 15.3). Using the presented model, the error is much smaller. Also, note that the initial data 2 produces a very fast result. For both initial data 1 and 2 by the presented model, the numerical solution $U(x) = \frac{V(x)}{|V(x)|}$ at steady state approximates very well the exact solution $\frac{x-x_0}{|x-x_0|}$ in $\Omega = [0,1]^2$, better than using the classical harmonic map scheme with the re-normalization at each step.

The results obtained with the presented model for $p = 2$ for both data are shown in Figure 15.4 together with the angle $\theta = \tan^{-1}\left(\frac{v}{u}\right)$.

Corresponding results obtained with the presented model for $p = 1$ are shown in Figures 15.5 and 15.6.

We show next a numerical result for maps with values in S^2, in the three-

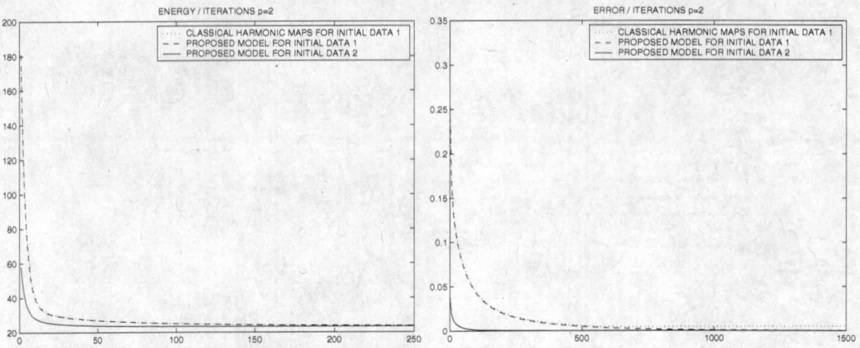

FIGURE 15.3: Energy and error versus iterations for the classical harmonic map scheme applied to the initial data 1 and for the presented model applied to initial data 1 and 2. Note a better accuracy obtained with the presented model compared with the classical formulation (we use the same $\triangle t = 0.0001$, $h = 1./21$, for both formulations).

dimensional case. Following [14], a test is performed which shows again that, for the Dirichlet boundary conditions, the numerical solution approximates well the exact solution, for $p = 2$. Indeed, in Figure 15.7, the initial data is to the left, and the result on the right. We see that the singularity has moved to the center of the domain and this is therefore an approximation of $\frac{x-x_0}{|x-x_0|}$, with $x_0 = (0.5, 0.5, 0.5)$.

15.4.2 Application to directional denoising and color image denoising

Next, we consider the case with Neumann boundary conditions. For the initial data in Figure 15.8, the results for $p = 1$ and $p = 2$ are presented in Figure 15.9. Note that for $p = 1$ (left), the "edges" are very well preserved, thanks to the total variation minimization [277], and the homogeneous regions are well denoised (we show the results at steady state).

Finally, we show applications more related to denoising of color RGB images. In the first test (Figure 15.10), we consider a map from $\mathbb{R}^2 \to S^2$, but instead of vectors we plot colors using the rectangular color space RGB. In Figure 15.10, left, we show an initial image of noisy directions (the components of the unit vector (u, v, w) are visualized as channels in a color RGB picture). We show at middle and right two numerical results in the case of directional diffusion, with $p = 1$ (middle) and $p = 2$ (right), with Neumann boundary conditions. As expected, in the case of the total variation [277], the edges are well preserved, while they are smeared out with the heat flow.

We end this chapter with an application to denoising color RGB images. We consider a color image $I = (I_R, I_G, I_B) \in \mathbb{R}^3$, from which we can extract

FIGURE 15.4: Left: numerical result approximating the exact solution and minimizer for $p = 2$ with Dirichlet boundary conditions ($\triangle t = 0.0001$, $h = 1/21$). Right: corresponding angle $\theta = \tan^{-1}\left(\frac{v}{u}\right)$.

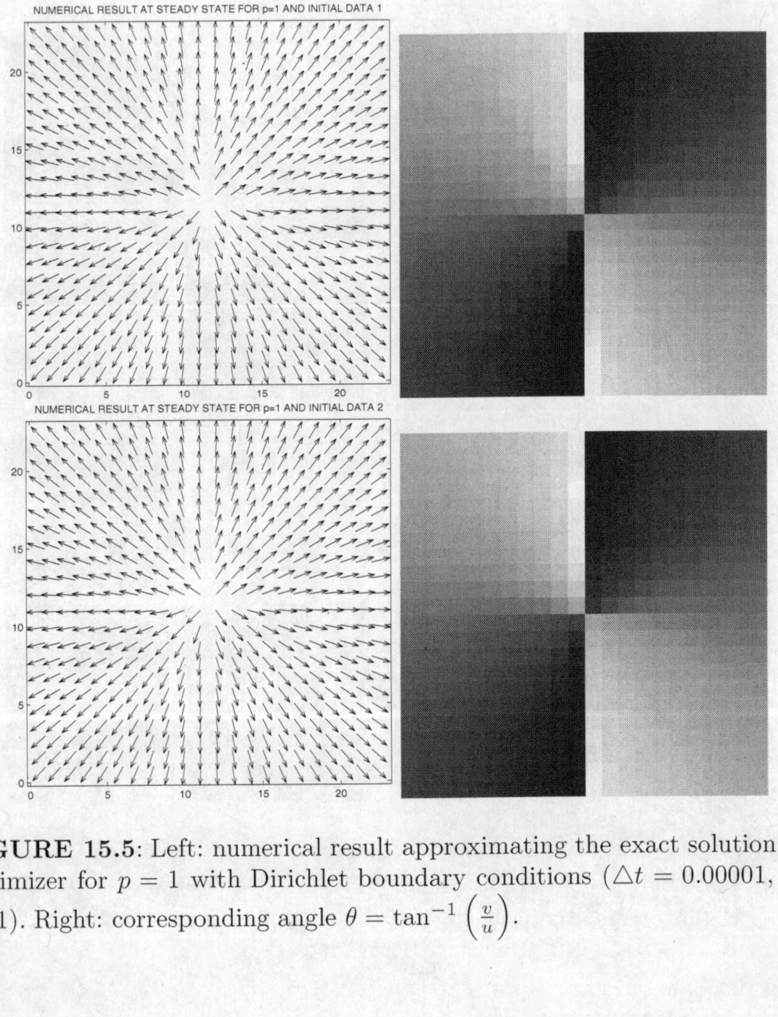

FIGURE 15.5: Left: numerical result approximating the exact solution and minimizer for $p = 1$ with Dirichlet boundary conditions ($\triangle t = 0.00001$, $h = 1/21$). Right: corresponding angle $\theta = \tan^{-1}\left(\frac{v}{u}\right)$.

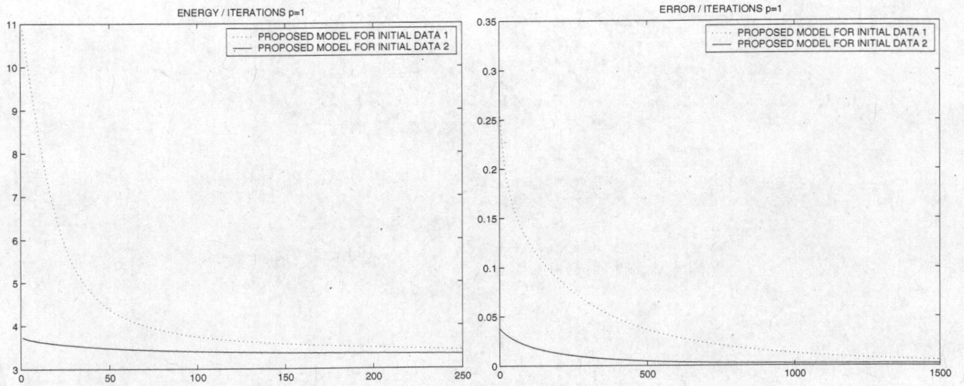

FIGURE 15.6: Energy and error versus iterations for $p = 1$ with Dirichlet boundary conditions corresponding to the results in Figure 15.5.

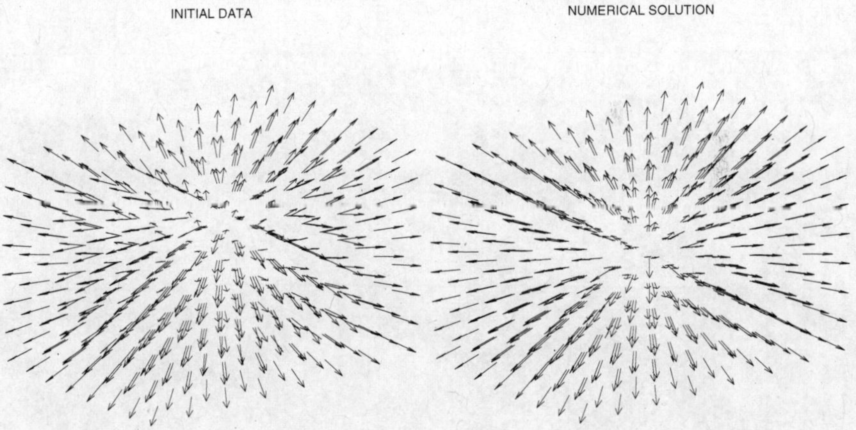

FIGURE 15.7: Left: initial flow $\frac{x-x_1}{|x-x_1|}$ from $(0,1)^3$ into S^2 with prescribed Dirichlet boundary conditions equal to $\frac{x-x_0}{|x-x_0|}$, where $x_0 = (0.5, 0.5, 0.5)$ and $x_1 = (0.64, 0.64, 0.64)$. Right: numerical solution obtained for $p = 2$. The singularity has moved to the center of the domain approximating the exact solution and minimizer ($\triangle t = 0.00001$, $h = 1/7$).

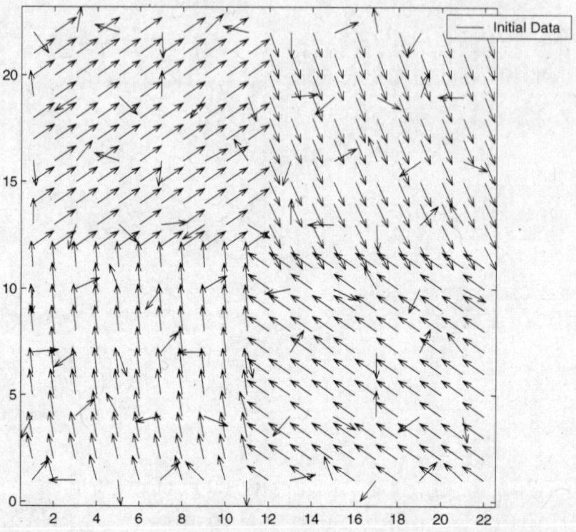

FIGURE 15.8: Initial noisy data for the case with Neumann boundary conditions.

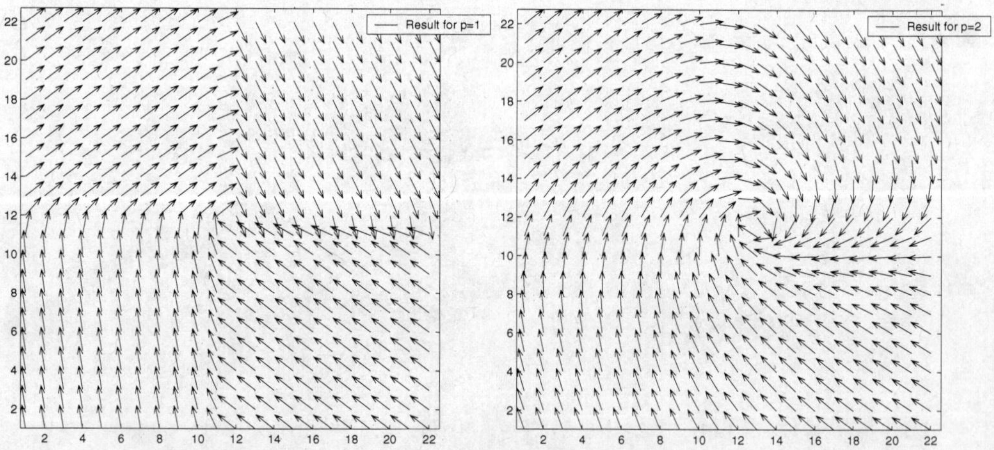

FIGURE 15.9: Numerical results with the initial noisy data from Figure 15.8, for Neumann boundary conditions and $p = 1$ (left), with $\triangle t = 0.00005$, $h = 1$, steady state, and $p = 2$ (right), with $\triangle t = 0.00005$, $h = 1$, steady state.

the intensity or brightness $|I| = \sqrt{I_R^2 + I_G^2 + I_B^2}$ and the chromaticity

$$\frac{I}{|I|} = \Big(\frac{I_R}{\sqrt{I_R^2 + I_G^2 + I_B^2}}, \frac{I_G}{\sqrt{I_R^2 + I_G^2 + I_B^2}}, \frac{I_B}{\sqrt{I_R^2 + I_G^2 + I_B^2}}\Big) \in S^2.$$

Let us assume that noise has been added to the image, but only to the chromaticity $\frac{I}{|I|}$. We can apply the above directional denoising method with $p = 1$ to the chromaticity (in this test case, we do not add noise to the brightness $|I|$). With the processed result, we obtain a denoised version of the image using the unchanged brightness. The idea of decomposing a color RGB image into its brightness and chromaticity and processing these two quantities separately has been already used in other works, for example in [325], [327], [326], [314], [315], [289], [296], [194], [295].

This type of application is illustrated in the last numerical example. In Figure 15.11, we show an original color RGB image $I = (I_R, I_G, I_B) \in \mathbb{R}^3$ (left), a noisy version (middle), where only the directions $\frac{I}{|I|}$ (the chromaticity) were noisy, keeping the brightness $|I|$ or magnitude of the vectors unchanged, and a denoised version obtained with $p = 1$ (right), where only the chromaticity or directions were denoised, keeping the brightness or magnitude unchanged from the original image, equal to $|I|$.

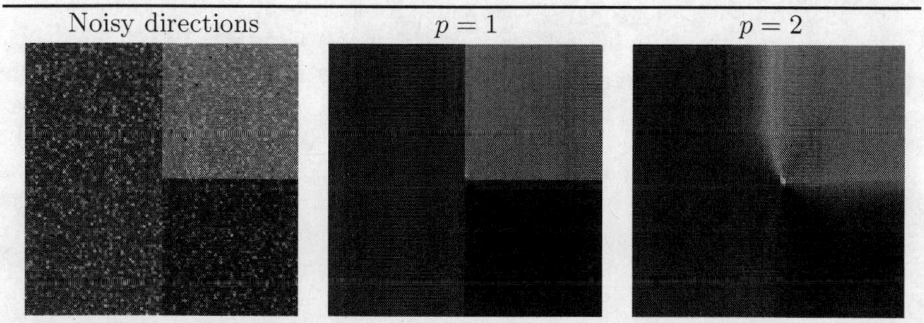

FIGURE 15.10: Directions denoising with $p = 1$ (middle) and $p = 2$ (right). The unit vectors are represented as RGB colors ($\triangle t = 0.01$, $h = 1$).

15.5 Concluding remarks and discussions for more general manifolds

In this chapter, we presented an alternative approach for computing harmonic maps and harmonic flows. We illustrated the presented methods by ex-

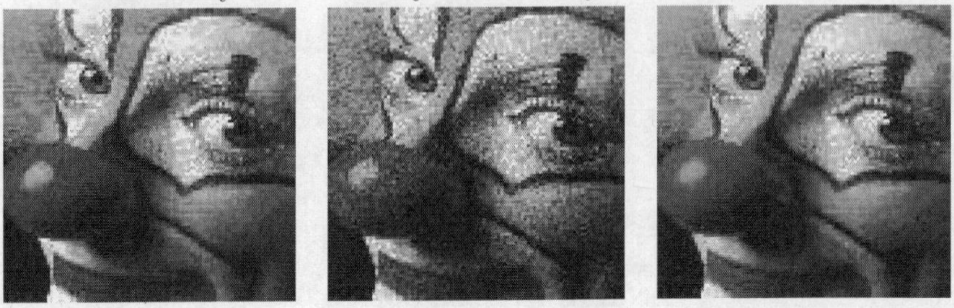

FIGURE 15.11: Chromaticity denoising with $p = 1$. The brightness is kept unchanged from the original image ($\triangle t = 0.01$, $h = 1$, 50 iterations).

perimental results and comparisons with classical schemes, and applications to directional diffusion and image processing.

It is easy to see that the minimization problems (15.1) and (15.4) have the same infimum and that solving one problem yields a minimizer for the other one and vice versa. Of course we cannot expect to have uniqueness of minimizers for (15.4), because λV is a minimizer for any non-zero constant λ if V is a minimizer. Showing the existence of minimizers for (15.4) may be a difficult problem because the energy is not convex. We also posed the question: given Dirichlet boundary conditions on $\partial\Omega$, what would be a good initial condition in Ω to guarantee a fast computation of a minimizer (to find a specific initial condition, we used the distance function to the boundary $\partial\Omega$, although perhaps other choices could also be constructed).

This method can be extended to more general manifolds. For instance, if we consider a manifold $\mathcal{M} \subset \mathbb{R}^N$, the associated constrained minimization problem can be formulated as follows:

$$\inf_{U:\Omega\to\mathcal{M}} F(U) = \int_\Omega |\nabla U|^p dx.$$

The proposed method for the case when $\mathcal{M} = S^{N-1}$ can be extended to such general cases if we assume for example that $\mathcal{M}$ can be represented implicitly, via a level set function given by the signed distance function to $\mathcal{M}$ from any other point in $\mathbb{R}^N$ (we refer to [254] for definitions and dynamics of closed hyper-surfaces defined implicitly, via level set functions and signed distance functions). Then we can write $\mathcal{M} = \{x \in \mathbb{R}^N : \ d(x) = 0\}$, where d is the signed distance function to $\mathcal{M}$ (in particular a Lipschitz continuous function, taking real values). To any $U : \Omega \to \mathcal{M}$, we associate $V : \Omega \to \mathbb{R}^N$, such that U is the projection of V on the manifold $\mathcal{M}$. This can be done using the closest point or the projection $U = V - d(V)\nabla_V d(V)$, and we have $d(U) = 0$. Then, we can associate the unconstrained minimization problem:

$$\inf_{V:\Omega\to\mathbb{R}^N} \int_\Omega |\nabla(V - d(V)\nabla_V d(V))|^p dx.$$

This is a generalization of the case $\mathcal{M} = S^{N-1}$, because in this case, we have $d(V) = |V| - 1$, and $V - d(V)\nabla_V d(V) = \frac{V}{|V|}$.

We would like to mention that the case of more general manifolds, and in particular of manifolds defined implicitly, has been considered in [46] for the manifold of origin and in [235] for the target manifold, but using different formulations.

15.6 Exercises

Exercise 15.1 *Prove the validity of the Euler–Lagrange equations associated with the minimization of the p–energy (15.1), as given in (15.2).*

Exercise 15.2 *Derive the Euler–Lagrange equations associated with the minimization of the Ginzburg–Landau functionals (15.3). Then implement and test them in two space dimensions for the S^1 case.*

Exercise 15.3 *Directly using the expressions for u_t and v_t, verify the claim given in Remark 18.*

Exercise 15.4 *Verify the claim given in Remark 20.*

Exercise 15.5 *Implement the algorithm described in Subsection 15.2.1 for both $p = 1$ and $p = 2$, in two dimensions for the S^1 case, and apply it to direction denoising.*

Exercise 15.6 *Directly using the equations (15.13), (15.14) and (15.15) for $p = 2$ (and similarly the corresponding ones for $p = 1$), verify that $u^2(t,x) + v^2(t,x) + w^2(t,x)$ remains constant in time for fixed x.*

Exercise 15.7 *Verify the claim given in Remark 23.*

Bibliography

[1] Compressive Sensing Resources. `http://dsp.rice.edu/cs`.

[2] R. Acar and C.R. Vogel. Analysis of bounded variation penalty methods for ill-posed problems. *Inverse Problems*, 10(6):1217–1229, 1994.

[3] R.A. Adams. *Sobolev Spaces*. Pure and Applied Mathematics Series. Academic Press, 1975.

[4] J.E. Adams Jr. and J.F. Hamilton Jr. Adaptive color plan interpolation in single sensor color electronic camera. U.S. Patent 5 629 734, 1996.

[5] M.N. Ahmed, S.M. Yamany, N. Mohamed, A.A. Farag, and T. Moriarty. A modified fuzzy C-means algorithm for bias field estimation and segmentation of MRI data. *Medical Imaging, IEEE Transactions on*, 21(3):193–199, 2002.

[6] O. Alexandrov and F. Santosa. A topology-preserving level set method for shape optimization. *Journal of Computational Physics*, 204(1):121–130, 2005.

[7] R. Alicandro, A. Braides, and J. Shah. Free-discontinuity problems via functionals involving the L^1-norm of the gradient and their approximations. *Interfaces and Free Boundaries*, 1:17–37, 1999.

[8] W.K. Allard. Total variation regularization for image denoising, I. geometric theory. *SIAM Journal on Mathematical Analysis*, 39(4):1150–1190, 2008.

[9] W.K. Allard. Total variation regularization for image denoising, II. examples. *SIAM Journal on Imaging Sciences*, 1(4):400–417, 2008.

[10] W.K. Allard. Total variation regularization for image denoising, III. examples. *SIAM Journal on Imaging Sciences*, 2(2):532–568, 2009.

[11] S. Alliney. Digital filters as L^1-norm regularizers. In *Multidimensional Signal Processing Workshop, 1989.*, page 105, 1989.

[12] S. Alliney. Digital filters as absolute norm regularizers. *Signal Processing, IEEE Transactions on*, 40(6):1548–1562, 1992.

[13] S. Alliney. Recursive median filters of increasing order: a variational approach. *Signal Processing, IEEE Transactions on*, 44(6):1346–1354, 1996.

[14] F. Alouges. A new algorithm for computing liquid crystal stable configurations: the harmonic mapping case. *SIAM Journal on Numerical Analysis*, 34(5):1708–1726, 1997.

[15] L. Alvarez, P.-L. Lions, and J.-M. Morel. Image selective smoothing and edge detection by nonlinear diffusion. *SIAM Journal on Numerical Analysis*, 29(3):845–866, 1992.

[16] O. Alvarez, P. Hoch, Y. Le Bouar, and R. Monneau. Dislocation dynamics: short-time existence and uniqueness of the solution. *Archive for Rational Mechanics and Analysis*, 181(3):449–504, 2006.

[17] O. Amadieu, E. Debreuve, M. Barlaud, and G. Aubert. Inward and outward curve evolution using level set method. In *Image Processing, 1999. Proceedings of International Conference*, volume 3, pages 188–192, 1999.

[18] L. Ambrosio. A compactness theorem for a special class of functions of bounded variation. *Bollettino dell'Unione Matematica Italina*, 3(B):857–881, 1989.

[19] L. Ambrosio. Existence theory for a new class of variational problems. *Archive for Rational Mechanics and Analysis*, 111(4):291–322, 1990.

[20] L. Ambrosio, N. Fusco, and D. Pallara. *Functions of Bounded Variation and Free Discontinuity Problems.* Clarendon Press, 2000.

[21] L. Ambrosio and V.M. Tortorelli. Approximation of functionals depending on jumps by elliptic functionals via Γ–convergence. *Communication on Pure and Applied Mathematics*, 43(8):999–1036, 1990.

[22] L. Ambrosio and V.M. Tortorelli. On the approximation of free discontinuity problems. *Bollettino dell'Unione Matematica Italina*, B7(6):105–123, 1992.

[23] F. Andreu-Vaillo, V. Caselles, and J.M. Mazon. *Parabolic Quasilinear Equations Minimizing Linear Growth Functionals.* Progress in Mathematics Series. Birkhäuser Basel, 2004.

[24] F. Arrate, J.T. Ratnanather, and L. Younes. Diffeomorphic active contours. *SIAM Journal on Imaging Sciences*, 3(2):176–198, 2010.

[25] J. Ashburner. A fast diffeomorphic image registration algorithm. *NeuroImage*, 38(1):95–113, 2007.

[26] K. Astala and L. Päivärinta. Calderón's inverse conductivity problem in the plane. *Annals of Mathematics*, 163(1):265–299, 2006.

[27] H. Attouch, G. Buttazzo, and G. Michaille. *Variational Analysis in Sobolev and BV Spaces: Applications to PDEs and Optimization.* MPS-SIAM Series on Optimization. Society for Industrial and Applied Mathematics, 2006.

[28] G. Aubert and J.-F. Aujol. Modeling very oscillating signals. Application to image processing. *Applied Mathematics and Optimization*, 51(2):163–182, 2005.

[29] G. Aubert and L. Blanc-Féraud. Some remarks on the equivalence between 2D and 3D classical snakes and geodesic active contours. *International Journal of Computer Vision*, 34(1):19–28, 1999.

[30] G. Aubert and P. Kornprobst. *Mathematical Problems in Image Processing: Partial Differential Equations and the Calculus of Variations.* Series Applied Mathematical Sciences. Springer, 2001.

[31] G. Aubert and L. Vese. A variational method in image recovery. *SIAM Journal on Numerical Analysis*, 34(5):1948–1979, 1997.

[32] J.-F. Aujol, G. Aubert, L. Blanc-Féraud, and A. Chambolle. Image decomposition into a bounded variation component and an oscillating component. *Journal of Mathematical Imaging and Vision*, 22(1):71–88, 2005.

[33] J.-F. Aujol and A. Chambolle. Dual norms and image decomposition models. *International Journal of Computer Vision*, 63(1):85–104, 2005.

[34] J.-F. Aujol, G. Gilboa, T. Chan, and S. Osher. Structure-texture image decomposition – modeling, algorithms, and parameter selection. *International Journal of Computer Vision*, 67(1):111–136, 2006.

[35] J.-F. Aujol, S. Ladjal, and S. Masnou. Exemplar-based inpainting from a variational point of view. *SIAM Journal on Mathematical Analysis*, 42(3):1246–1285, 2010.

[36] J. M. Ball. Convexity conditions and existence theorems in nonlinear elasticity. *Archive for Rational Mechanics and Analysis*, 63(4):337–403, 1976.

[37] C. Ballester, M. Bertalmio, V. Caselles, G. Sapiro, and J. Verdera. Filling-in by joint interpolation of vector fields and gray levels. *Image Processing, IEEE Transactions on*, 10(8):1200–1211, 2001.

[38] L. Bar, A. Brook, N. Sochen, and N. Kiryati. Deblurring of color images corrupted by impulsive noise. *Image Processing, IEEE Transactions on*, 16(4):1101–1111, 2007.

[39] L. Bar, N. Kiryati, and N. Sochen. Image deblurring in the presence of impulsive noise. *International Journal of Computer Vision*, 70(3):279–298, 2006.

[40] L. Bar, N. Sochen, and N. Kiryati. Variational pairing of image segmentation and blind restoration. In *Computer Vision - ECCV 2004*, Lecture Notes in Computer Science, pages 166–177. Springer, 2004.

[41] L. Bar, N. Sochen, and N. Kiryati. Image deblurring in the presence of salt-and-pepper noise. In *Scale Space and PDE Methods in Computer Vision*, Lecture Notes in Computer Science, pages 107–118. Springer, 2005.

[42] L. Bar, N. Sochen, and N. Kiryati. Semi-blind image restoration via Mumford–Shah regularization. *Image Processing, IEEE Transactions on*, 15(2):483–493, 2006.

[43] B.E. Bayer. Color imaging array, 1976. U.S. Patent 3 971 065.

[44] M.F. Beg, M.I. Miller, A. Trouvé, and L. Younes. Computing large deformation metric mappings via geodesic flows of diffeomorphisms. *International Journal of Computer Vision*, 61(2):139–157, 2005.

[45] M. Bertalmio, A.L. Bertozzi, and G. Sapiro. Navier–Stokes, fluid dynamics, and image and video inpainting. In *Computer Vision and Pattern Recognition. Proceedings of IEEE Computer Society Conference on*, volume 1, pages I–355–I–362, 2001.

[46] M. Bertalmío, L.-T. Cheng, S. Osher, and G. Sapiro. Variational problems and partial differential equations on implicit surfaces. *Journal of Computational Physics*, 174(2):759–780, 2001.

[47] M. Bertalmio, G. Sapiro, V. Caselles, and C. Ballester. Image inpainting. In *Proceedings of 27th Annual Conference on Computer Graphics and Interactive Techniques*, pages 417–424. ACM Press/Addison-Wesley, 2000.

[48] M. Bertalmio, L. Vese, G. Sapiro, and S. Osher. Simultaneous structure and texture image inpainting. *Image Processing, IEEE Transactions on*, 12(8):882–889, 2003.

[49] F. Bethuel, H. Brezis, and J.M. Coron. Relaxed energies for harmonic maps. In *Variational Methods*, volume 4 of *Progress in Nonlinear Differential Equations and Their Applications Series*, pages 37–52. Birkhäuser, 1990.

[50] F. Bethuel, H. Brezis, and F. Hélein. Singular limit for the minimization of Ginzburg-Landau functionals. *Comptes Rendus de l'Académie des Sciences Paris - Séries I - Mathematics*, 314:891–895, 1992.

[51] F. Bethuel, H. Brezis, and F. Hélein. Asymptotics for the minimization of a Ginzburg–Landau functional. *Calculus of Variations and Partial Differential Equations*, 1(2):123–148, 1993.

[52] A. Blake and A. Zisserman. *Visual Reconstruction.* MIT Press, 1987.

[53] P. Blomgren. *Total Variation Methods for Restoration of Vector-Valued Images.* PhD thesis, University of California, Los Angeles, 1998.

[54] P. Blomgren and T.F. Chan. Color TV: total variation methods for restoration of vector-valued images. *Image Processing, IEEE Transactions on*, 7(3):304–309, 1998.

[55] L. Borcea. Electrical impedance tomography. *Inverse Problems*, 18(6):R99–R136, 2002.

[56] L. Borcea, J.G. Berryman, and G.C. Papanicolaou. High-contrast impedance tomography. *Inverse Problems*, 12(6):835–858, 1996.

[57] L. Borcea, G.A. Gray, and Y. Zhang. Variationally constrained numerical solution of electrical impedance tomography. *Inverse Problems*, 19(5):1159–1184, 2003.

[58] B. Bourdin. Image segmentation with a finite element method. *ESAIM: Mathematical Modelling and Numerical Analysis*, 33:229–244, 1999.

[59] B. Bourdin and A. Chambolle. Implementation of an adaptive finite-element approximation of the Mumford-Shah functional. *Numerische Mathematik*, 85(4):609–646, 2000.

[60] S.P. Boyd and L. Vandenberghe. *Convex Optimization.* Cambridge University Press, 2004.

[61] A. Braides. *Approximation of Free-Discontinuity Problems.* Lecture Notes in Mathematics. Springer, 1998.

[62] A. Braides and G. Dal Maso. Non-local approximation of the Mumford–Shah functional. *Calculus of Variations and Partial Differential Equations*, 5(4):293–322, 1997.

[63] L.M. Bregman. The relaxation method for finding common points of convex sets and its application to the solution of problems in convex programming. *USSR Computational Mathematics and Mathematical Physics*, (7):200–217, 1967.

[64] X. Bresson and T.F. Chan. Fast minimization of the vectorial total variation norm and applications to color image processing. *Inverse Problems and Imaging*, 2(4):455–484, 2008.

[65] H. Brézis. *Analyse fonctionnelle. Théorie et Applications.* Collection Mathématiques appliquées pour la maîtrise. Dunod, 2005.

[66] H. Brezis, J.-M. Coron, and E.H. Lieb. Harmonic maps with defects. *Communications in Mathematical Physics*, 107(4):649–705, 1986.

[67] C. Broit. *Optimal Registration of Deformed Images.* PhD thesis, University of Pennsylvania, Philadelphia, 1981.

[68] R.M. Brown and G.Uhlmann. Uniqueness in the inverse conductivity problem for nonsmooth conductivities in two dimensions. *Communications in Partial Differential Equations*, 2:1009–1027, 1997.

[69] C.C. Brun, N. Leporé, X. Pennec, Y.-Y. Chou, A.D. Lee, G. de Zubicaray, K.L. McMahon, M.J. Wright, J.C. Gee, and P.M. Thompson. A nonconservative Lagrangian framework for statistical fluid registration - SAFIRA. *Medical Imaging, IEEE Transactions on*, 30(2):184–202, 2011.

[70] C. Brune, A. Sawatzky, and M. Burger. Bregman-EM-TV methods with application to optical nanoscopy. In *Scale Space and Variational Methods in Computer Vision*, volume 5567 of *Lecture Notes in Computer Science*, pages 235–246. Springer, 2009.

[71] C. Brune, A. Sawatzky, and M. Burger. Primal and dual Bregman methods with application to optical nanoscopy. Technical Report 09-47, UCLA, 2009.

[72] A. Buades, B. Coll, and J.-M. Morel. A review of image denoising algorithms, with a new one. *Multiscale Modeling & Simulation*, 4(2):490–530, 2005.

[73] A. Buades, B. Coll, J.-M. Morel, and C. Sbert. Non local demosaicing. Technical Report Preprint 2007-15, CMLA, 2007.

[74] R.L. Burden, D.J. Faires, and A.M. Burden. *Numerical Analysis.* Brooks Cole, 2015.

[75] M. Burger. A level set method for inverse problems. *Inverse Problems*, 17(5):1327–1355, 2001.

[76] M. Burger, J. Modersitzki, and L. Ruthotto. A hyperelastic regularization energy for image registration. *SIAM Journal on Scientific Computing*, 35(1):B132–B148, 2013.

[77] A.P. Calderón. On an inverse boundary value problem. *Computational & Applied Mathematics*, 25:133–138, 2006.

[78] E.J. Candes, J. Romberg, and T. Tao. Robust uncertainty principles: exact signal reconstruction from highly incomplete frequency information. *Information Theory, IEEE Transactions on*, 52(2):489–509, 2006.

[79] E.J. Candes and T. Tao. Near-optimal signal recovery from random projections: universal encoding strategies? *Information Theory, IEEE Transactions on*, 52(12):5406–5425, 2006.

[80] V. Caselles, R. Kimmel, and G. Sapiro. Geodesic active contours. *International Journal of Computer Vision*, 22(1):61–79, 1997.

[81] T. Cecil. *Numerical Methods for Partial Differential Equations Involving Discontinuities.* PhD thesis, University of California, Los Angeles, 2003.

[82] A. Chambolle. Un théorème de Γ–convergence pour la segmentation des signaux. *Comptes Rendus de l'Académie des Sciences Paris - Séries I - Mathematics*, 314(3):191–196, 1992.

[83] A. Chambolle. Image segmentation by variational methods: Mumford and Shah functional, and the discrete approximation. *SIAM Journal on Applied Mathematics*, 55(3):827–863, 1995.

[84] A. Chambolle. Finite-differences discretizations of the Mumford-Shah functional. *ESAIM: Mathematical Modelling and Numerical Analysis*, 33:261–288, 3 1999.

[85] A. Chambolle and G. Dal Maso. Discrete approximation of the Mumford-Shah functional in dimension two. *ESAIM: Mathematical Modelling and Numerical Analysis*, 33(4):651–672, 1999.

[86] A. Chambolle and P.-L. Lions. Image recovery via total variation minimization and related problems. *Numerische Mathematik*, 76(2):167–188, 1997.

[87] T. Chan and S. Esedoḡlu. Aspects of total variation regularized L^1 function approximation. *SIAM Journal on Applied Mathematics*, 65(5):1817–1837, 2005.

[88] T. Chan, S. Esedoḡlu, and M. Nikolova. Algorithms for finding global minimizers of image segmentation and denoising models. *SIAM Journal on Applied Mathematics*, 66(5):1632–1648, 2006.

[89] T. Chan and J. Shen. Mathematical models for local nontexture inpaintings. *SIAM Journal on Applied Mathematics*, 62(3):1019–1043, 2002.

[90] T. Chan and J. Shen. *Image Processing and Analysis. Variational, PDE, Wavelet, and Stochastic Methods.* SIAM, Philadelphia, 2005.

[91] T. Chan and X.-C. Tai. Level set and total variation regularization for elliptic inverse problems with discontinuous coefficients. *Journal of Computational Physics*, 193(1):40–66, 2004.

[92] T. Chan and L. Vese. An active contour model without edges. *Lecture Notes in Computer Science*, 1682:141–151, 1999.

[93] T. Chan and L. Vese. An efficient variational multiphase motion for the Mumford–Shah segmentation model. In *Asilomar Conference on Signals, Systems, and Computers*, volume 1, pages 490–494, 2000.

[94] T. Chan and L. Vese. Active Contours Without Edges. *Image Processing, IEEE Transactions on*, 10:266–277, 2001.

[95] T. Chan and L. Vese. A level set algorithm for minimizing the Mumford–Shah functional in image processing. In *IEEE/Computer Society Proceedings of the first IEEE Workshop on Variational and Level Set Methods in Computer Vision*, pages 161–168, 2001.

[96] T. Chan and C.-K. Wong. Total variation blind deconvolution. *Image Processing, IEEE Transactions on*, 7(3):370–375, 1998.

[97] Y. Chen and F.H. Lin. Remarks on approximate harmonic maps. *Commentarii Mathematici Helvetici*, 70(1):161–169, 1995.

[98] Y. Chen and M. Struwe. Existence and partial regularity results for the heat flow for harmonic maps. *Mathematische Zeitschrift*, 201(1):83–103, 1989.

[99] Y. Chen, H. Thiruvenkadam, K. Gopinath, and R. Brigg. Image registration using the Mumford–Shah functional and shape information. In *World Multiconference on Systems, Cybernetics and Informatics*, pages 580–583, 2002.

[100] Y. Chen, S. Thiruvenkadam, H.D. Tagare, F. Huang, D. Wilson, and E.A. Geiser. On the incorporation of shape priors into geometric active contours. In *Variational and Level Set Methods in Computer Vision. Proceedings of IEEE Workshop*, pages 145–152, 2001.

[101] G.E. Christensen, R.D. Rabbitt, and M.I. Miller. Deformable templates using large deformation kinematics. *Image Processing, IEEE Transactions on*, 5(10):1435–1447, 1996.

[102] E.T. Chung, T. Chan, and X.-C. Tai. Electrical impedance tomography using level set representation and total variational regularization. *Journal of Computational Physics*, 205(1):357–372, 2005.

[103] G. Chung and L.A. Vese. Energy minimization based segmentation and denoising using a multilayer level set approach. In *Energy Minimization Methods in Computer Vision and Pattern Recognition*, volume 3757 of *Lecture Notes in Computer Science*, pages 439–455. Springer, 2005.

[104] G. Chung and L.A. Vese. Image segmentation using a multilayer level-set approach. *Computing and Visualization in Science*, 12(6):267–285, 2009.

[105] P.G. Ciarlet. *Élasticité tridimensionnelle*. Recherches en mathématiques appliquées. Masson, 1986.

[106] P.G. Ciarlet. *Mathematical Elasticity: Three-Dimensional Elasticity*, volume 1. North-Holland, 1993.

[107] P.G. Ciarlet and G. Geymonat. Sur les lois de comportement en élasticité non linéaire compressible. *Comptes Rendus de l'Académie des Sciences Paris - Séries II*, 295:423–426, 1982.

[108] L. Cohen, E. Bardinet, and N. Ayache. Surface reconstruction using active contour models. In *SPIE '93 Conference on Geometric Methods in Computer Vision*, 1993.

[109] L.D. Cohen. Avoiding local minima for deformable curves in image analysis. In *Curves and Surfaces with Applications in CAGD*, pages 77–84, 1997.

[110] R. Cohen, R. Hardt, D. Kinderlehrer, S.-Y. Lin, and M. Luskin. Minimum energy configurations for liquid crystals: computational results. In *Theory and Applications of Liquid Crystals*, volume 5 of *The IMA Volumes in Mathematics and Its Applications*, pages 99–121. Springer, 1987.

[111] J.-M. Coron and R. Gulliver. Minimizing p–harmonic maps into spheres. *Journal für die reine und angewandte Mathematik*, 401:82–100, 1989.

[112] P. Courilleau and F. Demengel. Heat flow for p–harmonic maps with values in the circle. *Nonlinear Analysis: Theory, Methods & Applications*, 41(5-6):689–700, 2000.

[113] A. Criminisi, P. Perez, and K. Toyama. Region filling and object removal by exemplar-based image inpainting. *Image Processing, IEEE Transactions on*, 13(9):1200–1212, 2004.

[114] B. Dacorogna. *Direct Methods in the Calculus of Variations, Second Edition.* Applied Mathematical Sciences. Springer, 2008.

[115] B. Dacorogna, I. Fonseca, J. Maly, and K. Trivisa. Manifold constrained variational problems. *Calculus of Variations and Partial Differential Equations*, 9(3):185–206, 1999.

[116] G. Dal Maso. *An Introduction to Γ–Convergence.* Birkhäuser, 1993.

[117] G. Dal Maso, J.-M. Morel, and S. Solimini. Variational approach in image processing - existence and approximation properties. *Comptes Rendus de l'Académie des Sciences Paris - Séries I - Mathematics*, 308(19):549–554, 1989.

[118] G. Dal Maso, J.-M. Morel, and S. Solimini. A variational method in image segmentation: existence and approximation properties. *Acta Mathematica*, 168(1-2):89–151, 1992.

[119] D. Datsenko and M. Elad. Example-based single document image super-resolution: a global MAP approach with outlier rejection. *Multidimensional Systems and Signal Processing*, 18(2-3):103–121, 2007.

[120] G. David. *Singular Sets of Minimizers for the Mumford-Shah Functional.* Progress in Mathematics. Birkhäuser, 2006.

[121] E. De Giorgi and L. Ambrosio. Un nuovo tipo di funzionale del Calcolo delle Variazioni. *Atti della Accademia Nazionale dei Lincei. Classe di Scienze Fisiche, Matematiche e Naturali*, 82:199 – 210, 1988.

[122] E. De Giorgi, M. Carriero, and A. Leaci. Existence theorem for a minimum problem with free discontinuity set. *Archive for Rational Mechanics and Analysis*, 108(4):195–218, 1989.

[123] L. Demanet, B. Song, and T. Chan. Image inpainting by correspondence maps: a deterministic approach. Technical report, University of California, Los Angeles, 2003.

[124] F. Demengel. Fonctions à hessien borné. *Annales de l'institut Fourier*, 34(2):155–190, 1984.

[125] F. Demengel and G. Demengel. *Functional Spaces for the Theory of Elliptic Partial Differential Equations.* Springer, 2012.

[126] F. Demengel and R. Temam. Convex functions of a measure and applications. *Indiana University Mathematics Journal*, 33:673–709, 1984.

[127] A.P. Dempster, N.M. Laird, and D.B. Rubin. Maximum likelihood from incomplete data via the EM algorithm. *Journal of the Royal Statistical Society, B*, 39(1):1–38, 1977.

[128] A. Desolneux, L. Moisan, and Morel J.-M. Variational snake theory. In *Geometric Level Set Methods in Imaging, Vision and Graphics.* Springer, 2003.

[129] D.L. Donoho. Compressed sensing. *Information Theory, IEEE Transactions on*, 52(4):1289–1306, 2006.

[130] M. Droske, W. Ring, and M. Rumpf. Mumford-Shah based registration: a comparison of a level set and a phase field approach. *Computing and Visualization in Science*, 12(3):101–114, 2009.

[131] M. Droske and M. Rumpf. A variational approach to non-rigid morphological registration. *SIAM Journal on Applied Mathematics*, 64(2):668–687, 2003.

[132] M. Droske and M. Rumpf. Multiscale joint segmentation and registration of image morphology. *Pattern Analysis and Machine Intelligence, IEEE Transactions on*, 29(12):2181–2194, 2007.

[133] V. Duay, N. Houhou, and J. Thiran. Atlas-based segmentation of medical images locally constrained by level sets. In *ICIP 2005*, Parallel Computing in Electrical Engineering. IEEE, 2005.

[134] S. Dumont and R. Hadiji. Regularity of minimizing maps with values in S^2 and some numerical simulations. *Advances in Mathematical Sciences and Applications*, 10(2):711–733, 2000.

[135] A.A. Efros and T.K. Leung. Texture synthesis by non-parametric sampling. In *Computer Vision, Proceedings of Seventh IEEE International Conference on*, volume 2, pages 1033–1038, 1999.

[136] I. Ekeland and R. Témam. *Convex Analysis and Variational Problems*. Classics in Applied Mathematics. Society for Industrial and Applied Mathematics, 1999.

[137] M. Elad, J.-L. Starck, P. Querre, and D.L. Donoho. Simultaneous cartoon and texture image inpainting using morphological component analysis (MCA). *Applied and Computational Harmonic Analysis*, 19(3):340–358, 2005.

[138] S. Esedoḡlu and J. Shen. Digital inpainting based on the Mumford–Shah–Euler image model. *European Journal of Applied Mathematics*, 13:353–370, 2002.

[139] L.C. Evans. *Partial Differential Equations*. Graduate Studies in Mathematics. American Mathematical Society, 1998.

[140] L.C. Evans and R.F. Gariepy. *Measure Theory and Fine Properties of Functions*. Studies in Advanced Mathematics. Taylor & Francis, 1991.

[141] J. Faires and R. Burden. *Numerical Methods, Fourth Edition*. Cengage Learning, 2012.

[142] S. Farsiu, D. Robinson, M. Elad, and P. Milanfar. Advances and challenges in super-resolution. *International Journal of Imaging Systems and Technology*, 14(2):47–57, 2004.

[143] B. Fischer and J. Modersitzki. Fast diffusion registration. *AMS Contemporary Mathematics, Inverse Problems, Image Analysis, and Medical Imaging*, 313:117–129, 2002.

[144] B. Fischer and J. Modersitzki. Curvature-based image registration. *Journal of Mathematical Imaging and Vision*, 18(1):81–85, 2003.

[145] B. Fischer and J. Modersitzki. A unified approach to fast image registration and a new curvature based registration technique. *Linear Algebra and its Applications*, 380:107–124, 2004.

[146] G.B. Folland. *Real Analysis: Modern Techniques and Their Applications*. Wiley, 1999.

[147] N. Forcadel. Dislocations dynamics with a mean curvature term: short time existence and uniqueness. *Differential Integral Equations*, 21(3-4):285–304, 2008.

[148] N. Forcadel and C. Le Guyader. A short time existence/uniqueness result for a nonlocal topology-preserving segmentation model. *Journal of Differential Equations*, 253(3):977 – 995, 2012.

[149] W.T. Freeman, T.R. Jones, and E.C. Pasztor. Example-based super-resolution. *Computer Graphics and Applications, IEEE*, 22(2):56–65, 2002.

[150] M. Gage and R.S. Hamilton. The heat equation shrinking convex plane curves. *Journal of Differential Geometry*, 23(1):69–96, 1986.

[151] J. Garnett, P. Jones, T. Le, and L. Vese. Modeling oscillatory components with the homogeneous spaces $\dot{BMO}^{-\alpha}$ and $\dot{W}^{-\alpha,p}$. *Pure and Applied Mathematics Quarterly*, 7(2):275–318, 2011.

[152] J.B. Garnett, T.M. Le, Y. Meyer, and L.A. Vese. Image decompositions using bounded variation and generalized homogeneous Besov spaces. *Applied and Computational Harmonic Analysis*, 23(1):25–56, 2007.

[153] M. Giaquinta, G. Modica, and J. Souček. Variational problems for maps of bounded variation with values in S^1. *Calculus of Variations and Partial Differential Equations*, 1(1):87–121, 1993.

[154] G. Gilboa and S. Osher. Nonlocal linear image regularization and supervised segmentation. *Multiscale Modeling & Simulation*, 6(2):595–630, 2007.

[155] G. Gilboa and S. Osher. Nonlocal operators with applications to image processing. *Multiscale Modeling & Simulation*, 7(3):1005–1028, 2009.

[156] E. Giusti. *Minimal Surfaces and Functions of Bounded Variation*. Birkhäuser, 1984.

[157] R. Glowinski and P. Le Tallec. *Augmented Lagrangian and Operator-Splitting Methods in Nonlinear Mechanics*. Society for Industrial and Applied Mathematics, 1989.

[158] C. Goffman and J. Serrin. Sublinear functions of measures and variational integrals. *Duke Mathematical Journal*, 31(1):159–178, 1964.

[159] R.C. Gonzalez and R.E. Woods. *Digital Image Processing*. Addison-Wesley, 1992.

[160] Y. Gousseau and J.-M. Morel. Are natural images of bounded variation? *SIAM J. Math. Anal.,*, 33(3):634–648, 2001.

[161] M. A. Grayson. The heat equation shrinks embedded plane curves to round points. *Journal of Differential Geometry*, 26(2):285–314, 1987.

[162] B.K. Gunturk, Y. Altunbasak, and R.M. Mersereau. Color plane interpolation using alternating projections. *Image Processing, IEEE Transactions on*, 11(9):997–1013, 2002.

[163] B.K. Gunturk, J. Glotzbach, Y. Altunbasak, R.W. Schafer, and R.M. Mersereau. Demosaicking: color filter array interpolation. *Signal Processing Magazine*, 22(1):44–54, 2005.

[164] E. Haber and J. Modersitzki. Numerical methods for volume preserving image registration. *Inverse Problems*, 20(5):1621–1638, 2004.

[165] E. Haber and J. Modersitzki. Image registration with guaranteed displacement regularity. *International Journal of Computer Vision*, 71(3):361–372, 2007.

[166] X. Han, C. Xu, U. Braga-Neto, and J.L. Prince. Topology correction in brain cortex segmentation using a multiscale, graph-based algorithm. *Medical Imaging, IEEE Transactions on*, 21(2):109–121, 2002.

[167] X. Han, C. Xu, U. Braga-Neto, and J.L. Prince. A topology preserving geometric deformable model and its application in brain cortical surface reconstruction. In *Geometric Level Set Methods in Imaging, Vision and Graphics*. Springer, 2003.

[168] X. Han, C. Xu, and J.L. Prince. A topology preserving level set method for geometric deformable models. *Pattern Analysis and Machine Intelligence, IEEE Transactions on*, 25(6):755–768, 2003.

[169] Z.T. Harmany, R.F. Marcia, and R.M. Willett. This is SPIRAL-TAP: Sparse Poisson Intensity Reconstruction Algorithms, Theory and Practice. *Image Processing, IEEE Transactions on*, 21(3):1084–1096, 2012.

[170] L. He, A. Marquina, and S.J. Osher. Blind deconvolution using TV regularization and Bregman iteration. *International Journal of Imaging Systems and Technology*, 15(1):74–83, 2005.

[171] G. Hermosillo and O. Faugeras. Well-posedness of two nonrigid multimodal image registration methods. *SIAM Journal on Applied Mathematics*, 64(5):1550–1587, 2004.

[172] M.R. Hestenes. Multiplier and gradient methods. *Journal of Optimization Theory and Applications*, 4(5):303–320, 1969.

[173] M.R. Hestenes and E. Stiefel. Methods of conjugate gradients for solving linear systems. *Journal of Research of the National Bureau of Standards*, 49(6):409–436, 1952.

[174] K. Hirakawa and T.W. Parks. Adaptive homogeneity-directed demosaicing algorithm. *Image Processing, IEEE Transactions on*, 14(3):360–369, 2005.

[175] S. Jafarpour, R. Willett, M. Raginsky, and R. Calderbank. Performance bounds for expander-based compressed sensing in the presence of Poisson noise. In *Proceedings of 43rd Asilomar Conference on Signals, Systems and Computers*, pages 513–517. IEEE Press, 2009.

[176] X. Jia, Y. Lou, R. Li, W.Y. Song, and S.B. Jiang. GPU-based fast cone beam CT reconstruction from undersampled and noisy projection data via total variation. *Medical Physics*, 37(4):1757–1760, 2010.

[177] N. Jifeng, W. Chengke, L. Shigang, and Y. Shuqin. NGVF: An improved external force field for active contour model. *Pattern Recognition Letters*, 28(1):58–63, 2007.

[178] H.J. Johnson and G.E. Christensen. Consistent landmark and intensity-based image registration. *Medical Imaging, IEEE Transactions on*, 21(5):450–461, 2002.

[179] M. Jung. PhD thesis, University of California, Los Angeles, 2010.

[180] M. Jung, X. Bresson, T.F. Chan, and L.A. Vese. Nonlocal Mumford–Shah regularizers for color image restoration. *Image Processing, IEEE Transactions on*, 20(6):1583–1598, 2011.

[181] M. Jung, G. Chung, G. Sundaramoorthi, L.A. Vese, and A.L. Yuille. Sobolev gradients and joint variational image segmentation, denoising, and deblurring. In *Proceedings of SPIE, Computational Imaging VII*, volume 7246, pages 72460I–72460I–13, 2009.

[182] M. Jung, E. Resmerita, and L.A. Vese. Dual norm based iterative methods for image restoration. CAM Report 09-88, University of California, Los Angeles, 2009.

[183] M. Jung and L.A. Vese. Nonlocal variational image deblurring models in the presence of Gaussian or impulse noise. In *Scale Space and Variational Methods in Computer Vision*, volume 5567 of *Lecture Notes in Computer Science*, pages 401–412. Springer, 2009.

[184] S.H. Kang. *Mathematical approaches to color denoising and image inpainting problems.* PhD thesis, University of California, Los Angeles, 2002.

[185] G. Kanizsa. *La Grammaire du Voir. Essais sur la perception.* Arts et Sciences. Diderot Editeur, 1997.

[186] W. Karush. Minima of functions of several variables with inequalities as side constraints. Master's thesis, University of Chicago, 1939.

[187] M. Kass, A. Witkin, and D. Terzopoulos. Snakes: active contour models. *International Journal of Computer Vision*, 1(4):321–331, 1988.

[188] S. Kichenassamy, A. Kumar, P. Olver, A. Tannenbaum, and A. Yezzi Jr. Conformal curvature flows: from phase transitions to active vision. *Archive for Rational Mechanics and Analysis*, 134(3):275–301, 1996.

[189] J. Kim, A. Tsai, M. Cetin, and A.S. Willsky. A curve evolution-based variational approach to simultaneous image restoration and segmentation. In *Image Processing. Proceedings of International Conference on*, volume 1, pages I–109–I–112, 2002.

[190] R. Kimmel. Demosaicing: image reconstruction from color CCD samples. *Image Processing, IEEE Transactions on*, 8(9):1221–1228, 1999.

[191] R. Kimmel. Fast edge integration. In *Geometric Level Set Methods in Imaging, Vision, and Graphics*, pages 59–77. Springer, 2003.

[192] R. Kimmel and A.M. Bruckstein. Regularized Laplacian zero crossings as optimal edge integrators. *International Journal of Computer Vision*, 53(3):225–243, 2003.

[193] R. Kimmel, R. Malladi, and N. Sochen. Images as embedded maps and minimal surfaces: movies, color, texture, and volumetric medical images. *International Journal of Computer Vision*, 39(2):111–129, 2000.

[194] R. Kimmel and N. Sochen. Orientation diffusion or how to comb a porcupine. *Journal of Visual Communication and Image Representation*, 13(1-2):238–248, 2002.

[195] G. Koepfler, C. Lopez, and J.-M. Morel. A multiscale algorithm for image segmentation by variational method. *SIAM Journal on Numerical Analysis*, 31(1):282–299, 1994.

[196] R. Kohn and M. Vogelius. Determining conductivity by boundary measurements. *Communications on Pure and Applied Mathematics*, 37(3):289–298, 1984.

[197] R.V. Kohn and M. Vogelius. Determining conductivity by boundary measurements II. interior results. *Communications on Pure and Applied Mathematics*, 38(5):643–667, 1985.

[198] B. Kortschak and B. Brandstätter. A FEM-BEM approach using levelsets in electrical capacitance tomography. *International Journal for Computation and Mathematics in Electrical and Electronic Engineering*, 24(2):591–605, 2005.

[199] B. Kortschak, H. Wegleiter, and B. Brandstätter. Formulation of cost functionals for different measurement principles in nonlinear capacitance tomography. *Measurement Science and Technology*, 18(1):71, 2007.

[200] H.W. Kuhn and A.W. Tucker. Nonlinear programming. In *Proceedings of Second Berkeley Symposium on Mathematical Statistics and Probability*, pages 481–492. University of California Press, 1951.

[201] Y. Gousseau L. Alvarez and J.-M. Morel. Scales in natural images and a consequence on their bounded variation norm. In *Scale-Space Theories in Computer Vision, Lecture Notes in Comput. Sci., Springer, New York*, volume 1862, pages 247–258, 1999.

[202] C. J. Larsen. A new proof of regularity for two-shaded image segmentations. *Manuscripta Mathematica*, 96(2):247–262, 1998.

[203] T. Le, R. Chartrand, and T.J. Asaki. A variational approach to reconstructing images corrupted by Poisson noise. *Journal of Mathematical Imaging and Vision*, 27(3):257–263, 2007.

[204] T. Le and L. Vese. Additive and multiplicative piecewise-smooth segmentation models in a functional minimization approach. In *Interpolation Theory and Applications: A Conference in Honor of Michael Cwikel*, volume 445, pages 207–223. American Mathematical Society, 2007.

[205] T.M. Le. *Image segmentation and decomposition models in a variational approach.* PhD thesis, University of California, Los Angeles, 2006.

[206] T.M. Le and L.A. Vese. Image decomposition using total variation and div(BMO). *Multiscale Modeling & Simulation*, 4(2):390–423, 2005.

[207] H. Le Dret. Méthodes mathématiques en élasticité. Notes de cours de DEA 2003–2004. Université Pierre et Marie Curie.

[208] C. Le Guyader and L.A. Vese. Self-repelling snakes for topology-preserving segmentation models. *Image Processing, IEEE Transactions on*, 17(5):767–779, 2008.

[209] C. Le Guyader and L.A. Vese. A combined segmentation and registration framework with a nonlinear elasticity smoother. In *Scale Space and Variational Methods in Computer Vision*, volume 5567 of *Lecture Notes in Computer Science*, pages 600–611. Springer, 2009.

[210] C. Le Guyader and L.A. Vese. A combined segmentation and registration framework with a nonlinear elasticity smoother. *Computer Vision and Image Understanding*, 115(12):1689–1709, 2011.

[211] F. Lecellier. *Les contours actifs basés régions avec a priori de bruit, de texture et de forme : application à l'échocardiographie.* PhD thesis, Université de Caen, 2009.

[212] G.P. Leonardi and I. Tamanini. On minimizing partitions with infinitely many components. *Annali dell'Università di Ferrara Sezione VII - Scienze Matematiche*, XLIV:41–57, 1998.

[213] G. Leoni and D. Spector. Characterization of Sobolev and BV spaces. *Journal of Functional Analysis*, 261(10):2926–2958, 2011.

[214] A. Leow, M.-C. Chiang, H. Protas, P. Thompson, L. Vese, and H.S.C. Huang. Linear and non-linear geometric object matching with implicit representation. *Pattern Recognition, International Conference on*, 3:710–713, 2004.

[215] C. Li, Ç.-Y. Kao, J.C. Gore, and Z. Ding. Implicit active contours driven by local binary fitting energy. In *Computer Vision and Pattern Recognition, IEEE Conference on*, pages 1–7, 2007.

[216] W.-H. Liao, A. Khuu, M. Bergsneider, L. Vese, Huang S.-C., and S. Osher. From landmark matching to shape and open curve matching: a level set approach. CAM Report 02-59, University of California, Los Angeles, 2002.

[217] W.-H. Liao, C.-L. Yu, M. Bergsneider, L. Vese, and Huang S.-C. A new framework of quantifying differences between images by matching gradient fields and its application to image blending. In *Nuclear Science Symposium Conference Record, IEEE*, volume 2, pages 1092–1096, 2002.

[218] L. Lieu and L.A. Vese. Image restoration and decomposition via bounded total variation and negative Hilbert-Sobolev spaces. *Applied Mathematics and Optimization*, 58(2):167–193, 2008.

[219] T. Lin, C. Le Guyader, I. Dinov, P. Thompson, A. Toga, and L. Vese. Gene Expression Data to Mouse Atlas Registration Using a Nonlinear Elasticity Smoother and Landmark Points Constraints. *Journal of Scientific Computing*, 50(3):586–609, 2012.

[220] T. Lin, E.-F. Lee, I. Dinov, C. Le Guyader, P. Thompson, A.W. Toga, and L.A. Vese. A landmark-based nonlinear elasticity model for mouse atlas registration. In *Biomedical Imaging: From Nano to Macro. Fifth IEEE International Symposium on*, pages 788–791, 2008.

[221] F. Liu, Y. Luo, and D. Hu. Adaptive level set image segmentation using the Mumford and Shah functional. *Optical Engineering*, 41(12):3002–3003, 2002.

[222] N. Lord, J. Ho, B. Vemuri, and S. Eisenschenk. Simultaneous registration and parcellation of bilateral hippocampal surface pairs for local asymmetry quantification. *Medical Imaging, Transactions on*, 26(4):471–478, 2007.

[223] Y. Lou, X. Zhang, S. Osher, and A. Bertozzi. Image recovery via nonlocal operators. *Journal of Scientific Computing*, 42(2):185–197, 2010.

[224] W. Lu and Y.-P. Tan. Color filter array demosaicking: new method and performance measures. *Image Processing, IEEE Transactions on*, 12(10):1194–1210, 2003.

[225] L.B. Lucy. An iterative technique for the rectification of observed distributions. *The Astronomical Journal*, 79(6):745–754, 1974.

[226] J. Cong M. Yan, A. Bui and L.A. Vese. General convergent expectation maximization (EM)-type algorithms for image reconstruction. *Inverse Problems and Imaging*, 7(3):1007–1029, 2013.

[227] F. Maddalena and S. Solimini. Lower semicontinuity properties of functionals with free discontinuities. *Archive for Rational Mechanics and Analysis*, 159(4):273–294, 2001.

[228] S. Maheswaran, H. Barjat, S.T. Bate, P. Aljabar, D.L.G. Hill, L. Tilling, N. Upton, M.F. James, J.V. Hajnal, and D. Rueckert. Analysis of serial magnetic resonance images of mouse brains using image registration. *NeuroImage*, 44(3):692–700, 2009.

[229] M. Mahmoudi and G. Sapiro. Fast image and video denoising via nonlocal means of similar neighborhoods. *Signal Processing Letters, IEEE*, 12(12):839–842, 2005.

[230] F. Malgouyres and F. Guichard. Edge direction preserving image zooming: a mathematical and numerical analysis. *SIAM Journal on Numerical Analysis*, 39(1):1–37, 2001.

[231] D. Marr and E. Hildreth. Theory of edge detection. *Proceedings of the Royal Society of London B: Biological Sciences*, 207(1167):187–217, 1980.

[232] S. Masnou. Disocclusion: a variational approach using level lines. *Image Processing, IEEE Transactions on*, 11(2):68–76, 2002.

[233] S. Masnou and J.-M. Morel. Level lines-based disocclusion. In *Image Processing, Proceedings of International Conference on*, volume 3, pages 259–263, 1998.

[234] U. Massari and I. Tamanini. On the finiteness of optimal partitions. *Annali dell'Università di Ferrara Sezione VII - Scienze Matematiche*, XXXIX:167–185, 1993.

[235] F. Mémoli, G. Sapiro, and S. Osher. Solving variational problems and partial differential equations mapping into general target manifolds. *Journal of Computational Physics*, 195(1):263–292, 2004.

[236] Y. Meyer. *Oscillating Patterns in Image Processing and Nonlinear Evolution Equations: The Fifteenth Dean Jacqueline B. Lewis Memorial Lectures.* American Mathematical Society, 2001.

[237] M.I. Miller, A. Trouvé, and L. Younes. On the metrics and Euler–Lagrange equations of computational anatomy. *Annual Review of Biomedical Engineering*, 4(1):375–405, 2002.

[238] J. Modersitzki. *Numerical Methods for Image Registration.* Numerical Mathematics and Scientific Computation Series. Oxford University Press, 2004.

[239] J. Modersitzki. *FAIR: Flexible Algorithms for Image Registration.* Society for Industrial and Applied Mathematics (SIAM), 2009.

[240] L. Modica. The gradient theory of phase transitions and the minimal interface criterion. *Archive for Rational Mechanics and Analysis*, 98(2):123–142, 1987.

[241] L. Modica and S. Mortola. Un esempio di Γ–convergenza. *Bollettino dell'Unione Matematica Italina*, B(5)(14):285–299, 1977.

[242] J.-M. Morel and S. Solimini. *Variational Methods in Image Segmentation: with Seven Image Processing Experiments.* Birkhäuser, 1995.

[243] K.W. Morton and D.F. Mayers. *Numerical Solution of Partial Differential Equations: An Introduction.* Cambridge University Press, 2005.

[244] D. Mumford and B. Gidas. Stochastic models for generic images. *Quart. Appl. Math.*, LIX(1):85–111, 2001.

[245] D. Mumford and J. Shah. Boundary detection by minimizing functionals. In *IEEE Conference on Computer Vision and Pattern Recognition*, pages 22–26, 1985.

[246] D. Mumford and J. Shah. Boundary detection by minimizing functionals. In *Image Understanding*, pages 19–43, 1989.

[247] D. Mumford and J. Shah. Optimal approximations by piecewise smooth functions and associated variational problems. *Communications on Pure and Applied Mathematics*, 42:577–685, 1989.

[248] A.I. Nachman. Global uniqueness for a two-dimensional inverse boundary value problem. *Annals of Mathematics*, 143(1):71–96, 1996.

[249] P. V. Negrón Marrero. A numerical method for detecting singular minimizers of multidimensional problems in nonlinear elasticity. *Numerische Mathematik*, 58(1):135–144, 1990.

[250] L. K. Nielsen, X.-C. Tai, S. I. Aanonsen, and M. Espedal. A binary level set model for elliptic inverse problems with discontinuous coefficients. *International Journal of Numerical Analysis and Modeling*, 4(1):74–99, 2007.

[251] M. Nikolova. Minimizers of cost-functions involving nonsmooth data-fidelity terms. Application to the processing of outliers. *SIAM Journal on Numerical Analysis*, 40(3):965–994, 2002.

[252] J. Nocedal and S. Wright. *Numerical Optimization*. Operations Research and Financial Engineering Series. Springer, 2006.

[253] S. Osher. Level set methods. In *Geometric Level Set Methods in Imaging, Vision and Graphics*. Springer, 2003.

[254] S. Osher and J.A. Sethian. Fronts propagating with curvature-dependent speed: Algorithms based on Hamilton-Jacobi formulations. *Journal of Computational Physics*, 79(1):12–49, 1988.

[255] S. Osher, A. Solé, and L. Vese. Image decomposition and restoration using total variation minimization and the H^1-norm. *Multiscale Modeling & Simulation*, 1(3):349–370, 2003.

[256] S.J. Osher and R.P. Fedkiw. *Level Set Methods and Dynamic Implicit Surfaces*. Springer, 2003.

[257] L. Päivärinta, A. Panchenko, and G. Uhlmann. Complex geometrical optics solutions for Lipschitz conductivities. *Revista Matemática Iberoamericana*, 19(1):57–72, 03 2003.

[258] W. Peckar, C. Schnörr, K. Rohr, and H.S. Stiehl. Parameter-free elastic deformation approach for 2D and 3D registration using prescribed displacements. *Journal of Mathematical Imaging and Vision*, 10(2):143–162, 1999.

[259] P. Perona. Orientation diffusions. *Image Processing, IEEE Transactions on*, 7(3):457–467, 1998.

[260] G. Peyré. Image processing with nonlocal spectral bases. *Multiscale Modeling & Simulation*, 7(2):703–730, 2008.

[261] G. Peyré, S. Bougleux, and L. Cohen. Non-local regularization of inverse problems. In *Computer Vision*, volume 5304 of *Lecture Notes in Computer Science*, pages 57–68. Springer, 2008.

[262] D. L. Phillips. A technique for the numerical solution of certain integral equations of the first kind. *Journal of the ACM*, 9(1):84–97, 1962.

[263] A.C. Ponce. A new approach to Sobolev spaces and connections to Γ-convergence. *Calculus of Variations and Partial Differential Equations*, 19(3):229–255, 2004.

[264] M.J.D. Powell. A method for nonlinear constraints in minimization problems. In *Optimization*, pages 283–298. Academic Press, 1969.

[265] M. Protter, M. Elad, H. Takeda, and P. Milanfar. Generalizing the nonlocal-means to super-resolution reconstruction. *Image Processing, IEEE Transactions on*, 18(1):36–51, 2009.

[266] R.D. Rabbitt, J.A. Weiss, G.E. Christensen, and M.I. Miller. Mapping of hyperelastic deformable templates using the finite element method. In *Proceedings of SPIE, Deformation Methods*, volume 2573, pages 252–265, 1995.

[267] A. Raoult. Non-polyconvexity of the stored energy function of a Saint Venant- Kirchhoff material. *Aplikace Matematiky*, 31:417–419, 1986.

[268] W.H. Richardson. Bayesian-based iterative method of image restoration. *Journal of the Optical Society of America*, 62(1):55–59, 1972.

[269] M. Rochery, I. Jermyn, and J. Zerubia. Higher order active contours and their application to the detection of line networks in satellite imagery. In *2nd IEEE Workshop on Variational, Geometric and Level Set Methods in Computer Vision*, 2003.

[270] M Rochery, I. Jermyn, and J. Zerubia. Phase field models and higher-order active contours. In *Computer Vision, Tenth IEEE International Conference on*, volume 2, pages 970–976, 2005.

[271] M. Rochery, I.H. Jermyn, and J. Zerubia. Higher order active contours. *International Journal of Computer Vision*, 69(1):27–42, 2006.

[272] R.T. Rockafellar. A dual approach to solving nonlinear programming problems by unconstrained optimization. *Mathematical Programming*, 5(1):354–373, 1973.

[273] L. Rondi and F. Santosa. Enhanced electrical impedance tomography via the Mumford–Shah functional. *ESAIM: Control, Optimization and Calculus of Variations*, 6:517–538, 2001.

[274] Y. Rouchdy, J. Pousin, J. Schaerer, and P. Clarysse. A nonlinear elastic deformable template for soft structure segmentation: application to the heart segmentation in MRI. *Inverse Problems*, 23(3):1017–1035, 2007.

[275] L. Rudin, P.-L. Lions, and S. Osher. Multiplicative denoising and deblurring: theory and algorithms. In *Geometric Level Set Methods in Imaging, Vision, and Graphics*, pages 103–119. Springer, 2003.

[276] L.I. Rudin and S. Osher. Total variation-based image restoration with free local constraints. In *Image Processing, Proceedings of IEEE International Conference on*, volume 1, pages 31–35, 1994.

[277] L.I. Rudin, S. Osher, and E. Fatemi. Nonlinear total variation-based noise removal algorithms. *Physica D: Nonlinear Phenomena*, 60(1-4):259–268, 1992.

[278] G. Russo and P. Smereka. A remark on computing distance functions. *Journal of Computational Physics*, 163(1):51–67, 2000.

[279] C. Samson, L. Blanc-Féraud, G. Aubert, and J. Zerubia. A level set model for image classification. *International Journal of Computer Vision*, 40(3):187–197, 2000.

[280] C. Samson, L. Blanc-Féraud, J. Zerubia, and G. Aubert. A level set model for image classification. In *Scale-Space Theories in Computer Vision*, volume 1682 of *Lecture Notes in Computer Science*, pages 306–317. Springer, 1999.

[281] G. Sapiro and D.L. Ringach. Anisotropic diffusion of multivalued images with applications to color filtering. *Image Processing, IEEE Transactions on*, 5(11):1582–1586, 1996.

[282] R. Schoen and K. Uhlenbeck. A regularity theory for harmonic maps. *Journal of Differential Geometry*, 17(2):307–335, 1982.

[283] R. Schoen and K. Uhlenbeck. Boundary regularity and the Dirichlet problem for harmonic maps. *Journal of Differential Geometry*, 18(2):253–268, 1983.

[284] R. Schoen and K. Uhlenbeck. Regularity of minimizing harmonic maps into the sphere. *Inventiones Mathematicae*, 78(1):89–100, 1984.

[285] J.A. Sethian. *Level Set Methods. Evolving Interfaces in Geometry, Fluid Mechanics, Computer Vision, and Materials Science.* Cambridge University Press, 1996.

[286] J.A. Sethian. *Level Set Methods and Fast Marching Methods: Evolving Interfaces in Computational Geometry, Fluid Mechanics, Computer Vision, and Materials Science.* Cambridge University Press, 1999.

[287] S. Setzer, G. Steidl, and T. Teuber. Deblurring Poissonian images by split Bregman techniques. *Journal of Visual Communication and Image Representation*, 21(3):193–199, 2010.

[288] J. Shah. A common framework for curve evolution, segmentation and anisotropic diffusion. In *Computer Vision and Pattern Recognition, Proceedings of IEEE Computer Society Conference on*, pages 136–142, 1996.

[289] J. Shen and T. Chan. Variational restoration of nonflat image features: models and algorithms. *SIAM Journal on Applied Mathematics*, 61(4):1338–1361, 2001.

[290] J. Shen and T.F Chan. Variational image inpainting. *Communications on Pure and Applied Mathematics*, 58(5):579–619, 2005.

[291] J. Shen, S.H. Kang, and T.F. Chan. Euler's elastica and curvature-based inpainting. *SIAM Journal on Applied Mathematics*, 63(2):564–592, 2003.

[292] L.A. Shepp and B.F. Logan. The Fourier reconstruction of a head section. *Nuclear Science, IEEE Transactions on*, 21(3):21–43, 1974.

[293] L.A. Shepp and Y. Vardi. Maximum likelihood reconstruction for emission tomography. *Medical Imaging, IEEE Transactions on*, 1(2):113–122, 1982.

[294] R.L. Siddon. Fast calculation of the exact radiological path for a three-dimensional CT array. *Medical Physics*, 12(2):252–255, 1985.

[295] N. Sochen, R. Kimmel, and R. Malladi. A general framework for low level vision. *Image Processing, IEEE Transactions on*, 7(3):310–318, 1998.

[296] N.A. Sochen and R. Kimmel. Combing a porcupine via stereographic direction diffusion. In *Proceedings of Third International Conference on Scale-Space and Morphology in Computer Vision*, pages 308–316. Springer, 2001.

[297] M. Soleimani, O. Dorn, and W.R.B. Lionheart. A narrow-band level set method applied to EIT in brain for cryosurgery monitoring. *Biomedical Engineering, IEEE Transactions on*, 53(11):2257–2264, 2006.

[298] C.O.S. Sorzano, P. Thevenaz, and M. Unser. Elastic registration of biological images using vector-spline regularization. *Biomedical Engineering, IEEE Transactions on*, 52(4):652–663, 2005.

[299] A. Sotiras, C. Davatzikos, and N. Paragios. Deformable medical image registration: a survey. *Medical Imaging, IEEE Transactions on*, 32(7):1153–1190, 2013.

[300] D. Strong and T. Chan. Edge-preserving and scale-dependent properties of total variation regularization. *Inverse Problems*, 19(6):S165, 2003.

[301] M. Struwe. On the evolution of harmonic mappings of Riemannian surfaces. *Commentarii Mathematici Helvetici*, 60(1):558–581, 1985.

[302] M. Struwe. The evolution of Harmonic maps: existence, partial regularity, and singularities. In *Nonlinear Diffusion Equations and Their Equilibrium States*, volume 7 of *Progress in Nonlinear Differential Equations and Their Applications*, pages 485–491. Birkhäuser, 1992.

[303] M. Struwe. Uniqueness of harmonic maps with small energy. *Manuscripta Mathematica*, 96(4):463–486, 1998.

[304] M. Struwe. *Variational Methods: Applications to Nonlinear Partial Differential Equations and Hamiltonian Systems, Third Edition.* Springer, 2000.

[305] G. Sundaramoorthi and A. Yezzi. More-than-topology-preserving flows for active contours and polygons. In *Tenth IEEE International Conference on Computer Vision*, pages 1276–1283, 2005.

[306] G. Sundaramoorthi and A. Yezzi. Global regularizing flows with topology preservation for active contours and polygons. *Image Processing, IEEE Transactions on*, 16(3):803–812, 2007.

[307] J. Sylvester and G. Uhlmann. A global uniqueness theorem for an inverse boundary value problem. *Annals of Mathematics*, 125(1):153–169, 1987.

[308] A. Szlam, M. Maggioni, and R.R. Coifman. A general framework for adaptive regularization based on diffusion processes on graphs. *Journal of Machine Learning Research*, 19:1711–1739, 2008.

[309] E. Tadmor, S. Nezzar, and L. Vese. Multiscale hierarchical decomposition of images with applications to deblurring, denoising, and segmentation. *Communications in Mathematical Sciences*, 6(2):281–307, 2008.

[310] H.D. Tagare, D. Groisser, and O. Skrinjar. A geometric theory of symmetric registration. In *Computer Vision and Pattern Recognition Workshop, Conference on*, pages 73–73, 2006.

[311] X.-C. Tai and C. Wu. Augmented Lagrangian method, dual methods and split Bregman iteration for ROF model. In *Scale Space and Variational Methods in Computer Vision*, volume 5567 of *Lecture Notes in Computer Science*, pages 502–513. Springer, 2009.

[312] I. Tamanini. Optimal approximation by piecewise constant functions. *Progress in Nonlinear Differential Equations and Their Applications*, 25:73–85, 1996.

[313] I. Tamanini and G. Congedo. Optimal segmentation of unbounded functions. *Rendiconti del Seminario Matematico della Università di Padova*, 95:153–174, 1996.

[314] B. Tang, G. Sapiro, and V. Caselles. Diffusion of general data on nonflat manifolds via harmonic maps theory: the direction diffusion case. *International Journal of Computer Vision*, 36(2):149–161, 2000.

[315] B. Tang, G. Sapiro, and V. Caselles. Color image enhancement via chromaticity diffusion. *Image Processing, IEEE Transactions on*, 10:701–707, 2002.

[316] N.M. Tanushev and L.A. Vese. A piecewise-constant binary model for electrical impedance tomography. *Inverse Problems and Imaging*, 1(2):423–435, 2007.

[317] R. Temam and A. Miranville. *Mathematical Modeling in Continuum Mechanics*. Cambridge University Press, 2000.

[318] R.C. Thompson and L.J. Freede. On the eigenvalues of sums of Hermitian matrices. *Linear Algebra and its Applications*, 4(4):369–376, 1971.

[319] A.N. Tikhonov. On the stability of inverse problems. *Doklady Akademii Nauk SSSR*, 39(5):195–198, 1943.

[320] A.N. Tikhonov. Regularization of incorrectly posed problems. *Soviet Mathematics, Doklady*, 4:1624–1627, 1963.

[321] A.N. Tikhonov. Solution of incorrectly formulated problems and the regularization method. *Soviet Mathematics, Doklady*, 4:1035–1038, 1963.

[322] A.N. Tikhonov and V.I.A. Arsenin. *Solutions of Ill-Posed Problems*. Scripta Series in Mathematics. Winston, 1977.

[323] A.N. Tikhonov, A.V. Goncharsky, V.V. Stepanov, and A.G. Yagola. *Numerical methods for the solution of ill-posed problems*. Mathematics and Its Applications. Springer, 1995.

[324] A.N. Tikhonov, A.S. Leonov, and A.G. Yagola. *Nonlinear Ill-Posed Problems*. Chapman & Hall, 1998.

[325] P.E. Trahanias. Generalized multichannel image-filtering structures. *Image Processing, IEEE Transactions on*, 6(7):1038–1045, 1997.

[326] P.E. Trahanias, D. Karakos, and A.N. Venetsanopoulos. Directional processing of color images: theory and experimental results. *Image Processing, IEEE Transactions on*, 5(6):868–880, 1996.

[327] P.E. Trahanias and A.N. Venetsanopoulos. Vector directional filters: a new class of multichannel image processing filters. *Image Processing, IEEE Transactions on*, 2(4):528–534, 1993.

[328] A. Tsai, A. Yezzi Jr., and A.S. Willsky. Curve evolution implementation of the Mumford-Shah functional for image segmentation, denoising, interpolation, and magnification. *Image Processing, IEEE Transactions on*, 10(8):1169–1186, 2001.

[329] D. Tschumperlé. *PDE's based regularization of multivalued images and applications*. PhD thesis, INRIA, Sophia-Antipolis, 2002.

[330] D. Tschumperlé and R. Deriche. Regularization of orthonormal vector sets using coupled PDE's. In *Variational and Level Set Methods in Computer Vision, Proceedings of IEEE Workshop on*, pages 3–10, 2001.

[331] D. Tschumperlé and R. Deriche. Vector-valued image regularization with PDEs: a common framework for different applications. *Pattern Analysis and Machine Intelligence, IEEE Transactions on*, 27(4):506–517, 2005.

[332] G. Unal and G. Slabaugh. Coupled PDEs for non-rigid registration and segmentation. In *Computer Vision and Pattern Recognition, IEEE Computer Society Conference on*, volume 1, pages 168–175, 2005.

[333] B. Vemuri and Y. Chen. Joint Image Registration and Segmentation. In *Geometric Level Set Methods in Imaging, Vision, and Graphics*, pages 251–269. Springer New York, 2003.

[334] B.C. Vemuri, J. Ye, Y. Chen, and C.M. Leonard. A level-set based approach to image registration. In *Mathematical Methods in Biomedical Image Analysis, Proceedings of IEEE Workshop on*, pages 86–93, 2000.

[335] L. Vese and T.F. Chan. A multiphase level set framework for image segmentation using the Mumford and Shah model. *International Journal of Computer Vision*, 50(3):271–293, 2002.

[336] L.A. Vese. A study in the BV space of a denoising-deblurring variational problem. *Applied Mathematics and Optimization*, 44:131–161, 2001.

[337] L.A. Vese and S.J. Osher. Numerical methods for p−harmonic flows and applications to image processing. *SIAM Journal on Numerical Analysis*, 40(6):2085–2104, 2002.

[338] L.A. Vese and S.J. Osher. Modeling textures with total variation minimization and oscillating patterns in image processing. *Journal of Scientific Computing*, 19(1-3):553–572, 2003.

[339] F. Wang and B.C. Vemuri. Simultaneous registration and segmentation of anatomical structures from brain MRI. In *Medical Image Computing and Computer-Assisted Intervention*, volume 3749 of *Lecture Notes in Computer Science*, pages 17–25. Springer, 2005.

[340] L.-Y. Wei and M. Levoy. Fast texture synthesis using tree-structured vector quantization. In *Proceedings of 27th Annual Conference on Computer Graphics and Interactive Techniques*, pages 479–488. ACM Press/Addison-Wesley, 2000.

[341] J. Weickert and G. Kühne. Fast methods for implicit active contour models. In *Geometric Level Set Methods in Imaging, Vision, and Graphics*, pages 43–57. Springer, 2003.

[342] E. Weinan and X.-P. Wang. Numerical methods for the Landau–Lifshitz equation. *SIAM Journal on Numerical Analysis*, 38(5):1647–1665, 2000.

[343] W.M. Wells III, W.E.L. Grimson, R. Kikinis, and F.A. Jolesz. Adaptive segmentation of MRI data. *Medical Imaging, IEEE Transactions on*, 15(4):429–442, 1996.

[344] R.M. Willett, Z.T. Harmany, and R.F. Marcia. Poisson image reconstruction with total variation regularization. In *Image Processing, 17th IEEE International Conference on*, pages 4177–4180, 2010.

[345] C. Xiaohua, M. Brady, and D. Rueckert. Simultaneous segmentation and registration for medical image. In *Medical Image Computing and Computer-Assisted Intervention*, volume 3216 of *Lecture Notes in Computer Science*, pages 663–670. Springer, 2004.

[346] C. Xu and J.L. Prince. Snakes, shapes, and gradient vector flow. *Image Processing, IEEE Transactions on*, 7(3):359–369, 1998.

[347] M. Yan and L. A. Vese. Expectation maximization and total variation-based model for computed tomography reconstruction from undersampled data. *Proceedings of SPIE, Medical Imaging*, 7961:79612X–79612X–8, 2011.

[348] I. Yanovsky, S. Osher, P. Thompson, and A. Leow. Log-unbiased large deformation image registration. In *International Conference on Computer Vision Theory and Applications*, volume 1, pages 272–279, 2007.

[349] I. Yanovsky, P.M. Thompson, S. Osher, and A.D. Leow. Topology preserving log-unbiased nonlinear image registration: theory and implementation. In *Computer Vision and Pattern Recognition, IEEE Conference on*, pages 1–8, 2007.

[350] A. Yezzi, L. Zollei, and T. Kapur. A variational framework for joint segmentation and registration. In *Mathematical Methods in Biomedical Image Analysis, IEEE Workshop on*, pages 44–51, 2001.

[351] H. Zhao and A.J. Reader. Fast ray-tracing technique to calculate line integral paths in voxel arrays. In *Nuclear Science Symposium Conference Record*, volume 4, pages 2808–2812, 2003.

[352] H.-K. Zhao, T. Chan, B. Merriman, and S. Osher. A variational level set approach to multiphase motion. *Journal of Computational Physics*, 127(1):179–195, 1996.

[353] H.-K. Zhao, S. Osher, B. Merriman, and M. Kang. Implicit and nonparametric shape reconstruction from unorganized data using a variational level set method. *Computer Vision and Image Understanding*, 80(3):295–314, 2000.

Index